Natural Computing Series

Series Editors: G. Rozenberg
Th. Bäck A.E. Eiben J.N. Kok H.P. Spaink
Leiden Center for Natural Computing

T0181478

Christian Blum · Daniel Merkle (Eds.)

Swarm Intelligence

Introduction and Applications

Dr. Christian Blum
ALBCOM
Dept. Llenguatges i Sistemes
Informàtics
Universitat Politècnica de Catalunya,
Jordi Girona 1-3
Omega 112 Campus Nord
08034 Barcelona, Spain
cblum@lsi.upc.edu

Dr. Daniel Merkle
Dept. of Mathematics and Computer
Science
University of Southern Denmark
Campusvej 55
5230 Odense M, Denmark
daniel@imada.sdu.dk

ISBN: 978-3-642-09343-2   e-ISBN: 978-3-540-74089-6

ACM Computing Classification (1998): I.2.8, I.2.9, I.2.11

© Springer-Verlag Berlin Heidelberg 2010

Cover design: KuenkelLopka GmbH

Printed on acid-free paper

9 8 7 6 5 4 3 2 1

springer.com

For María and Marc
(Christian Blum)

For Janine
(Daniel Merkle)

Preface

Swarm intelligence is a modern artificial intelligence discipline that is concerned with the design of multiagent systems with applications, e.g., in optimization and in robotics. The design paradigm for these systems is fundamentally different from more traditional approaches.

Instead of a sophisticated controller that governs the global behavior of the system, the swarm intelligence principle is based on many unsophisticated entities that cooperate in order to exhibit a desired behavior. Inspiration for the design of these systems is taken from the collective behavior of social insects such as ants, termites, bees, and wasps, as well as from the behavior of other animal societies such as flocks of birds or schools of fish. Colonies of social insects have mesmerized researchers for many years. However, the principles that govern their behavior remained unknown for a long time. Even though the single members of these societies are unsophisticated individuals, they are able to achieve complex tasks in cooperation. Coordinated behavior emerges from relatively simple actions or interactions between the individuals.

For example, ants, termites and wasps are able to build sophisticated nests in cooperation, without any of the individuals having a global master plan of how to proceed. Another example is the foraging behavior that ants or bees exhibit when searching for food. While ants employ an indirect communication strategy via chemical pheromone trails in order to find shortest paths between their nest and food sources, bee colonies are very efficient in exploiting the richest food sources based on scouts that communicate information about new food sources by means of a so-called waggle dance. For more examples and a more detailed description of the fascinating biological role models that inspired swarm intelligence applications see Chaps. 1 and 2 of this book.

Scientists have applied these principles to new approaches, for example, in optimization and the control of robots. Characterizing properties of the resulting systems include robustness and flexibility. The field of research that is concerned with collective behavior in self-organized and decentralized systems is now referred to as *swarm intelligence*. The term swarm intelligence was first used by Beni and colleagues in the context of cellular robotic sys-

tems where simple agents organize themselves through nearest neighbor interactions. Meanwhile, the term swarm intelligence is used for a much broader
research field, as documented in the seminal book *Swarm Intelligence—From
Natural to Artificial Systems* by Dorigo, Theraulaz, and Bonabeau, published
by Oxford University Press. However, since the appearance of the above-
mentioned book in 1999, the literature on swarm intelligence topics has grown
significantly. This was the motivation for editing this book, whose intention
is to provide an overview of swarm intelligence to novices of the field, and
to provide researchers from the field with a collection of some of the most
interesting recent developments. In order to achieve this goal we were able
to convince some of the top researchers in their respective domains to write
chapters on their work.

Introductory chapters in the first part of the book are on biological
foundations of swarm intelligence, optimization, swarm robotics, and applications in new-generation telecommunication networks. Optimization and
swarm robotics are nowadays two of the domains where swarm intelligence
principles have been applied very successfully. A third and very popular application domain concerns routing and loadbalancing in telecommunication
networks. The second part of the book contains chapters on more specific
topics of swarm intelligence research such as the evolution of robot behavior,
the use of particle swarms for dynamic optimization, organic computing, and
the decentralized traffic flow in production networks.

Finally, we hope that the readers enjoy reading this book, and, most importantly, that they learn something new by seeing things from a new perspective.

Barcelona, Odense   *Christian Blum*
April 2008   *Daniel Merkle*

Contents

Part I Introduction

Biological Foundations of Swarm Intelligence
Madeleine Beekman, Gregory A. Sword, Stephen J. Simpson 3

Swarm Intelligence in Optimization
Christian Blum, Xiaodong Li 43

Swarm Robotics
Erol Şahin, Sertan Girgin, Levent Bayındır, Ali Emre Turgut 87

Routing Protocols for Next-Generation Networks Inspired by Collective Behaviors of Insect Societies: An Overview
Muddassar Farooq, Gianni A. Di Caro 101

Part II Applications

Evolution, Self-organization and Swarm Robotics
Vito Trianni, Stefano Nolfi, Marco Dorigo 163

Particle Swarms for Dynamic Optimization Problems
Tim Blackwell, Jürgen Branke, Xiaodong Li 193

An Agent-Based Approach to Self-organized Production
Thomas Seidel, Jeanette Hartwig, Richard L. Sanders, Dirk Helbing 219

Organic Computing and Swarm Intelligence
Daniel Merkle, Martin Middendorf, Alexander Scheidler 253

Part I

Introduction

Biological Foundations of Swarm Intelligence

Madeleine Beekman[1], Gregory A. Sword[2], and Stephen J. Simpson[2]

[1] Behaviour and Genetics of Social Insects Lab, School of Biological Sciences,
University of Sydney, Sydney, Australia
`mbeekman@bio.usyd.edu.au`
[2] Behaviour and Physiology Research Group, School of Biological Sciences,
University of Sydney, Sydney, Australia
`{greg.sword,stephen.simpson}@bio.usyd.edu.au`

Summary. Why should a book on swarm intelligence start with a chapter on biology? Because swarm intelligence is biology. For millions of years many biological systems have solved complex problems by sharing information with group members. By carefully studying the underlying individual behaviours and combining behavioral observations with mathematical or simulation modeling we are now able to understand the underlying mechanisms of collective behavior in biological systems. We use examples from the insect world to illustrate how patterns are formed, how collective decisions are made and how groups comprised of large numbers of insects are able to move as one. We hope that this first chapter will encourage and inspire computer scientists to look more closely at biological systems.

1 Introduction

"He must be a dull man who can examine the exquisite structure of a comb so beautifully adapted to its end, without enthusiastic admiration."

Charles Darwin (1872)

When the Egyptians first started to keep honeybees 5,000 years ago, they surely must have marveled on the beauty of the bees' comb. Not only is the honeycomb beautiful to look at, but how did the bees decide to build hexagonal cells and not cells of another form? Initially it was suggested that hexagonal cells hold the most honey, but the French physicist R.A.F. de Réaumur realized that it was not the content of the cells that counts, but the amount of material, wax, that is needed to divide a given area into equal cells. Obviously at that time it was assumed that the bees were "blindly using the highest mathematics by divine guidance and command" (Ball 1999). It was not until Darwin that the need for divine guidance was removed and the hexagonal cells were thought to be the result of natural selection. In this view the bees' ancestors 'experimented' with different shaped cells, but the bees that by chance 'decided' to build hexagonal cells did better and, as a result, the building of hexagonal cells spread. In Darwin's words, *"Thus, as I believe, the most wonderful of all known instincts, that of the hive-bee, can be explained by natural selection having taken advantage of numerous, successive, slight modifications of simpler instincts; natural selection having by slow degrees, more and more perfectly, led the bees to sweep equal spheres at a given distance from each other in a double layer, and to build up and excavate the wax along the planes of intersection."* (Chapter 7, Darwin 1872).

It was exactly such 'Darwinian fables' that inspired the biologist and mathematician D'Arcy Wentworth Thompson to write his book *On Growth and Form* (Thompson 1917). The central thesis of this book is that biologists overemphasize the role of evolution and that many phenomena can be more parsimoniously explained by applying simple physical or mathematical rules. Thompson argued that the bees' hexagonal cells are a clear example of a pattern formed by physical forces that apply to all layers of bubbles that are pressed into a two-dimensional space. Bees' wax is not different, the soft wax forms bubbles that are simply pulled into a perfect hexagonal array by physical forces. Hence, the pattern forms spontaneously and no natural selection or divine interference needs to be invoked (Ball 1999).

In fact, many instances of spontaneous pattern formation can be explained by physical forces, and given the almost endless array of patterns and shapes found around us, it is perhaps not surprising that such patterns are an inspiration for many people, scientists and non-scientists alike. Upon closer examination, amazing similarities reveal themselves among patterns and shapes of very different objects, biological as well as innate objects. As we already alluded above, the characteristic hexagonal pattern found on honeycombs are not unique; the same pattern can be obtained by heating a liquid uniformly from below. Autocatalytic reaction-diffusion systems will lead to Turing patterns (think stripes on tigers) in both chemical and biological mediums (Kondo and Asai 1995; Ball 1999), and minerals form patterns that have even been mistaken for extra-terrestrial fossils (McKay et al. 1996).

The similarity of patterns found across a huge range of systems suggests that there are underlying principles that are shared by both biological and innate objects. Such similarities have been nicely illustrated by work on pattern formation in bacterial colonies. When one manipulates the amount of food available to bacteria and the viscosity of their medium, patterns emerge that are remarkably similar to those found in, for example, snowflakes (Ben-Jacob et al. 2000). In fact, the growth of bacterial colonies has proven to be an important playground for testing ideas on non-living branching systems (Ball 1999; Ben-Jacob and Levine 2001; Levine and Ben-Jacob 2004). As it turns out, many branching patterns found across nature can be explained by the same process, known as diffusion-limited aggregation, resulting from the interactions of the particles, be they molecules or individual bacteria (Ball 1999).

All patterns described above have been explained by approaching the systems from the bottom up: how do the particles interact with each other and with their immediate environment? One may not really be surprised by the fact that the same approach helps one to understand bacteria as well as molecules. After all, bacteria aren't really that different from molecules, are they? In the following we will illustrate how such a bottom-up approach can explain another remarkable feature of honeybees: the typical pattern of honey, pollen and brood found on combs.

The honeybee's comb is not only a marvel because of its almost perfect hexagonal cells, the bees also seem to fill the cells with brood (eggs that develop into larvae and then pupae and finally emerge as young workers or males), pollen (to feed the brood) and nectar (which will be converted into honey) in a characteristic pattern. This pattern consists of three distinct concentric regions: a central brood area, a surrounding rim of pollen, and a large peripheral region of honey (Fig. 1). If we envision the honeybee colony as a three-dimensional structure, this pattern is most pronounced in the central combs which intersect a large portion of the almost spherical volume of brood. How does this pattern come about? The storage of pollen close to the brood certainly makes sense as it reduces the time needed to get the pollen to the brood. But how do the bees know this? Do they use a blueprint (or template) to produce this characteristic pattern, implying that there are particular locations specified for the deposition of pollen, nectar and brood? Or is the pattern self-organized and emerges spontaneously from the dynamic interactions between the honeybee queen, her workers and the brood? Scott Camazine set out to determine which of these two hypotheses is the most parsimonious (Camazine 1991).

The beauty of working on macroscopic entities such as insects is that you can individually mark them. Honeybees are particularly suitable because we can then house them in what we call an observation hive, a glass-walled home for the bees. This means that we can study the interactions of the individually marked bees without taking them out of their natural environment (see Fig. 1).

Camazine did just that. He monitored the egg-laying behavior of the queen, of foragers that returned with pollen or nectar, and of nurse workers, those that feed the brood. The first thing that he observed was that the queen is rather sloppy in her egg-laying behavior, moving about in a zig-zag-like manner, often missing empty cells and retracing her own steps. Camazine further noticed that she has a clear preference to lay a certain distance from the periphery of the comb and never more than a few cell lengths of the nearest brood-containing cell. Interestingly, even though the queen somewhat has a preference for at least the middle of the comb and the vicinity of brood, bees returning with pollen or nectar did not seem to have a preference for specific cells at all. When an empty comb was left in the colony and the deposition of nectar and pollen observed, both could be found in any cell. Even though such absence of a preference clearly refutes the blueprint hypothesis, it does not explain how the characteristic pattern ultimately arises.

Fig. 1. Because of their relatively large size, we can easily mark individual bees in a colony. In this particular colony we marked 5,000 bees by combining numbered plates and different paint colors. This allowed us to study their behavior at an individual level. Photograph taken by M. Beekman.

As it turns out, bees do have a clear preference when they *remove* pollen or honey from cells. Both honey and pollen are preferentially removed from cells closest to the brood. By following the pattern of cell emptying during a period in which foraging activity was low (overnight or during rain), Camazine observed that all the cells that were emptied of their pollen or nectar were located within two cells or less from a cell containing brood. No cells were emptied that were further from brood cells. It is easy to see why the bees would have a preference for the removal (through use) of pollen that is found closest to the brood, as it is the brood that consumes the pollen. In addition, nurse bees are the younger bees which restrict most of their activity to the brood area (Seeley 1982).

The preferential removal of pollen and nectar from cells closest to cells containing brood and the queen's preference for laying eggs in cells close to brood made Camazine realism that this might explain the honeybee's characteristic comb pattern. But how to prove this? This is where the physicist's approach comes in. By constructing a simulation model based on his behavioral observations, Camazine was able to closely follow the emergence of the pattern. Initially, both pollen and nectar were deposited randomly throughout the frame with the queen wandering over the comb from her initial starting point. Despite the random storage of pollen and nectar, the queen's tendency to lay eggs in the vicinity of cells that already contain brood rapidly results in an area in which mostly brood is found. This is enhanced by the bees' pref-

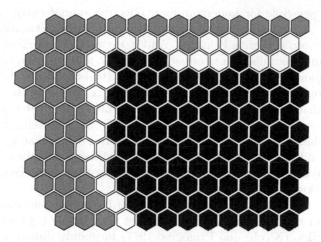

Fig. 2. The typical pattern of honey (grey cells), pollen (white cells), and brood (black cells) as seen on a honeybee's comb. Shown is the top-left corner of the comb

erence to remove honey and pollen from cells close to brood, which increases the availability to the queen of cells to lay eggs in. This further reduces the number of cells available for storage of honey or pollen. Thus, the brood area is continually freed of honey and pollen and filled with eggs resulting in a compact brood structure. But how do the pollen and nectar get separated (Fig. 2)?

Because initially both are deposited randomly, both pollen and nectar will be present in the periphery of the comb. However, most pollen that gets collected on a daily basis is consumed that same day. This means that given the normal fluctuations in pollen availability, there is often a net loss of pollen, with pollen present in the periphery being consumed at nearly the same rate as pollen being stored elsewhere. At the same time, these empty cells are most likely to be filled with nectar, as the nectar intake is much higher, and soon there is no longer space to store pollen. Where is pollen stored then?

Eventually the only place left for pollen to be stored is the band of cells adjacent to the brood. The developmental time from egg to adult is 21 days, meaning that for three weeks a brood cell cannot be used for anything else. But in the interface zone between the brood and the stores of honey at the periphery, the preferential removal of honey and pollen continuously provides a region in which cells are being emptied at a relatively high rate. And it is these cells that are available for pollen. Other cells that become available because bees emerge from them are found in the middle of the brood nest, but these will then be preferentially emptied and again filled with eggs.

Without his computer simulation Camazine would not have been able to fully understand how the behavior of the individual bees resulted in the organized pattern of brood, honey and pollen on the comb of the bees. And this is a general principle for understanding collective animal behavior: without tools such as simulations or mathematics, it is impossible to translate individual behavior into collective behavior. And it is exactly with those tools that originally came from disciplines outside of biology, and with the view that interactions among individuals yield insights into the behavior of the collective, that we biologists have learned from physics. In fact, we began this chapter by illustrating that even biological phenomena can often more parsimoniously be understood using physical explanations, and that many systems, both innate and living, share the same physical principles. And it has exactly been these similarities and the wide applicability of the mathematical rules that govern diverse behaviors that have led to the field of Swarm Intelligence (e.g. Dorigo et al. 1996; Dorigo and Di Caro 1999).

However, it is important to realise that our biological 'particles' are more complex than molecules and atoms and that the 'simple rules of thumb' of self-organization (Nicolis and Prigogine 1977) have only limited explanatory power when it comes to biological systems (Seeley 2002). Bacterial colonies may grow in a similar pattern as minerals, Turing patterns may be found on fish, in shells and in chemical reactions, and we can understand the bees' hexagonal cells using physics, but when it comes to biological systems, an extra layer of complexity needs to be added. Besides the complexity of the individuals, we cannot ignore natural selection acting on, for example, the foraging efficiency of our ant colony, or the building behavior of our termites. If the underlying principles that govern the building behavior of termites results in colony-level behavior that is far from functional, this would be rapidly selected against. Moreover, it is of no use to assume that certain systems must behave similarly simply because they 'look' similar. It is true that if the same mathematical model or behavioral algorithm captures the behavior of different systems, then we can talk about similarities between systems that go beyond simple analogy (Sumpter 2005). However, as we will explain in the concluding section of this chapter, true biological inspiration needs to come not from the superficial similarities between systems, but from the intricate and often subtle differences between them. We shall illustrate this standpoint by drawing examples from our own study systems: decentralized decision making in social insects and the coordinated movement of animal groups.

2 Decentralized Decision Making

The evolution of sociality, the phenomenon where individuals live together within a nest such as is found in many bees and wasps, and all ants and termites, has created the need for information transfer among group members. No longer can each individual simply behave as if solitary, but actions by different group members need to be carefully tuned to achieve adaptive behavior at the level of the whole group. Insect colonies need to make many collective decisions, for example where to forage, which new nest to move to, when to reproduce, and how to divide the necessary tasks among the available work-force. It is by now well known that such group-level decisions are the result of the individual insects acting mainly on local information obtained from interactions with their peers and their immediate environment (Bonabeau et al. 1997; Camazine et al. 2001). In other words, decision making in insect societies is decentralized. To illustrate how insect colonies achieve this, we will describe foraging and nest site selection in ants and honeybees.

2.1 Where to Forage?

In order to organize foraging, social insects need a form of recruitment. Recruitment is a collective term for any behavior that results in an increase in the number of individuals at a particular place (Deneubourg et al. 1986), and allows insect societies to forage efficiently in an environment in which food sources are patchily distributed or are too large to be exploited by single individuals (Beckers et al. 1990; Beekman and Ratnieks 2000; Detrain and Deneubourg 2002). In addition, social insects that transfer information about the location of profitable food sources can exploit an area much larger than those that lack such a sophisticated recruitment mechanism. Honeybees are a prime example. Their sophisticated dance language (von Frisch 1967) allows them to forage food sources as far as 10 km from the colony (Beekman and Ratnieks 2000).

Exact recruitment mechanisms vary greatly among the social insects but can be divided into two main classes: direct and indirect mechanisms. Mass recruitment via a chemical trail is a good example of indirect recruitment. The recruiter and recruited are not physically in contact with each other; communication is instead via modulation of the environment: the trail. The recruiter deposits a pheromone on the way back from a profitable food source and recruits simply follow that trail. In a way such a recruitment mechanism is comparable to broadcasting: simply spit out the information without controlling who receives it. The other extreme is transferring information, figuratively speaking, mouth to mouth: direct recruitment. The best-known example of such a recruitment mechanism is the honeybees' dance language. Successful foragers, the recruiters, perform a stylized 'dance' which encodes information about the direction and distance of the food source found and up to seven dance followers (Tautz and Rohrseitz 1998), potential recruits, are able to extract this information based upon which they will leave the colony and try to locate the advertised food source. Recruitment trails and the honeybee dance language can be seen as the two extremes of a whole range of different mechanisms used by social insects to convey information about profitable food sources.

Many computer scientists are familiar with the double bridge experiment as an example of the means by which foraging is organized in ant colonies. In this experiment a colony of trail-laying ants is offered two equal food sources located at the end of two paths of different lengths. After some time the vast majority of foragers converges on the shorter path (Beckers et al. 1993). This collective choice for the nearest source is the result of a positive feedback process: Ants finding food mark the environment with pheromone trails during their return to the nest, and ants searching for food probabilistically follow these trails.

The same trail-following behavior allows an ant colony to choose the best food source out of several possibilities without the individual ants directly comparing the quality of the food sources on offer. Experiments performed on several species of ants have shown that ants modulate the amount of pheromone deposited depending on the quality of the food source, such that the better the quality, the more pheromone is left and the more likely other ants are to follow the trail to the best food source (Beckers et al. 1990; Beckers et al. 1993; Sumpter and Beekman 2003).

The success of the pheromone trail mechanism is likely to be due, at least in part, to the non-linear response of ants to pheromone trails where, for example, the distance that an ant follows a trail before leaving it is a saturating function of the concentration of the pheromone (Pasteels et al. 1986). In other words, the probability an ant will follow a trail is a function of trail strength (expressed as concentration of pheromone), but ants never have a zero probability of losing a trail, irrespective of the strength of the trail. Mathematically, non-linearity in response means an increase in the number and complexity of solutions of the model equations that may be thought of as underlying foraging (linear equations have only a single solution). Biologically, a solution to a differential equation corresponds to a distribution of ants between food sources and an increase in solutions implies more flexibility as the ants 'choose' between possible solutions. Such an allocation of workers among food sources, which assigns nearly all trail-following foragers to the best food source, is optimal provided the food source has unlimited capacity. When the food source does not have unlimited capacity, the result is that trail-following ants will be directed to a food source at which they cannot feed. In a way the colony gets 'stuck' in a suboptimal solution and can only get out of this solution by adding some layers of complexity, such as negative pheromones signalling 'don't go there', or individual memory so that the individual remembers that following that particular trail does not yield anything. A second drawback of relying on pheromone trails is that it may be difficult to compete with an existing trail, even if a better food source is found. If, due to initial conditions, a mediocre food source is discovered first, ants that have found a better quality food source after the first trail has been established will not be able to build up a trail strong enough to recruit nest mates to the newly discovered bonanza (Sumpter and Beekman 2003). Again, the ants are stuck in a sub-optimal solution.

Because of their fundamentally different recruitment mechanism, honeybees cannot get stuck in a sub-optimal solution. This direct recruitment behavior, the dance, encodes two main pieces of spatial information: the direction and the distance to the target. Both are necessary as, unlike ants, honeybees need to deal with a three-dimensional space. During a typical dance the dancer strides forward about 1.5 times her length vigorously shaking her body from side to side (Tautz et al. 1996). This is known as the 'waggle phase' of the dance. After the waggle phase the bee makes an abrupt turn to the left or right, circling back to start the waggle phase again. This is known as the 'return phase'. At the end of the second waggle, the bee turns in the opposite direction so that with every second circuit of the dance she will have traced the famous figure-of-eight pattern of the waggle dance (von Frisch 1967).

The most information rich phase of the dance is the waggle phase. During the waggle phase the bee aligns her body so that the angle of deflection from vertical is similar to the angle of the goal from the sun's current azimuth. Distance information is encoded in the duration of the waggle phase. Dances for nearby targets have short waggle phases, whereas dances for distant targets have protracted waggle phases.

Dance followers need to be in close contact with the dancer in order to be able to decode the directional information (Rohrseitz and Tautz 1999). Hence, directional information is transferred to a limited number of individuals. Moreover, more than one dance can take place at the same time, and these dances can be either for the same site or for different sites. This means that there is no direct competition between dances, provided the number of bees available to 'read' a dance is infinite (a likely assumption). Dances are only performed for food sources that are really worthwhile.

In order to assess the quality of the food encountered, a forager uses an internal gauge to assess the profitability of her source, based on the sugar content of the nectar and the distance of the patch from the colony, as well as the ease with which nectar (or pollen) can be collected. A bee's nervous system, even at the start of her foraging career, has a threshold calibrated into it which she uses to weigh these variables when deciding whether a patch is firstly worth foraging for at all, and secondly worth advertising to her fellow workers (Seeley 1995).

Dancing bees also adjust both the duration and the vigor of their dancing as a function of profitability of their current source (Seeley et al. 2000). The duration of the dance is measured by the number of waggle phases that the dancer performs in a particular dance, and the vigor is measured by the time interval between waggle phases (the return phase). The larger the number of waggle phases, and the smaller the return phase, the more profitable the source is and the more nest mates will be recruited to it. This means that when two dances are performed simultaneously, one for a mediocre and one for a superb site, the dance for the superb site is more likely to attract dance followers than the one for the mediocre site. At the same time, however, the dance for the mediocre site will still attract some dance followers, because potential dance followers do not compare dances before deciding which one to follow (Seeley and Towne 1992). The result is that a honeybee colony can not only focus on the best food sources to the extent that most foragers will collect food at the best sites (Seeley et al. 1991), but is also able to swiftly refocus its foraging force in response to day-to-day, or even hour-to-hour, variation in available forage (Visscher and Seeley 1982; Schneider and McNally 1993; Waddington et al. 1994; Beekman and Ratnieks 2000; Beekman et al. 2004).

2.2 Exploration Versus Exploitation

Most, if not all, studies on the allocation of foragers over food sources have used stable environments in which the feeders or forage sites were kept constant (e.g. Beckers et al. 1990; Seeley et al. 1991; Sumpter and Beekman 2003). When conditions are stable the optimal solution from the colony's point of view is to focus solely on the best food source (provided this food source is so large that it allows an infinite number of individuals to forage it) and this is exactly what many species of ants do that collect stable food sources such as honeydew (a sugary secretion produced by aphids) (Quinet et al. 1997), or leaves (Darlington 1982; Wetterer et al. 1992). These species construct long lasting trails (trunk trails) that connect the nest to foraging locations. In some species trails are more or less permanent due to the workers actively changing the environment by removing vegetation (Rosengren and Sundström 1987; Fewell 1988). As soon as conditions are not stable, however, which is mostly the case in nature, it becomes important to have a mechanism that allows the change-over to another food source or food sources when they have become more profitable or when the initial food source has been depleted. This means that in order to do well in a dynamically changing environment, insect colonies should allow the storage of information about food patches which are currently exploited but at the same time allow the exploration for new sites.

The key to keeping track of changing conditions is the trade-off between exploitation and exploration: the use of existing information (exploitation) versus the collection of new information (exploration). How do the two extreme recruitment mechanisms, trail-based foraging and the honeybee's dance language, allow for the discovery of new food sources?

As mentioned earlier, trail-following ants never have a zero probability of losing a trail, irrespective of the trail resulting in some ants getting lost even when the trail is at its strongest. Assuming that these 'lost' ants are able to discover new food sources and thus serve as the colony's explorers or scouts, this 'strategy of errors' (Deneubourg et al. 1983; Jaffe and Deneubourg 1992) allows the colony to fine-tune the number of scouts depending on the profitability of the food source that has already been exploited. This is because the weaker the trail (indicating the presence of a mediocre food source), the more the number of ants that get 'lost' and hence become scouts. When the trail is very strong (indicating that a high-quality food source has been found) a smaller number of ants will lose the trail, resulting in a smaller number of scouts.

The regulation of scouts in honeybees similarly assures the correct balance between the number of individuals allocated to exploration and exploitation. An unemployed forager (an individual that wants to forage but does not know where to forage) will first attempt to locate a dance to follow. If this fails because the number of dancers is low, she will leave the colony and search the surroundings, thereby becoming a scout (Beekman et al. 2007). As a result, the number of scouts is high when the colony has not discovered many profitable forage sites, as dancing will then be low, whereas the number of scouts will be low when forage is plentiful and the number of bees performing recruitment dances high (Seeley 1983; Beekman et al. 2007). This so-called 'failed follower mechanism' (Sumpter 2000; Beekman et al. 2007) provides the colony with the means to rapidly adjust its number of scouts depending on the amount of information available about profitable forage sites. Even when the colony is exploiting profitable patches, there may still be other, undiscovered, profitable sites that are not yet exploited. As soon as there is a reduction in the number of dances occurring in the colony, the probability that some unemployed foragers are unable to locate a dance increases, and the colony therefore sends out some scouts. Such fluctuations in the number of dances regularly occur in honeybee colonies, even when there is plenty of forage (Beekman et al. 2004).

2.3 Where to Live?

Amazing as an insect colony's collective food collection is, even more amazing is that the same communication mechanisms are often used to achieve a very different goal: the selection of a new nest. A colony needs to select a new home under two conditions. Either the whole colony needs to move after the old nest has been destroyed, or part of the colony requires a new nest site in the case of reproductive swarming (where the original colony has grown so much that part of it is sent off with one or more new queens to start a new colony). This means that colonies of insects need to address questions very similar to the questions we ask ourselves when changing homes (Franks et al. 2002). What alternative potential new homes are available? How do their attributes compare? Has sufficient information been collected or is more needed? House hunting by social insects is even more piquant, as it is essential for the colony that the decision be unanimous. Indecisiveness and disagreement are fatal (Lindauer 1955). House hunting has been studied in detail in two species of social insect only, in the ant and the honeybee. Both study systems have been selected for ease with which this process can be studied. The ant *Temnothorax albipennis* forms small colonies (often containing about 100 workers) and lives in thin cracks in rocks (Partridge et al. 1997) and can easily be housed in the laboratory. By simply destroying their old nest, the ants are forced to select a new one (Sendova-Franks and Franks 1995). Moreover, because of their small colony size, it is not that hard to uniquely mark all individuals (and they don't sting!), which greatly facilitates the study of their behavior.

Honeybee swarms normally have many more individuals (approximately 15,000 bees (Winston 1987)), but researchers often work with swarms that contain 4,000 to 5,000 bees (Seeley et al. 1998; Camazine et al. 1999; Seeley and Buhrman 1999; Seeley and Buhrman 2001; Seeley 2003; Seeley and Visscher 2004; Beekman et al. 2006). The great benefit of honeybees is that we can artificially make swarms by simply taking the queen out of the colony and shaking the 5,000 or so bees needed to produce an experimental swarm. And, if necessary, we label the bees individually in the same manner as when we study foraging.

By offering our homeless insects nest sites that differ in quality, we can carefully study which attributes the ants or bees value in their new home. At the same time we can get a clear picture of the behavioral repertoire that underlies collective house hunting. These behaviors can then be used to construct individual-based models aimed at understanding precisely how the actions of the individuals result in collective choice.

House Hunting in Honeybees

Tom Seeley was the first to systematically study house hunting in social insects using the honeybee as his model organism. Seeley started out by determining what attributes the bees look for when judging the suitability of a potential new nest site. By working on a tree-less island (this species of honeybee normally inhabits tree hollows but happily lives in man-made hives when no tree hollows are available), Seeley and his colleague Buhrman could manipulate the kind of nest sites the bees could choose from. By manipulating the nest site's main attributes, such as content and size of the entrance, they could determine what constitutes a 'mediocre' and a 'superb' site from the bees' point of view. This further allowed them to study how good a bee swarm is at choosing the best nest site out of several possibilities (Seeley and Buhrman 2001). And, not surprisingly, they are pretty good at it because when offered an array of five nest boxes, four of which were mediocre because they were too small (bees like large nest sites with a small nest entrance), in four out of five trials the bees chose the superb nest site. How do they do it?

As with recruitment to forage sites, bees that have found a nest site that is considered worthwhile will perform a dance upon returning to the swarm. By filming the dances of bees returning from both mediocre and superb sites, Seeley and Buhrman (2001) could study how the dances differed between the two. What they observed was that bees tune their dance in three ways. Firstly, bees returning from a superb site dance longer than bees returning from a mediocre site. Secondly, a dance for a superb site contains more waggle phases (the part of the dance that encodes the distance to the site). Lastly, dances for superb sites are 'livelier', meaning that the period between two subsequent waggle phases is shorter. Hence, there is a clear difference in dance behavior between bees returning from mediocre and superb sites, but this is not different from bees dancing for high and mediocre quality food sources which leads to most, but not all, bees focusing on the best food source. Dances for forage will never converge; instead there will always be different sites advertised simultaneously. However, a swarm needs to select just one nest site unanimously and this suggests that dances for nest sites do converge.

Scout bees, those bees that search the environment for suitable new nest sites, fly out in every direction and return to the swarm with information about nest sites found. Initially, many dances will take place on the swarm, advertising all sites that have been judged to be good enough. Within a few hours, however, many sites are no longer danced for, and just before the swarm takes to the air to fly to its new home, most, if not all, dances will be for a single site (Seeley and Buhrman 1999). Such a unanimous decision is reached without scouts comparing multiple nest sites (Visscher and Camazine 1999) or potential dance followers selecting dances for the best nest sites. The most likely reason why the swarm is ultimately able to select one nest site that is mostly the best is dance attrition. In contrast to the dances for forage, where bees will keep dancing for a forage site provided it remains profitable, bees returning from a potential new nest site ultimately cease dancing (Seeley and Buhrman 2001; Seeley 2003) even when their discovered nest site is of superb quality.

The process goes like this. A bee that has returned from the best site possible will perform, say, 100 waggle phases during the first dance that she performs for that nest site. After she has finished her dance, she returns to the nest site to confirm that it is still superb. Upon returning to the swarm, she will advertise her site again, but will now reduce the number of waggle phases to 80. After this dance she flies off again to her site and the process repeats itself. This means that after five trips, this bee will not perform a dance upon her return (as she will have reduced the number of waggle phases after each return trip), but in the meantime she will have performed protracted and lively dances for her site. Compare this with a bee that has found a mediocre site. This bee will perform, say, only 60 waggle phases during her first dance, 40 on her second dance, etc. until she ceases dancing altogether. She not only dances for a shorter period, but the number of dances performed for her site is also less than the number of dances performed by the bee that found the superb site. Hence, the 'length of advertising' differs significantly between the mediocre and superb site and, as a result, more bees will be recruited to the superb site than to the mediocre site, and those bees, provided they also rate the site as superb, will perform lengthy dances and recruit more bees. It has been suggested, based on a mathematical model, that this dance attrition is crucial to the swarm's ability to decide on one site (Myerscough 2003), but this assumption awaits empirical testing.

Even though many behaviors of the bees involved in the swarm's decision-making process have been described in great detail (for a nice overview see Visscher 2007), without the use of a mathematical or simulation model it is not immediately obvious how individual behavior is translated into collective behavior and the swarm's ability to choose the best nest site. Understanding in more detail the swarm's decision-making process led one of us, Madeleine Beekman, together with two computer scientists, Stefan Janson and Martin

Middendorf, to construct an individual-based model of a honeybee swarm choosing a new home (Janson et al. 2007). Not only did we want to construct a model that would behave in a realistic way, we wanted also to use that model to get an idea about two aspects in particular which are hard to study using real honeybee swarms: how a swarm regulates the number of individuals that explore the surroundings for nest sites (as opposed to recruiting individuals to nest sites that have already been discovered), and how scouts search their environment. Both questions address the trade-off between using existing information (exploitation) and acquiring new information (exploration) and how this trade-off affects the quality of the decision made at the level of the swarm.

We assumed that the bees would use the same behaviors and decision rules both when foraging and when deciding on a new home. We therefore started by applying the same exploration decision rule as had been applied in the context of foraging: an individual bee that has not yet decided where to search will always start by attempting to locate a dance to follow. The longer it takes to find a dance, the more likely this bee is to fly off and search independently (explore). This simple decision rule gave the following result (see Fig. 3): when the nest site known to the swarm is only of mediocre quality, not many bees will dance for that nest site and many bees will search independently because they have a low chance of finding a dance to follow; the reverse is true when a superb nest site has been found, as now most returning bees will dance for this nest site (note that we included individual variation in our bees such that an individual has a probability of dancing that increases with increasing quality of nest site). Clearly, applying the failed follower mechanism also works very well in house hunting and ensures an elegant balance between the number of bees recruiting to a known nest site and the need to search the environment for a better nest site.

All experimental work done so far on nest site selection in honeybee swarms used nest sites which were located at equal distance from the swarm, a situation which is highly unlikely under natural conditions where nest sites are present at all distances. Imagine a situation in which the swarm has discovered a nest site nearby, but this nest site is only of very mediocre quality. We now know that under this condition the swarm allows for more bees to explore the surroundings in case a better nest site is found; but how should the swarm distribute its scouts over the environment to allow the discovery of such a further site in the first place? The great benefit of using models is that one can manipulate the experiment. Hence, in our simulation model we were able to control which nest site was discovered first by the swarm by simply sending the first scout to that particular nest site. At the same time we could give our scouts different 'search rules' to investigate how these rules affect the swarm's decision. The search rules we used were the following: scouts were sent out such that all nest sites irrespective of their distance were equally likely to be discovered (*uniform* $P_u=1/250$); the chance of discovery decreased with increasing distance from the swarm (*distance* $P_d=1/$distance); sites nearby had

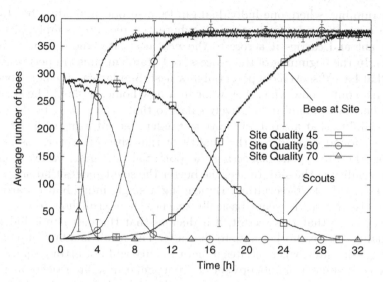

Fig. 3. Average number of bees scouting for a nest site and average number of bees assessing a nest site (e.g., the number of bees that already know about the location of a potential new nest site) when nest sites of different quality (70: good site, 50: mediocre site, 45: poor site) were offered to the swarm. The vertical bars indicate the standard variation (50 runs per experiment). (Fig. 6 in Janson et al. (2007).)

a much higher probability of being discovered than sites present further from the swarm (*distance squared* P_s =250/distance^2). Our results showed that the best search strategy from the swarm's point of view would be to focus on nearby sites without ignoring possible sites further away. Hence, most scouts are expected to search in the vicinity of the swarm, whereas some are likely to fly out further. This prediction can relatively easily be tested using real honeybee swarms.

Ants Moving House

The ant *Temnothorax albipennis* is rather different from the honeybee. Not only are its colonies much smaller, decision making seems much more dependent on individual decisions. For example, when offered a choice between two nest sites, about half of the ants directly compare the quality of those sites and can therefore make an informed choice (Mallon et al. 2001). At the same time, however, the other half does not directly compare the different options but these poorly informed ants still contribute to the colony's overall decision. How does their decision-making mechanism work?

Individual behavior of *T. albipennis* during nest site selection has been described in great detail (Mallon et al. 2001; Pratt et al. 2002). *T. albipennis* does not use pheromone trails to recruit nest mates but instead relies on tandem running, where one individual guides a second individual by staying in close contact, and social carrying, where the recruiter simply picks up another ant and carries it across to the new nest site (the queen is always carried!). In the beginning of the process, only tandem runs are performed by scouts that have discovered a potential new nest. Similarly to the honeybee, *T. albipennis* scouts know what they want in a new home: it should be spacious and the entrance should be relatively small so that it can easily be defended. The probability that a scout will initiate tandem recruitment to the site that it has just found depends on the quality of that site. Moreover, assessment duration (the time spent inspecting the potential new nest) varies inversely with the quality of the site. Hence, the better the scout judges that site to be, the faster it will start recruiting. By leading a single individual towards the nest site discovered, the scout basically teaches the recruit how to get to the new nest site so that this recruit, if it decides that the nest site is indeed of good quality, can lead other ants to that site. The result is a build-up of ants at good quality sites, whereas sites of poor quality will not attract many ants.

When the number of ants present at a particular nest site reaches a certain level, the quorum threshold, no more tandem recruitment takes place but instead ants still present in the old nest are picked up and simply carried to the new nest site. Brood items (eggs, larvae and pupae) will also be moved in this way. Why does *T. albipennis* have two recruitment methods, one slow (tandem runs) and one fast (social carrying)?

During the period in which tandem recruitment takes place, the quality of the nest site discovered is assessed independently by each ant that either discovered that site via scouting or was led to it via tandem recruitment. This ensures that the 'opinions' of many ants about the site's quality are pooled, thereby increasing the likelihood that that site is indeed of sufficient quality. At the same time, the slow build-up of ants at the discovered site allows for a better site to be discovered, as recruitment to this site will be faster and hence the number of ants will rapidly increase. Because of the different recruitment to sites depending on their quality, the quorum will be reached faster at the better site, after which social carrying will be initiated. This last phase enables the colony to move into the chosen site rapidly (remember that these ants move when their old nest site has been destroyed).

The above is a verbal description of the ants' collective choice based on observations of their individual behaviors. But can this sequence of behaviors really account for the ants' collective choice? To answer this question Stephen Pratt and his colleagues (Pratt et al. 2005) incorporated everything they knew about individual behavior into an agent-based model of collective nest choice. They then used this model to simulate emigrations and compared the outcomes of these *in silico* emigrations with those performed by real ant colonies. When the simulated ants were presented with a single site, the time course of the emigration generally conformed to experimental data. More interesting, however, was what the simulated ants did when confronted with two potential sites that differed in quality. The model predicted that about 10% of each colony should typically be carried to the site of lower quality by the time the old nest is completely empty, a result of many individuals basing their decision on information on one nest site only. This prediction was confirmed by using real colonies and offering them the same choice as the *in silico* ants. The agent-based model therefore provides strong support for the interpretation of the ants' individual behavior.

3 Moving in Groups

In many animal species, individuals move in groups as they perform seasonal migrations, travel to food sources and return to safe havens, often over considerable distances (Boinski and Garber 2000; Krause and Ruxton 2002; Couzin and Krause 2003). The movement of these groups is commonly self-organized, arising from local interactions between individuals rather than from a hierarchical command center. Self-organized group movement is not restricted to groups of relatively 'simple' creatures such as insect swarms or schools of fish, but may even include 'intelligent' species like us. One of the most disastrous examples of collective human group movement is crowd stampede induced by panic, often leading to fatalities as people are crushed or trampled (Helbing et al. 2000).

There are two extreme ways in which groups can 'decide' on a direction of movement. Either all individuals within the group contribute to a consensus, or else relatively few individuals (for convenience we will call these 'leaders') have information about the group's travel destination and guide the uninformed majority. Thus, in some species, all individuals within a group share a genetically determined propensity to travel in a certain direction (Berthold and Querner 1981; Berthold et al. 1992) or all are involved in choosing a particular travel direction (Neill 1979; Grünbaum 1998). In contrast, a few informed individuals within a fish school can determine the foraging movements of the group and can steer a group towards a target (Reebs 2000; Swaney et al. 2001). Similarly, very few individuals (approx. 5%) within a honeybee swarm can guide the group to a new nest site (Seeley et al. 1979).

When leaders are present, the question arises as to how these informed individuals transfer directional information to the uninformed majority. Similarly, in the absence of leaders how is a consensus reached about travel direction? Such questions are almost impossible to address without having first developed a theoretical framework that explores possible mechanisms.

Recently, two theoretical studies have addressed the issue of information transfer from informed to uninformed group members. Stefan Janson, Martin Middendorf and Madeleine Beekman (2005) modeled a situation in which the informed individuals make their presence known by moving at a higher speed than the average group member in the direction of travel. Guidance of the group is achieved by uninformed individuals aligning their direction of movement with that of their neighbors. Because the informed individuals initially move faster, they have a larger influence on the directional movement of the uninformed individuals, thereby steering the group.

A second model by Iain Couzin and colleagues (2005) shows that the movement of a group can be guided by a few informed individuals without these individuals providing explicit guidance signals and even without any individual in the group 'knowing' which individuals possess information about travel direction. Only the informed members of the group have a preferred direction, and it is their tendency to go in this direction that steers the group. The main difference between the two models lies in the presence or absence of cues or signals from the informed individuals to the uninformed majority. Janson et al.'s (2005) leaders clearly make their presence known, whereas Couzin et al.'s (2005) model suggests that leadership can arise simply as a function of information difference between informed and uninformed individuals, without the uninformed individuals being able to tell which ones have more information. It seems likely that the exact guidance mechanism is species-dependent. When the group needs to move fast, for example a swarm of honeybees that cannot run the risk of losing its queen during flight, the presence of leaders that clearly signal their presence might be essential, as the group otherwise takes a long time to start moving into the preferred direction. However, when the speed of movement is less important than group cohesion, for example because being in a group reduces the chance of predation, leaders do not need to signal their presence.

If there are no leaders, the essential first step before a group can start to move cohesively is some level of consensus among the individuals in their alignment. How is this achieved when there are no leaders? Most likely there are a minimum number of individuals that need to be aligned in the same direction before the group can start to move in a particular direction without breaking up. If the number of equally aligned individuals is below this threshold, the group does not move cohesively. As soon as this threshold is exceeded, coordinated movement is achieved. Such a non-linear transition at a threshold is known in theoretical physics and mathematics as a phase transition. Interestingly, we have recently discovered, for the first time, similar transitions in biological systems (Beekman et al. 2001). Theoretical physicists have developed a suite of models, termed self-propelled particle (SPP) models, which attempt to capture phase transitions in collective behavior (Vicsek et al. 1995). SPP models aim to explain the intrinsic dynamics of large groups of individuals. Later we shall show how this theoretical framework can be applied to the collective movement of locusts. But first we will describe some experimental results on group movement in honeybee swarms, locusts and Mormon crickets.

3.1 Honeybees on the Move

Deciding where to live is only one part of a honeybee swarm's problem. The second problem arises once that decision has been made: how does the small number of informed bees (about 5%) convey directional information to the majority of the uninformed bees in such a way that the swarm moves in unison? In the previous section we already described two theoretical possibilities: either leaders signal their presence to the uninformed majority, or they do not but simply move in their preferred direction. In fact, the model by Janson and colleagues was inspired by a suggestion made in the early 1950s by Martin Lindauer (1955). Lindauer observed in airborne swarms that some bees fly through the swarm cloud at high speed and in the correct travel direction, seemingly 'pointing' the direction to the new nest site. He suggested that these fast-flying bees, later named 'streakers' (Beekman et al. 2006), are the informed individuals or scouts. Lindauer got this idea while working in war-ravaged Munich where he used to run with his honeybee swarms in an attempt to find out where they were going. Like every scientist who takes himself or herself seriously, at least at that time, Lindauer used to wear a white lab coat, even when he was out in the field with the bees. One of his field sites was near a mental hospital and rumor has it that one day he was mistaken for an escaped mentally ill patient (Tom Seeley, personal communication). Luckily, Lindauer ran faster than the guards who tried to catch him, which gives one an indication of how fast a swarm of bees flies!

An alternative to Lindauer's hypothesis (which we will refer to as the 'vision' hypothesis) is the olfaction hypothesis of Avitabile et al. (1975). They proposed that the scouts provide guidance by releasing assembly pheromone from their Nasanov glands (a gland found between the last two tergites of the bee's abdomen) on one side of the swarm cloud, thereby creating an odor gradient that can guide the other bees in the swarm. Until very recently neither the vision hypothesis nor the olfaction hypothesis had been tested empirically, though other investigators have confirmed Lindauer's report that there are streakers in flying swarms (Seeley et al. 1979; Dyer 2000).

Madeleine Beekman, Rob Fathke and Tom Seeley (2006) decided that it was time to shed some light on this issue. In that study they did two things. They studied in detail the flights of normal honeybee swarms (containing approximately 15,000 bees) and smaller (4,000–5,000 bees) swarms in which the bee's Nasanov gland was sealed shut by applying paint to every single bee in the swarm. This meant that sealed-bee swarms could not emit the Nasanov pheromone (they had to apply paint to all bees in the swarm because they had no means of knowing which bees would be the scouts). By using a 'bait-nest site' that they made extremely attractive to a bee swarm, they could be almost certain that their swarms would select that nest site. This allowed them to follow the swarm (as Lindauer did through Munich, though they used an open field), measure its speed and the time it took the swarm to settle in its new home. Using this procedure and several sealed-bee swarms allowed them to show that even if every single bee in the swarm was unable to produce the Nasanov pheromone, the swarm was still able to fly more or less directly towards the new nest site. From this they concluded that scouts do not use pheromones to guide the swarm.

Proving the vision hypothesis was more difficult. They decided that a first step would be to show that there is variation in flight speed and flight direction among the individual bees within a flying swarm by taking photographs of a large swarm during its flight to the bait hive such that individual bees appear as small, dark streaks on a light background. The faster the flight speed of a bee, the longer the streak it produced using this technique (provided the bee flew in the plane of vision). Each photograph was analyzed by projecting it onto a white surface to create an enlarged image. They then measured the length (in mm) and the angle (in degrees, relative to horizontal) of each dark streak that was in focus in the enlarged image. Because a size reference was present in each photograph, and because each photograph recorded the bees' movements during a known time interval (1/30 s), they were able to calculate for each photograph the conversion factor between streak length and flight speed. Using this procedure they could quantify what they saw while running with the swarms: that a portion of the bees fly much faster in the direction of travel while the majority of the bees seem to fly much slower and with curved flight paths. Moreover, the fast-flying bees, the streakers, appeared to be most common in the upper region of a swarm. For humans, and probably also for bees, streakers are much more easily seen against the bright sky rather than the dark ground or vegetation, so by flying above most of the bees in a swarm, the streakers may be facilitating the transfer of their direction information to the other bees. Future work should focus on determining if it is indeed the streakers that are the scouts, those with information about the location of the new nest site.

3.2 Locusts

To this point we have considered examples of self-organization and swarm intelligence in highly structured social groups, in which there is a distinction between reproductive individuals and more or less sterile workers and pronounced division of labor among workers. But not all cohesively behaving animal groups are so structured. Some consist of individuals that are essentially all the same. And, as we shall see in the next section, the forces that bind and propel such groups may be very sinister indeed.

Of the approximately 13,000 described species of grasshopper that exist across the world, 20 or so are particularly notorious. For much of the time they are just like any other harmless grasshopper — but, occasionally, and catastrophically, they change and instead of living solitary lives, produce massive, migrating aggregations. As juveniles they form marching bands that may extend for kilometers. Once they become winged adults, they take to the air as migrating swarms that may be hundreds of square kilometers in area and travel hundreds of kilometers each day. More than one fifth of the earth's land surface is at risk from such plagues and the livelihood of one in ten people on the planet may be affected. These grasshoppers are called locusts.

Fig. 4. The two extreme forms of juvenile desert locusts. When reared in a crowd, locusts develop into the gregarious phase, whereas the same individual if reared alone would develop into the solitarious phase (photo by S. Simpson).

Phase Polyphenism: The Defining Feature of Locust Biology

Unlike other grasshoppers, locusts express an extreme form of density dependent phenotypic plasticity, known as 'phase polyphenism'. Individuals reared under low population densities (the harmless, non-migratory 'solitarious' phase) differ markedly in behavior, physiology, color and morphology from locusts reared under crowded conditions (the swarm-forming, migratory 'gregarious' phase) (Pener and Yerushalmi 1998; Simpson et al. 1999; Simpson and Sword 2007). In some species, such as the infamous migratory locust of Africa, Asia and Australia (*Locusta migratoria*), the phenotypic differences are so extreme that the two phases were once considered to be separate species (Uvarov 1921; Fig. 4). In fact, not only are the two phases not different species, they are not even different genotypes: the same animal can develop into the solitarious or the gregarious phase depending on its experience of crowding. The genetic instructions for producing the two phases are, therefore, packaged within a single genome, with expression of one or other gene set depending on cues associated with crowding.

At the heart of swarm formation and migration is the shift from the shy, cryptic behavior of solitarious phase locusts, which are relatively sedentary and avoid one another, to the highly active behavior and tendency to aggregate typical of gregarious phase insects. In the African desert locust, *Schistocerca gregaria*, this behavioral shift occurs after just one hour of crowding (Simpson et al. 1999). In recent years progress has been made towards understanding the physiological and neural mechanisms controlling behavioral phase change in locusts. In the desert locust the key stimulus evoking behavioral gregarization is stimulation of touch-sensitive receptors on the hind (jumping) legs. These receptors project via identified neural pathways to the central nervous system and cause release of a suite of neuro-modulators, among which serotonin initiates phase transition through its action on neural circuits controlling behavior (Simpson et al. 2001; Rogers et al. 2003, 2004; Anstey et al., unpublished).

Phase characteristics, including behavior, not only change within the life of an individual, they also accumulate epigenetically across generations (Simpson et al. 1999; Simpson and Miller 2007). Solitarious females produce hatchlings that are behaviorally gregarized to an extent that reflects the degree and recency of maternal crowding. If crowded for the first time at the time of laying her eggs, the mother will produce fully gregariously behaving offspring. In contrast, if a gregarious female finds herself alone when laying eggs, she will produce partially behaviorally solitarized young (Islam et al. 1994a,b; Bouaïchi et al. 1995). The gregarizing effect is mediated by a chemical which the mother produces in her reproductive accessory glands and adds to the egg foam in which she lays her eggs in the soil (McCaffery et al. 1998). In effect, female locusts use their own experience of being crowded to predict the population density that their young will experience upon emerging from the egg and predispose them to behave appropriately. As a result phase changes accumulate across generations.

Group Formation

Behavioral phase change within individuals sets up a positive feedback loop, which under appropriate environmental conditions promotes the rapid transition of a population from the solitarious to the gregarious phase. If they can, solitarious locusts will avoid each other. However, if the environment forces them to come together, close contact between individuals will rapidly induce the switch from avoidance to active aggregation, which will in turn promote further gregarization and lead to formation of groups. Given that gregarious phase locusts are migratory and move together, either as marching bands of juveniles or swarms of winged adults, there is the likelihood that local groups coalesce, ultimately seeding the formation of massive regional swarms. In contrast, when previously aggregated individuals become separated, they will begin to solitarize, hence reducing their tendency to aggregate and so promoting further solitarization. If the habitat tends to keep locusts apart, then this will ultimately lead to resolitarization of a gregarious population. Interestingly, the switch from solitarious to gregarious occurs more rapidly than the reverse transition (Roessingh and Simpson 1994), indicating a hysteresis effect.

Small-scale features of resource distribution determine the extent to which phase change occurs in a local population of desert locusts. Clumping of resources such as food plants, roosting sites, and areas of favorable microclimate encourages solitarious locusts to come together and as a consequence to gregarize and aggregate (Bouaïchi et al. 1996). The degree of clumping of food plants in the parental environment in turn influences the phase state of the offspring (Despland and Simpson 2000a).

The relationships between resource distribution, resource abundance, and locust population size have been explored using individual-based computer simulations, parameterized using experimental data from locusts (Collett et al. 1998) . The extent of gregarization within a simulated population increases with rising locust population density and increasing clumping of food resources. Critical zones at which solitarious populations gregarize precipitously appear in the model across particular combinations of resource abundance, resource distribution and population size. Subsequent experimental data support the predictions from the simulation model (Despland et al. 2000).

The spatial pattern of food distribution interacts with the nutritional quality of foods to determine the spread of phase change within local populations (Despland and Simpson 2000b). Nutritional effects are mediated through differences in locust movement (Simpson and Raubenheimer 2000). Insects provided with poor quality food patches are highly active and are likely to contact one another and gregarize even when food patches are not clumped. In contrast, locusts with nutritionally optimal food patches do not move far after feeding, resulting in limited physical interactions between individuals, even when food patches are highly clumped.

It is clear that small-scale features of the habitat such as resource abundance, quality and distribution either promote or impede phase change within local populations. The same pattern seems to apply at intermediate scales of a small number of kilometers (Babah and Sword, 2004) but at higher spatial scales the relationship between vegetation distribution and desert locust outbreaks changes as different ecological processes come into play. At the scale of individual plants, a fragmented habitat with multiple dispersed patches encourages solitarization, whereas at the landscape scale the pattern is reversed: habitat fragmentation brings migrating locusts together and encourages outbreaks (Despland et al. 2004).

Understanding patterns of collective movement across local to landscape scales requires answering two questions: what causes bands of marching hoppers (the juvenile stages) and flying adults to remain as cohesive groups, and what causes them to move synchronously and collectively between patches at different scales?

Collective Movement

Locust aggregations will build into major outbreaks only if locally gregarized populations remain together and move collectively into neighboring areas of habitat, where they can recruit further locusts to the growing band. Unless such cohesive movement occurs, local aggregations will disband and individuals will return to the solitarious phase.

Within marching bands of juvenile locusts, individuals tend to synchronize and align their directions of travel with those of near neighbors (Despland and Simpson 2006). It had been shown in the laboratory that marching begins only at high locust densities (Ellis, 1951), but these experiments did not measure how and why alignment increases with density to the point that an aggregation of locusts suddenly commences collective marching.

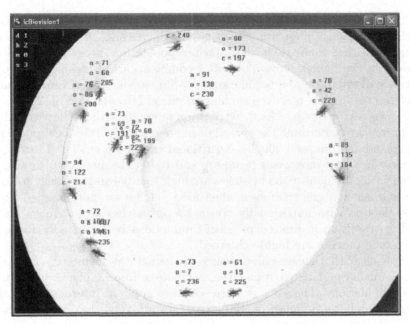

Fig. 5. An image from the Mexican hat marching arena and tracking software used in Buhl et al.'s (2006) study of collective marching in gregarious locusts. For a movie see http://www.sciencemag.org/cgi/content/full/312/5778/1402/DC1.

This problem has recently been studied by Jerome Buhl and colleagues (2006) by modeling locusts as self-propelled particles (SPP), each 'particle' adjusting its speed and/or direction in response to near neighbors. The model developed by Vicsek et al. (1995) was used because of its small number of underlying assumptions and the strength of the universal features it predicts. A central prediction from the model is that as the density of animals in the group increases, a rapid transition occurs from disordered movement of individuals within the group to highly aligned collective motion. Since SPP models underlie many theoretical predictions about how groups form complex patterns, avoid predators, forage, and make decisions, confirming such a transition for real animals has fundamental implications for understanding all aspects of collective motion. It is also particularly important in the case of locusts as it could explain the sudden appearance of mobile swarms.

Buhl et al.'s experiments involved studying marching in the laboratory in a ring-shaped arena, rather like a Mexican hat in shape, with a central dome to restrict optical flow in the direction opposite to that of individual motion. For data analysis, Iain Couzin developed an automated digital tracking system, allowing the simultaneous analysis of group-level and individual-level properties, which is technically extremely challenging but essential for discovering the link between these levels of organization (Fig. 5).

Juvenile locusts readily formed highly coordinated marching bands under laboratory conditions when placed in the Mexican hat arena. Individuals selected collectively either a clockwise or counter-clockwise direction of travel (the choice of which was random) and maintained this for extended periods. Experiments were conducted in which the numbers of locusts in the arena ranged from 5 to 120 insects (densities of 13 to 295 m^2). The locusts' motion was recorded for eight hours and the resulting data were processed using the tracking software to compute the position and orientation of each locust.

Coordinated marching behavior depended strongly on locust density (Figs. 6, 7). At low densities (2 to 7 locusts in the arena, equating to 5 to 17 locusts per m^2) there was a low incidence of alignment among individuals. In trials where alignment did occur it did so only sporadically and after long initial periods of disordered motion. Intermediate densities (10 to 25 locusts; 25 to 62 per m^2) were characterized by long periods of collective marching with rapid, spontaneous reversals in rotational direction. At densities higher than 74 per m^2 (30 or more locusts in the arena) spontaneous changes in direction did not occur, with the locusts quickly adopting and maintaining a common rotational direction.

Hence Buhl et al.'s experiments confirmed the theoretical prediction from the SPP model of a rapid transition from disordered to ordered movement (Figs. 6, 7) and identified a critical density for the onset of coordinated marching in juvenile locusts. In the field, small increases in density past this threshold would be predicted to result in a sudden transition to a highly unpredictable collective motion, making control measures difficult to implement. The experiments also demonstrated a dynamic instability in motion at densities typical of locusts in the field, whereby groups can switch direction without external perturbation, potentially facilitating rapid transfer of directional information. Buhl et al.'s data and model also suggest that predicting the motion of very high densities is easier than predicting that of intermediate densities.

Of course, it cannot be assumed that all of the collective behavior seen in laboratory experiments translates directly to that observed in the field. However, the wealth of mathematical and simulation-based understanding of SPP models provides tools for performing such scaling. In combination with the detailed understanding of the role of the environment in behavioral phase change, as discussed above, SPP models could now form the basis of prediction to improve control of locust outbreaks.

3.3 Mormon Crickets

As we have noted, superficial similarities in group-level characteristics of biological systems may mask subtle, but important, underlying differences among them. This scenario rings true for mass-migrating Mormon crickets (*Anabrus simplex*). Just like locusts, Mormon crickets form cohesive migratory bands during outbreak periods that march en masse across the landscape (Fig. 8a).

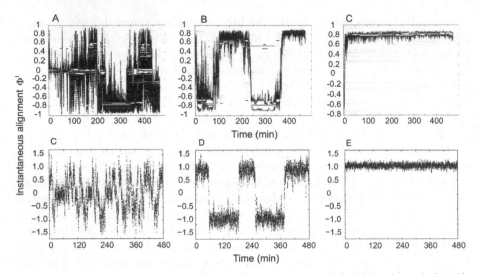

Fig. 6. Similarity between the self-propelled particles model of Vicsek et al. (1995) and experimental data as density of locusts in the arena was manipulated: (A) 7, (B) 20 and (C) 60 individuals in the arena (from Buhl et al. 2006). See text for explanation.

These bands can be huge, spanning over ten kilometers in length, several in width, containing dozens of insects per square meter, and capable of traveling up to 2.0 km per day (Cowan 1929; Lorch et al. 2005). Mormon cricket bands can cause serious damage when they enter crop systems and usually elicit prompt chemical control measures when they appear.

Although studied far less than locusts, laboratory and field analyses of Mormon cricket migratory behavior have provided important insights into the mechanisms underlying group formation and subsequent collective movement patterns. In addition, Mormon crickets have served as a key study system in the development of the nascent field of insect radiotelemetry in which the movement patterns of individual insects can be tracked across the landscape using small radiotransmitters. The use of this technology has enabled the study of landscape-scale collective movement to move beyond descriptions of observed patterns and into the realm of empirical hypothesis testing using manipulative field experiments.

Despite their name, Mormon crickets are not true crickets, but rather are classified as katydids or bush-crickets. They are flightless throughout their lives and possess small vestigial wings used by males for sound production and mate attraction (Gwynne 2001). As a result, they are incapable of forming flying swarms and travel on the ground as both juveniles and adults. Their religious name originates from a now legendary incident that occurred in the spring of 1848 involving the first Mormon settlers to arrive in the Great Salt Lake Valley in the western US. After surviving a difficult westward jour-

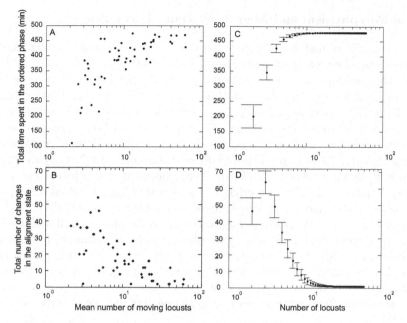

Fig. 7. The relation between the average number of moving locusts and the mean total time spent in the aligned state (A and C) and the mean number of changes in the alignment state (B and D) are displayed on a semi-log scale. Error bars, standard deviation. The 'ordered phase' refers to periods where the insects exhibited high alignment (> 0.3), and thus were moving collectively in one direction (either clockwise or anti-clockwise). From (Buhl et al. 2006).

ney and ensuing winter, the pioneers were enjoying what appeared to be a bountiful first spring in their newly established homeland. This serenity was shattered when their fields, planted with over 5,000 acres of wheat, corn and vegetables were invaded by marching hoards of large black 'crickets' that set upon their standing young crops (Hartley 1970). The devout surely interpreted this assault as an act of God analogous to the well-known Biblical plagues of Old World locusts. The settlers' attempts to battle the crickets using sticks, shovels, brooms, fire and trenches were futile, but their prayers for relief were answered by the arrival of seagulls that flew in from the Great Salt Lake and began to devour the marauding crickets. The gulls reportedly gorged themselves on crickets in the fields, often to the point of regurgitation, after which they would return to feast again (Hartley 1970). The gulls were credited with saving the remaining crops, and by extension the first settlers; a multi-trophic level interaction that resulted in the California Sea Gull being selected as the state bird of Utah. The Miracle of the Gulls was also commemorated by the erection of a monument at the headquarters of the Mormon Church in Salt Lake City, one of the few monuments, if not the only one, in the world dedicated to an insect predator (Gwynne, 2001).

Phase Polyphenism and Migratory Band Formation

Until recently, it had been widely assumed that Mormon crickets express density-dependent phase polyphenism similar to that known to occur in locusts. This assumption was due in large part to the similarities between migratory bands of locusts and those of Mormon crickets. The possibility of phase polyphenism in Mormon crickets was further supported by observed phenotypic differences in migratory behavior, coloration and body size between individuals from low-density, non-outbreak populations and their counterparts in high-density, band-forming populations (MacVean, 1987; Gwynne, 2001; Lorch and Gwynne, 2000). MacVean (1987) noted that the formation of migratory bands in the Mormon cricket "bears a striking resemblance to phase transition in the African plague locusts," and Cowan (1990) described the Mormon cricket as having gregarious and solitarious phases similar to locusts. Mormon crickets and locusts also share phase-related terminology in the scientific literature with Mormon crickets in non-outbreak populations, commonly referred to as inactive solitary forms (i.e. solitarious phase), whereas those in band-forming populations are referred to as gregarious forms (e.g. Wakeland 1959; MacVean 1987, 1990; Lorch and Gwynne 2000; Gwynne 2001; Bailey et al. 2005).

Two lines of recent evidence suggest that the expression density-dependent phase polyphenism in Mormon crickets plays little if any role in either the initial formation of migratory bands or the observed phenotypic differences between insects from high-density band-forming and low-density non-band-forming populations. Sword (2005) failed to find an endogenous effect of rearing density on Mormon cricket movement behavior in the lab, but rather demonstrated that individual movement was induced simply by the short-term presence of other nearby conspecifics. Although the lack of a behavioral phase change does not rule out the possibility of density-dependent changes in other traits, a recent phylogeographic analysis of genetic population structure suggests considerable divergence between the migratory and non-migratory forms (Bailey et al., 2005). Thus, the differences between crickets in migratory and non-migratory populations could primarily be due to genetic differences rather than the expression of phase polyphenism mediated by differences in population density.

Taken together, these studies strongly suggest, in contrast to the case with locusts, that the expression of phase polyphenism is not involved in the formation of Mormon cricket migratory bands. In other words, the expression phase polyphenism in not a prerequisite for migratory band formation.

Collective Movement

The initial formation of migratory bands in Mormon crickets and locusts appears to have convergently evolved via different underlying behavioral mechanisms. Is the same true for the mechanisms governing patterns of collective movement once these groups have formed? Are there general rules applicable to the movement patterns of both Mormon crickets and locust bands (not to mention other organisms), or do these differ as well? The answers to these questions have important implications for the broader understanding of collective animal movement as well as considerable practical implications for the development of predictive movement models that can aid in the management of these and other migratory pests.

Given that the frequency of contact among individuals will increase with local population density, the finding that Mormon cricket movement is induced by immediate behavioral interactions among nearby individuals predicts that there should be some threshold population density above which mass movement is induced (Sword 2005). Although this remains to be demonstrated in Mormon crickets, the recent application of SPP models by Buhl et al. (2006) to explain the induction of mass movement in locusts with increasing local density stands as a promising general framework to explain the onset of mass movement in Mormon cricket bands as well. Furthermore, as we shall discuss in detail later, understanding how individual insects contend with the ecological costs and benefits of living in a group has provided considerable insight into the general mechanisms that may drive migratory band movement.

Radiotelemetry is an extremely valuable tool available to biologists for tracking the movement patterns of individual animals in the wild. The approach has traditionally been limited to larger vertebrates capable of carrying the extra weight of a radiotransmitter. However, technological advances have reduced the size of transmitters such that they can be used to track the movements of individual insects on the ground (e.g. Lorch and Gwynne 2000; Lorch et al. 2005) (Fig. 8b) as well as in flight (Wikelski et al., 2006). Lorch and Gwynne (2000) first demonstrated the utility of small radiotransmitters to track individual Mormon crickets. Their study was followed by a similar, but much more rigorous analysis by Lorch et al. (2005) who compared the individual movement patterns of insects from several different band-forming and non-band-forming populations. These studies confirmed that Mormon crickets in migratory bands cover much greater distances (up to 2 km/day) and tend to move collectively in the same direction relative to insects from low-density, non-band-forming populations (Fig. 8c, d).

In addition to consistent group directionality within as well as across days, migratory bands also exhibit group-level turns in which similar direction changes are made by individuals regardless of their position in the band (Lorch et al. 2005). Two possible explanations for these synchronous turns are that either (i) group movement direction is determined by orientation towards some landscape-scale environmental cue such as wind direction that can be detected and responded to by all group members, or (ii) they are similar to turns in bird flocks or fish schools in which individuals adjust their direction in response to the movement of near neighbors and these turns are propagated through the group like a wave (Couzin and Krause 2003). Although the Lorch et al. (2005) experiment was not designed to examine the effect of wind

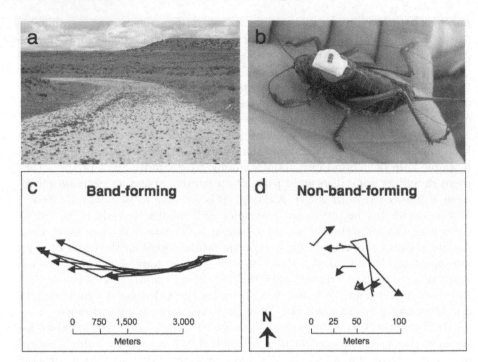

Fig. 8. Collective movement in Mormon cricket migratory bands. (a) A large migratory band crossing a dirt road in northeastern Utah, USA (photo by G. Sword). (b) A female Mormon cricket affixed with a small radiotransmitter (photo by D. Gwynne). (c) Example of individual movement patterns by radiotracked Mormon crickets in a high-density, migratory-band-forming population. Each line represents a single individual and each line segment depicts one day of movement. (d) Examples of radiotracked Mormon cricket movement patterns in a low-density, non-band-forming population. Note the differences in group directionality and scale of movement between the band-forming and non-band-forming populations. Radiotracking examples are from Lorch et al. (2005).

direction on migratory band movement, local wind direction data collected concurrently with the radiotracking data hinted that wind directions early in the day might correlate with migratory band directions. However, no effect whatsoever of wind direction on the movement of individuals within migratory bands was found in a follow-up study specifically designed to test the wind direction hypothesis. Migratory bands simultaneously tracked at three nearby sites in the same vicinity were found to travel in distinctly different directions despite experiencing very similar wind directions and other weather conditions (Sword et al., unpublished).

So what cues determine the direction in which a migratory band will move? One possible answer provided by simulation models of collective animal movement patterns is that nothing is responsible. A variety of movement models

in which individuals modify their direction and movement rate in response
to others have shown that group directionality can arise from inter-individual
interactions as a result of self-organization (Krause and Ruxton 2002; Couzin
and Krause 2003). The hypothesis that Mormon cricket migratory band movement direction and distance are collectively determined was tested by conducting a manipulative transplant experiment in the field as originally described in
Sword et al. (2005). Insects traveling in naturally occurring migratory bands
were captured and radiotracked. Half of these insects were released back into
the band while the other half were transported and released at a nearby site
where bands had previously been, but were absent at the time. The resulting differences in movement patterns between the crickets released into the
migratory band versus those that were isolated from the band were dramatic
and closely resembled the previously documented differences between crickets from band-forming versus non-band-forming populations shown in Figs. 8c
and d. Insects isolated from the band moved shorter distances, and were much
less directional as a group relative to the insects released back into the band
(Sword et al., unpublished). These findings quite clearly show that the distance and direction traveled by insects in a migratory band are group-level
properties that differ considerably from the movement patterns of individuals
when they are removed from the social context of the band.

A Forced March Driven by Cannibalism

Mormon crickets provide a unique model system in which understanding the
costs and benefits of migratory band formation has provided a unifying framework that explains both how and why inter-individual interactions can lead
to landscape-scale mass movement. The evolution and maintenance of migratory band formation in insects necessarily requires the benefits of such a
strategy to outweigh its costs in terms of individual survival and reproduction.
The radiotelemetry-based transplant study of Sword et al. (2005) mentioned
above was originally designed to study collective movement, but it unexpect-
edly yielded a critical insight into the benefits and selection pressures that
favor the formation of migratory bands. Individual band members were much
less likely to be killed by predators than were crickets that had been separated
from the group. The precise mechanisms by which individuals in bands gain
protection from predators were not identified (see Krause and Ruxton 2002 for
potential mechanisms), but 50–60% of the crickets removed from migratory
bands were killed by predators in just two days while none within the bands
were harmed during the same period. Thus, migratory bands form as part of
an anti-predator strategy and there is a very strong adaptive advantage to
staying in the group.

   Although migratory bands confer anti-predator benefits, living in a huge
group of conspecifics also has a variety of potential costs (see Krause and
Ruxton 2002). It is precisely the interplay between these costs and benefits
that promotes cohesive and coordinated mass movement among individual

Mormon crickets living in bands. Recent field experiments revealed that individual band members are subject to increased intraspecifc competition for nutritional resources. Individual crickets within migratory bands were shown to be deprived of specific nutrients, namely protein and salt (Simpson et al. 2006). When provided with augmented dietary protein, individual crickets spent less time walking; a response that was not found when crickets had ample carbohydrate. Thus, group movement results in part from locomotion induced by protein deprivation and should act to increase the probability that individual band members will encounter new resources and redress their nutritional imbalances.

An additional cost of group formation is that Mormon crickets are notoriously cannibalistic (MacVean 1987; Gwynne 2001). Their propensity to cannibalize is a function of the extent to which they are nutritionally deprived. Given that Mormon crickets are walking packages of protein and salt, the insects themselves are often the most abundant source of these nutrients in the habitat. As a result, individuals within the band that fail to move risk being attacked and cannibalized by other nutritionally deprived crickets approaching from the rear (Simpson et al. 2006). Thus, the mass movement of individuals in migratory bands is a forced march driven by cannibalism due to individuals responding to their endogenous nutritional state. The fact that migratory bands are maintained as cohesive groups despite these seemingly dire conditions suggests that the risk of predation upon leaving the band must outweigh the combined costs of intraspecific competition for resources and cannibalism. Importantly, ongoing experimental work strongly implicates the threat of cannibalism as a general mechanism that mediates migratory band movement in locusts as well as Mormon crickets (Couzin et al., unpublished).

4 Concluding Remarks

We have discussed in detail three cases of collective movement of large groups of animals: honeybees, locusts and Mormon crickets. A description of their movement would yield striking similarities: individuals in the group seem to keep an almost fixed distance from their neighbors; they tend to align themselves with their nearest neighbor, and show a clear tendency to stay with the group. In fact, this description can easily be extended to many other animals that move in groups, such as to schools of fish and flocks of birds (Couzin et al. 2005). However, the *reasons* for their collective movement are fundamentally different. For individual bees in a swarm it is critically important to stay with the swarm, as an individual bee cannot live. Cohesive movement in locusts is induced by the fine-scale structure of the environment they find themselves in. And if you are a Mormon cricket, moving faster than the ones behind you is essential to prevent yourself from becoming your neighbor's next meal.

We as biologists are fascinated by nature's diverse tapestry. Often, biologists tend to argue that nature is too diverse to allow its manifestations to be captured by generalist models. This is not the message that we want to convey in this chapter. As we have illustrated, many behaviors can only be understood by constructing models, which, by definition, are an abstract representation of reality. It is helpful to think about unifying theories that have the power to explain behaviors across a range of biological systems. We encourage computer scientists and mathematicians to look at biological systems and to become inspired, see patterns, and seek applications beyond biological systems. But in doing so, we hope that researchers will be awed not by the superficial similarities between natural systems but by the intricate and often subtle differences that distinguish them.

References

1. Avitabile, A., Morse, R. A. and Boch, R. (1975) Swarming honey bees guided by pheromones. *Annals of the Entomological Society of America*, 68:1079–1082.
2. Babah, M. A. O. and Sword, G. A. (2004) Linking locust gregarization to local resource distribution patterns across a large spatial scale. *Environmental Entomology*, 33:1577–1583.
3. Bailey, N. W., Gwynne, D. T. and Ritchie, M. G. (2005) Are solitary and gregarious Mormon crickets (*Anabrus simplex*, Orthoptera, Tettigoniidae) genetically distinct? *Heredity*, 95:166–173.
4. Ball, P. (1999) The self-made tapestry. Pattern formation in nature. Oxford University Press, Oxford UK, 287 pp.
5. Beckers, R., Deneubourg, J. L. and Goss, S. (1993) Modulation of trail laying in the ant *Lasius niger* (Hymeoptera: Formicidae) and its role in the collective selection of a food source. *Journal of Insect Behavior*, 6:751–759.
6. Beckers, R., Deneubourg, J. L., Gross, S. and Pasteels, J. M. (1990) Collective decision making through food recruitment. *Insectes Sociaux*, 37:258–267.
7. Beekman, M., Fathke, R. L. and Seeley, T. D. (2006) How does an informed minority of scouts guide a honey bee swarm as it flies to its new home? *Animal Behaviour*, 71:161–171.
8. Beekman, M., Gilchrist, A. L., Duncan, M. and Sumpter, D. J. T. (2007) What makes a honeybee scout? *Behavioral Ecology and Sociobiology*, in press.
9. Beekman, M. and Ratnieks, F. L. W. (2000) Long range foraging by the honeybee *Apis mellifera* L. *Functional Ecology*, 14:490–496.
10. Beekman, M., Sumpter, D. J. T. and Ratnieks, F. L. W. (2001) Phase transition between disorganised and organised foraging in Pharaoh's ants. *Proceedings of the National Academy of Science of the United States of America*, 98:9703–9706.
11. Beekman, M., Sumpter, D. J. T., Seraphides, N. and Ratnieks, F. L. W. (2004) Comparing foraging behaviour of small and large honey bee colonies by decoding waggle dances made by foragers. *Functional Ecology*, 18:829–835.
12. Ben-Jacob, E., Cohen, I. and Levine, H. (2000) Cooperative self-organization of microorganisms. *Advances in Physics*, 49:395-554.

13. Ben-Jacob, E. and Levine, H. (2001) The artistry of nature. *Nature*, 409:985–986.

14. Berthold, P., Helbig, A. J., Mohr, G. and Querner, U. (1992) Rapid microevolution of migratory behaviour in a wild bird species. *Nature*, 360:668–670.

15. Berthold, P., Querner, U. (1981) Genetic basis of migratory behaviour in European warblers. *Science*, 212:77–79.

16. Boinski, S., Garber, P. A. (2000) On the move: how and why animals travel in groups. The University of Chicago Press, Chicago.

17. Bonabeau, E., Theraulaz, G., Deneubourg, J.-L., Aron, S. and Camazine, S. (1997) Self-organization in social insects. *Trends in Ecology and Evolution*, 12:188–193.

18. Bouaichi, A., Roessingh, P. and Simpson, S. J. (1995) An analysis of the behavioural effects of crowding and re-isolation on solitary-reared adult desert locusts (*Schistocerca gregaria*, Forskal) and their offspring. *Physiological Entomology*, 20:199–208.

19. Bouaichi, A., Simpson, S.J. and Roessingh, P. (1996) The influence of environmental microstructure on the behavioural phase state and distribution of the desert locust. *Physiological Entomology*, 21:247-256.

20. Buhl, J., Sumpter, D. J. T., Couzin, I. D., Hale, J. J., Despland, E., Miller, E. R. and Simpson, S. J. (2006) From disorder to order in marching locusts. *Science*, 312:1402–1406.

21. Camazine, S. (1991) Self-organizing pattern formation on the combs of honey bee colonies. *Behavioral Ecology and Sociobiology*, 28:61–76.

22. Camazine, S., Deneubourg, J.L., Franks, N. R., Sneyd, J., Theraulaz, G. and Bonabeau, E. (2001) Self-organization in biological systems. Princeton University Press, Princeton and Oxford.

23. Camazine, S., Visscher, P. K., Finley, J. and Vetter, R. S. (1999) House-hunting by honey bee swarms: collective decisions and individual behaviors. *Insectes Sociaux*, 46:348–362.

24. Collett, M., Despland, E., Simpson, S. J. and Krakauer, D. C. (1998) Spatial scales of desert locust gregarization. *Proceedings of the National Academy of Sciences of the United States of America*, 95:13052–13055.

25. Couzin, I. D., Krause, J., Franks, N. R. and Levin, S. A. (2005) Effective leadership and decision making in animal groups on the move. *Nature*, 455:513–516.

26. Couzin, I. D. and Krause, J. K. (2003) Self-organization and collective behavior in vertebrates. *Advances in the Study of Behavior*, 32:1–75.

27. Cowan, F. T. (1929) Life history, habits, and control of the Mormon cricket. *USDA Technical Bulletin*, 161:1–28.

28. Cowan, F. T. (1990) The Mormon Cricket Story. Montana State University, *Agricultural Experiment Station Special Report*, 31:7–42.

29. Darlington, J. P. E. C. (1982) The underground passages and storage pits using in foraging by a nest of the termite Macrotermes michaelseni in Kajiado, Kenya. *Journal of Zoology, London*, 198:237–247.

30. Darwin, C. (1872) The origin of species, 6 edn. John Murray, London.

31. Deneubourg, J.-L., Aron, S., Goss, S., Pasteels, J. M. and Duerinck, G. (1986) Random behaviour, amplification processes and number of participants: how they contribute to the foraging properties of ants. *Physica D*, 22:176–186.

32. Deneubourg, J.-L., Pasteels, J. M. and Verhaeghe, J. C. (1983) Probabilistic behaviour in ants: a strategy of errors? *Journal of Theoretical Biology*, 105:259–271.

33. Despland, E. and Simpson, S. J. (2000a) Small-scale vegetation patterns in the parental environment influence the phase state of hatchlings of the desert locust. *Physiological Entomology*, 25:74–81.

34. Despland, E. and Simpson, S. J. (2000b) The role of food distribution and nutritional quality in behavioural phase change in the desert locust. *Animal Behaviour*, 59:643–652.

35. Despland, E. and Simpson, S. J. (2006) Resource distribution mediates synchronization of physiological rhythms in locust groups. *Proceedings of the Royal Society of London, Series B*, 273:1517–1522.

36. Despland, E., Collett, M. and Simpson, S. J. (2000) Small-scale processes in Desert Locust swarm formation: how vegetation patterns influence gregarization. *Oikos*, 88:652–662.

37. Despland, E., Rosenberg, J. and Simpson, S. J. (2004) Landscape structure and locust swarming: a satellite's view. *Ecography*, 27:381–391.

38. Detrain, C. and Deneubourg, J.-L. (2002) Complexity of environment and parsimony of decision rules in insect societies. *Biological Bulletin*, 202:268–274.

39. Dorigo, M. and Di Caro, G. (1999) The ant colony optimization meta-heuristic. In: D. Corne MDFG (ed), *New Ideas in Optimization*, McGraw-Hill, pp 11–32.

40. Dorigo, M., Maniezzo, V. and Colorni, A. (1996) The ant system: optimization by a colony of cooperating agents. *IEEE Transactions on Systems, Man and Cybernetics – Part B*, 26:29–41.

41. Dyer, F. C. (2000) Group movement and individual cognition: lessons from social insects. In: Boinski S, Garber PA (eds), *On the Move: How and Why Animals Travel in Groups*, The University of Chicago Press, Chicago, pp 127–164.

42. Ellis, P. E. (1951) The marching behaviour of hoppers of the African migratory locust (*Locusta migratoria migratorioides* R. & F.) in the laboratory. *Anti-Locust Bulletin*, 7, 46 pp.

43. Fewell, J. H. (1988) Energetic and time costs of foraging in harvester ants, *Pogonomyrmex occidentalis*. *Behavioral Ecology and Sociobiology*, 22:401–408.

44. Franks, N. R., Pratt, S. C., Mallon, E. B., Britton, N. F. and Sumpter, D. J. T. (2002) Information flow, opinion polling and collective intelligence in house-hunting social insects. *Philosophical Transactions of the Royal Society of London Series B*, 357:1567–1583.

45. Frisch von, K. (1967) The dance language and orientation of bees. Harvard University Press, Cambridge, MA.

46. Grünbaum, D. (1998) Schooling as a strategy for taxis in a noisy environment. *Evolutionary Ecology*, 12:503–522.

47. Gwynne, D. T. (2001) Katydids and Bush-Crickets: reproductive behavior and evolution of the Tettigoniidae. Cornell University Press, Ithaca.

48. Hartley, W. G. (1970) Mormons, crickets, and gulls: A new look at an old story. *Utah Historical Quarterly*, 38:224–239.

49. Helbing, D., Farkas, I. and Vicsek, T. (2000) Simulating dynamical features of escape panic. *Nature*, 407:487–490.

50. Islam, M. S., Roessingh, P., Simpson, S. J. and McCaffery, A. R. (1994a) Effects of population density experienced by parents during mating and oviposition

on the phase of hatchling desert locusts. *Proceedings of the Royal Society of London B*, 257:93–98.

51. Islam, M. S., Roessingh, P., Simpson, S. J. and McCaffery, A. R. (1994b) Parental effects on the behaviour and coloration of nymphs of the desert locust, *Schistocerca gregaria. Journal of Insect Physiology*, 40:173–181.

52. Jaffe, K. and Deneubourg, J.-L. (1992) On foraging, recruitment systems and optimum number of scouts in eusocial colonies. *Insectes Sociaux*, 39:201–213.

53. Janson, S., Middendorf, M. and Beekman, M. (2005) Honey bee swarms: How do scouts guide a swarm of uninformed bees? *Animal Behaviour*, 70:349–358.

54. Janson, S., Middendorf, M., Beekman, M. (2007) Searching for a new home — scouting behavior of honeybee swarms. *Behavioral Ecology*, 18:384–392.

55. Kondo, S. and Asai, R. (1995) A reaction-diffusion wave on the skin of the marine angelfish Pomacanthus. *Nature*, 376:765–768.

56. Krause, J. and Ruxton, G. D. (2002) Living in groups. Oxford University Press, Oxford.

57. Levine, H. and Ben-Jacob, E. (2004) Physical schemata underlying biological pattern formation — examples, issues and strategies. *Physical Biology*, 1:14–22.

58. Lindauer, M. (1955) Schwarmbienen auf Wohnungssuche. *Zeitschrift für vergleichende Physiologie*, 37:263–324.

59. Lorch, P. D. and D. T. Gwynne (2000) Radio-telemetric evidence of migration in the gregarious but not the solitary morph of the Mormon cricket (*Anabrus simplex*: Orthoptera: Tettigoniidae). *Naturwissenschaften*, 87:370–372.

60. Lorch, P. D., Sword, G. A., Gwynne, D. T. and Anderson, G. L. (2005) Radiotelemetry reveals differences in individual movement patterns between outbreak and non-outbreak Mormon cricket populations. *Ecological Entomology*, 30:548–555.

61. MacVean, C. M. (1987) Ecology and management of Mormon cricket, *Anabrus simplex* Haldeman. In J. L. Capinera (ed), *Integrated pest management on rangeland: a shortgrass prarie perspective*, Westview Press, Boulder, pp 116–136.

62. MacVean, C. M. (1990) Mormon crickets: A brighter side. *Rangelands*, 12:234–235.

63. Mallon, E. B., Pratt, S. C. and Franks, N. R. (2001) Individual and collective decision-making during nest site selection by the ant *Leptothorax albipennis*. *Behavioral Ecology and Sociobiology*, 50:352–359.

64. McCaffery, A. R., Simpson, S. J., Islam, M. S. and Roessingh, P. (1998) A gregarizing factor present in egg pod foam of the desert locust *Schistocerca gregaria. Journal of experimental Biology*, 201:347–363.

65. McKay, D. S., Gibson, E. K., Thomas-Keprta, K. L., Vali, H., Romanek, C. S., Clemett, S. J., Chillier, X. D. F., Maechling, C. R. and Zare, R. N. (1996) Search for past life on Mars: possible relic biogenic activity in Martian meteorite ALH84001. *Science*, 273:924–930.

66. Myerscough, M. R. (2003) Dancing for a decision: a matrix model for nest-site choice by honeybees. *Proceedings of the Royal Society of London Series B*, 270:577–582.

67. Neill, W. H. (1979) Mechanisms of fish distribution in heterothermal environments. *American Zoologist*, 19:305–317.

68. Nicolis, G. and Prigogine, I. (1977) Self-organization in nonequilibrium systems. From dissipative structures to order through fluctuations. John Wiley & Sons, Inc.

69. Partridge, L. W., Partridge, K. A. and Franks, N. R. (1997) Field survey of a monogynous leptohoracine ant (Hymenoptera, Formicidae): evidence of seasonal polydomy? *Insectes Sociaux*, 44:75–83.

70. Pasteels, J. M., Deneubourg, J.-L., Verhaeghe, J.-C., Boevé, J.-L. and Quinet, Y. (1986) Orientation along terrestrial trails by ants. In Payne, T. and Birch, M. (eds), *Mechanisms in Insect Olfaction*, Oxford University Press, Oxford, pp 131–138.

71. Pener, M. P. and Yerushalmi, Y. (1998) The physiology of locust phase polymorphism: an update. *Journal of Insect Physiology*, 44:365–377.

72. Pratt, S. C., Mallon, E. B., Sumpter, D. J. T. and Franks, N. R. (2002) Quorum sensing, recruitment, and collective decision-making during colony emigration by the ant *Leptothorax albipennis*. *Behavioral Ecology and Sociobiology*, 52:117–127.

73. Pratt, S. C., Sumpter, D. J. T., Mallon, E. B. and Franks, N. R. (2005) An agent-based model of collective nest choice by the ant *Temnothorax albipennis*. *Animal Behaviour*, 70:1023–1036.

74. Quinet, Y., de Biseau, J. C. and Pasteels, J. M. (1997) Food recruitment as a component of the trunk-trail foraging behaviour of *Lasius fuliginosus* (Hymenoptera: Formicidae). *Behavioural Processes*, 40:75–83.

75. Reebs, S. G. (2000) Can a minority of informed leaders determine the foraging movements of a fish shoal? *Animal Behaviour*, 59:403–409.

76. Roessingh, P. and Simpson, S. J. (1994) The time-course of behavioural phase change in nymphs of the desert locust, *Schistocerca gregaria*. *Physiological Entomology*, 19:191–197.

77. Rogers, S. M., Matheson, T., Despland, E., Dodgson, T., Burrows, M. and Simpson, S. J. (2003) Mechanosensory-induced behavioural gregarization in the desert locust, *Schistocerca gregaria*. *Journal of Experimental Biology*, 206:3991–4002.

78. Rogers, S. M., Matheson, T., Sasaki, K., Kendrick, K., Simpson, S. J. and Burrows, M. (2004) Substantial changes in central nervous neurotransmitters and neuromodulators accompany phase change in the locust. *Journal of Experimental Biology*, 207:3603–3617.

79. Rohrseitz, K. and Tautz, J. (1999) Honey bee dance communication: waggle run direction coded in antennal contacts? *Journal of Comparative Physiology A*, 184:463–470.

80. Rosengren, R. and Sundström, L. (1987) The foraging system of a red wood ant colony (Formica s. str) — collecting and defending food through an extended phenotype. In: Pasteels J.M., Deneubourg J.L. (eds), *From Individual to Colelctive Behavior in Social Insects*, vol 54, Birkhäuser, Basel, pp 117–137.

81. Schneider, S. S. and McNally, L. C. (1993) Spatial foraging patterns and colony energy status in the African honey bee, *Apis mellifera scutellata*. *Journal of Insect Behaviour*, 6:195–210.

82. Seeley, T. D. (1982) Adaptive significance of the age polyethism schedule in honeybee colonies. Behavioral Ecology and Sociobiology, 11:287–293.

83. Seeley, T. D. (1983) Division of labor between scouts and recruits in honeybee foraging. *Behavioral Ecology and Sociobiology*, 12:253–259.

84. Seeley, T. D. (1995) The wisdom of the hive. Harvard University Press, Cambridge, MA.

85. Seeley, T. D. (2002) When is self-organization used in biological systems? *Biological Bulletin*, 202:314–318.

86. Seeley, T. D. (2003) Consensus building during nest-site selection in honey bee swarms: the expiration of dissent. *Behavioral Ecology and Sociobiology*, 53:417–424.
87. Seeley, T. D. and Buhrman, S. C. (1999) Group decision making in swarms of honeybees. *Behavioral Ecology and Sociobiology*, 45:19–31.
88. Seeley, T. D. and Buhrman, S. C. (2001) Nest-site selection in honey bees: how well do swarms implement the "best-of-N" decision rule? *Behavioral Ecology and Sociobiology*, 49:416–427.
89. Seeley, T. D., Camazine, S. and Sneyd, J. (1991) Collective decision-making in honey bees: how colonies choose among nectar sources. *Behavioral Ecology and Sociobiology*, 28:277–290.
90. Seeley, T. D., Mikheyev, A. S. and Pagano, G. J. (2000) Dancing bees tune both duration and rate of waggle-run production in relation to nectar-source profitability. *Journal of Comparative Physiology A*, 186:813–819.
91. Seeley, T. D., Morse, R. A., Visscher, P. K. (1979) The natural history of the flight of honey bee swarms. *Psyche*, 86:103–113.
92. Seeley, T. D. and Towne, W. F. (1992) Tactics of dance choice in honey bees: do foragers compare dances? *Behavioral Ecology and Sociobiology*, 30:59–69.
93. Seeley, T. D. and Visscher, P. K. (2004) Group decision making in nest-site selection by honey bees. *Apidologie*, 35:101–116.
94. Seeley, T. D., Weidenmüller, A. and Kühnholz, S. (1998) The shaking signal of the honey bee informs workers to prepare for greater activity. *Ethology*, 104:10–26.
95. Sendova-Franks, A. B. and Franks, N. R. (1995) Division of labour in a crisis: task allocation during colony emigration in the ant *Leptothorax unifasciatus*. *Behavioral Ecology and Sociobiology*, 36:269–282.
96. Simpson, S. J. and Miller, G. A. (2007) Maternal effects on phase characteristics in the desert locust, *Schistocerca gregaria*: an appraisal of current understanding. *Journal of Insect Physiology*, in press.
97. Simpson, S. J. and Raubenheimer, D. (2000) The Hungry Locust. *Advances in the Study of Behavior*, 29:1–44.
98. Simpson, S. J. and Sword, G. A. (2007) Phase polyphenism in locusts: mechanisms, population consequences, adaptive significance and evolution. In: Whitman, D. and Ananthakrishnan, T.N. (eds), *Phenotypic Plasticity of Insects: Mechanisms and Consequences*.
99. Simpson, S. J., Despland, E., Haegele, B. F. and Dodgson, T. (2001) Gregarious behaviour in desert locusts is evoked by touching their back legs. *Proceedings of the National Academy of Sciences, USA*, 98:3895–3897.
100. Simpson, S. J., Sword, G. A., Lorch, P. D. and Couzin, I. D. (2006) Cannibal crickets on a forced march for protein and salt. *Proceedings of the National Academy of Sciences, USA*, 103:4152–4156.
101. Sumpter, D. J. T. (2000) From bee to society: an agent-based investigation of honey bee colonies. PhD thesis, University of Manchester, Manchester.
102. Sumpter, D. J. T. (2005) The principles of collective animal behaviour. *Philosophical Transactions of the Royal Society of London, series B*, 361:5–22.
103. Sumpter, D. J. T. and Beekman, M. (2003) From non-linearity to optimality: pheromone trail foraging by ants. *Animal Behaviour*, 66:273–280.
104. Swaney, W., Kendal, J., Capon, H., Brown, C. and Laland, K. N. (2001) Familiarity facilitates social learning of foraging behaviour in the guppy. *Animal Behaviour*, 62:591–598.

105. Sword, G. A. (2005) Local population density and the activation of movement in migratory band-forming Mormon crickets. *Animal Behaviour*, 69:437–444.
106. Sword, G. A., Lorch, P. D. and Gwynne, D. T. (2005) Migratory bands give crickets protection. *Nature*, 433:703.
107. Tautz, J. and Rohrseitz, K. (1998) What attracts honeybees to a waggle dancer? *Journal of Comparative Physiology A*, 183:661–667.
108. Tautz, J., Rohrseitz, K. and Sandeman, D.C. (1996) One-strided waggle dance in bees. *Nature*, 382:32.
109. Thompson, D. W. (1917) On growth and form. University Press, Cambridge.
110. Uvarov, B. P. (1921) A revision of the genus *Locusta* (L.) (=*Patchytylus* Fieb.), with a new theory as to the periodicity and migrations of locusts. *Bulletin of Entomological Research*, 12:135–163.
111. Vicsek, T., Czirók, A., Ben-Jacob, E., Cohen, I. and Shochet, O. (1995) Novel type of phase transition in a system of self-driven particles. *Physical Review Letters*, 75:1226–1229.
112. Visscher, P. K. (2007) Nest-site selection and group decision-making in social insects. *Annual Review of Entomology*, 52, in press.
113. Visscher, P. K. and Camazine, S. (1999) Collective decisions and cognition in bees. *Nature*, 397:400.
114. Visscher, P. K. and Seeley, T. D. (1982) Foraging strategy of honeybee colonies in a temperate deciduous forest. *Ecology*, 63:1790–1801.
115. Waddington, K. D., Visscher, P.K., Herbert, T. J. and Raveret Richter, M. (1994) Comparisons of forager distributions from matched honey bee colonies in suburban environments. *Behavioral Ecology and Sociobiology*, 35:423-429.
116. Wakeland, C. (1959) Mormon crickets in North America. *USDA Technical Bulletin*, 1202:1–77.
117. Wetterer, J., Shafir, S., Morrison, L., Lips, K., Gilbert, G., Cipollini, M. and Blaney, C. (1992) On- and off-trail orientation in the leaf-cutting ant, *Atta cephalotes* (L.) (Hymenoptera: Formicidae). *Journal of the Kansas Entomological*, Society 65:96–98.
118. Winston, M. L. (1987) The Biology of the Honey Bee. Harvard University Press, Cambridge, MA.
119. Wikelski, M., Moskowitz, D., Adelman, J., Cochran, J., Wilcove, D. and May, M. (2006) Simple rules guide dragonfly migration. *Biology Letters*, 2:325–329.

Swarm Intelligence in Optimization

Christian Blum[1] and Xiaodong Li[2]

[1] ALBCOM Research Group
Universitat Politècnica de Catalunya, Barcelona, Spain
cblum@lsi.upc.edu
[2] School of Computer Science and Information Technology
RMIT University, Melbourne, Australia
xiaodong@cs.rmit.edu.au

Summary. Optimization techniques inspired by swarm intelligence have become increasingly popular during the last decade. They are characterized by a decentralized way of working that mimics the behavior of swarms of social insects, flocks of birds, or schools of fish. The advantage of these approaches over traditional techniques is their robustness and flexibility. These properties make swarm intelligence a successful design paradigm for algorithms that deal with increasingly complex problems. In this chapter we focus on two of the most successful examples of optimization techniques inspired by swarm intelligence: ant colony optimization and particle swarm optimization. Ant colony optimization was introduced as a technique for combinatorial optimization in the early 1990s. The inspiring source of ant colony optimization is the foraging behavior of real ant colonies. In addition, particle swarm optimization was introduced for continuous optimization in the mid-1990s, inspired by bird flocking.

1 Introduction

Swarm intelligence (SI), which is an artificial intelligence (AI) discipline, is concerned with the design of intelligent multi-agent systems by taking inspiration from the collective behavior of social insects such as ants, termites, bees, and wasps, as well as from other animal societies such as flocks of birds or schools of fish. Colonies of social insects have fascinated researchers for many years, and the mechanisms that govern their behavior remained unknown for a long time. Even though the single members of these colonies are non-sophisticated individuals, they are able to achieve complex tasks in cooperation. Coordinated colony behavior emerges from relatively simple actions or interactions between the colonies' individual members. Many aspects of the collective activities of social insects are self-organized and work without a central control. For example, leafcutter ants cut pieces from leaves, bring them back to their nest, and grow fungi used as food for their larvae. Weaver

Fig. 1. Ants cooperate for retrieving a heavy prey. (Photographer: Christian Blum)

ant workers build chains with their bodies in order to cross gaps between two leaves. The edges of the two leaves are then pulled together, and successively connected by silk that is emitted by a mature larva held by a worker. Another example concerns the recruitment of other colony members for prey retrieval (see, for example, Fig. 1).

Other examples include the capabilities of termites and wasps to build sophisticated nests, or the ability of bees and ants to orient themselves in their environment. For more examples and a more detailed description see Chap. 1 of this book, as well as [21, 92]. The term swarm intelligence was first used by Beni in the context of cellular robotic systems where simple agents organize themselves through nearest-neighbor interaction [4]. Meanwhile, the term swarm intelligence is used for a much broader research field [21]. Swarm intelligence methods have been very successful in the area of optimization, which is of great importance for industry and science. This chapter aims at giving an introduction to swarm intelligence methods in optimization.

Optimization problems are of high importance both for the industrial world as well as for the scientific world. Examples of practical optimization problems include train scheduling, timetabling, shape optimization, telecommunication network design, and problems from computational biology. The research community has simplified many of these problems in order to obtain scientific test cases such as the well-known traveling salesman problem (TSP) [99]. The TSP models the situation of a traveling salesman who is required to pass through a number of cities. The goal of the traveling salesman is to traverse these cities (visiting each city exactly once) so that the total traveling distance is minimal. Another example is the problem of protein folding, which is one of the most challenging problems in computational biology, molecular biology, biochemistry, and physics. It consists of finding the functional shape or conformation of a protein in two- or three-dimensional space, for example, under simplified lattice models such as the hydrophobic-polar model [169]. The TSP and the protein folding problem under lattice models belong to an important class of optimization problems known as combinatorial optimization (CO).

In general, any optimization problem $\mathcal{P}$ can be described as a triple $(\mathcal{S}, \Omega, f)$, where

1. $\mathcal{S}$ is the search space defined over a finite set of decision variables X_i, $i = 1, \ldots, n$. In the case where these variables have discrete domains we deal with discrete optimization (or combinatorial optimization), and in the case of continuous domains $\mathcal{P}$ is called a continuous optimization problem. Mixed variable problems also exist. Ω is a set of constraints among the variables;

2. $f : \mathcal{S} \rightarrow I\!R^+$ is the objective function that assigns a positive cost value to each element (or solution) of $\mathcal{S}$.

The goal is to find a solution $s \in \mathcal{S}$ such that $f(s) \leq f(s')$, $\forall s' \in \mathcal{S}$ (in case we want to minimize the objective function), or $f(s) \geq f(s')$, $\forall s' \in \mathcal{S}$ (in case the objective function must be maximized). In real-life problems the goal is often to optimize several objective functions at the same time. This form of optimization is labelled multiobjective optimization.

Due to the practical importance of optimization problems, many algorithms to tackle them have been developed. In the context of combinatorial optimization (CO), these algorithms can be classified as either *complete* or *approximate* algorithms. Complete algorithms are guaranteed to find for every finite size instance of a CO problem an optimal solution in bounded time (see [133, 128]). Yet, for CO problems that are NP-hard [65], no polynomial time algorithm exists, assuming that $\mathcal{P} \neq \mathcal{NP}$. Therefore, complete methods might need exponential computation time in the worstcase. This often leads to computation times too high for practical purposes. In approximate methods such as SI-based algorithms we sacrifice the guarantee of finding optimal solutions for the sake of getting good solutions in a significantly reduced amount of time. Thus, the use of approximate methods has received more and more attention in the last 30 years. This was also the case in continuous optimization, due to other reasons: Approximate methods are usually easier to implement than classical gradient-based techniques. Moreover, generally they do not require gradient information. This is convenient for optimization problems where the objective function is only implicitly given (e.g., when objective function values are obtained by simulation), or where the objective function is not differentiable.

Two of the most notable swarm intelligence techniques for obtaining approximate solutions to optimization problems in a reasonable amount of computation time are ant colony optimization (ACO) and particle swarm optimization (PSO). These optimization methods will be explained in Sects. 2

and 3 respectively. In Sect. 4 we will give some further examples of algorithms for which swarm intelligence was the inspiring source.

2 Ant Colony Optimization

Ant colony optimization (ACO) [52] was one of the first techniques for approximate optimization inspired by swarm intelligence. More specifically, ACO is inspired by the foraging behavior of ant colonies. At the core of this behavior is the indirect communication between the ants by means of chemical pheromone trails, which enables them to find short paths between their nest and food sources. This characteristic of real ant colonies is exploited in ACO algorithms in order to solve, for example, discrete optimization problems.[3]

Seen from the operations research (OR) perspective, ACO algorithms belong to the class of metaheuristics [18, 68, 80]. The term *metaheuristic*, first introduced in [67], derives from the composition of two Greek words. *Heuristic* derives from the verb *heuriskein* ($\epsilon\upsilon\rho\iota\sigma\kappa\epsilon\iota\nu$) which means "to find", while the suffix *meta* means "beyond, in an upper level". Before this term was widely adopted, metaheuristics were often called *modern heuristics* [144]. In addition to ACO, other algorithms, such as evolutionary computation, iterated local search, simulated annealing, and tabu search, are often regarded as metaheuristics. For books and surveys on metaheuristics see [144, 68, 18, 80].

This section on ACO is organized as follows. First, in Sect. 2.1 we outline the origins of ACO algorithms. In particular, we present the foraging behavior of real ant colonies and show how this behavior can be transfered into a technical algorithm for discrete optimization. In Sect. 2.2 we provide a description of ACO in more general terms, outline some of the most successful current ACO variants, and list some representative examples of ACO applications. In Sect. 2.3, we shortly describe some recent trends in ACO.

2.1 The Origins of Ant Colony Optimization

Marco Dorigo and colleagues introduced the first ACO algorithms in the early 1990s [46, 50, 51]. The development of these algorithms was inspired by the observation of ant colonies. Ants are social insects. They live in colonies and their behavior is governed by the goal of colony survival rather than being focused on the survival of individuals. The behavior that provided the inspiration for ACO is the ants' foraging behavior, and in particular, how ants

---

[3] Even though ACO algorithms were originally introduced for the application to discrete optimization problems, the class of ACO algorithms also comprises methods for the application to problems arising in networks, such as routing and load balancing (see, for example, [44]), and continuous optimization problems (see, for example, [159]). In Sect. 2.3 we will shortly deal with ACO algorithms for continuous optimization.

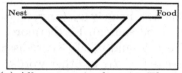

(a) All ants are in the nest. There is no pheromone in the environment

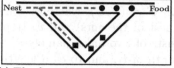

(b) The foraging starts. In probability, 50% of the ants take the short path (see the circles), and 50% take the long path to the food source (see the rhombs)

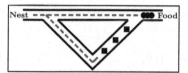

(c) The ants that have taken the short path have arrived earlier at the food source. Therefore, when returning, the probability that they again take the short path is higher

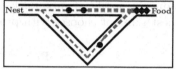

(d) The pheromone trail on the short path receives, in probability, a stronger reinforcement, and the probability of taking this path grows. Finally, due to the evaporation of the pheromone on the long path, the whole colony will, in probability, use the short path

Fig. 2. An experimental setting that demonstrates the shortest path finding capability of ant colonies. Between the ants' nest and the only food source exist two paths of different lengths. In the four graphics, the pheromone trails are shown as dashed lines whose thickness indicates the trails' strength

can find shortest paths between food sources and their nest. When searching for food, ants initially explore the area surrounding their nest in a random manner. While moving, ants leave a chemical pheromone trail on the ground. Ants can smell pheromone. When choosing their way, they tend to choose, in probability, paths marked by strong pheromone concentrations. As soon as an ant finds a food source, it evaluates the quantity and the quality of the food and carries some of it back to the nest. During the return trip, the quantity of pheromone that an ant leaves on the ground may depend on the quantity and quality of the food. The pheromone trails will guide other ants to the food source. It has been shown in [42] that the indirect communication between the ants via pheromone trails—known as *stigmergy* [70]—enables them to find shortest paths between their nest and food sources. This is explained in an idealized setting in Fig. 2.

As a first step towards an algorithm for discrete optimization we present in the following a discretized and simplified model of the phenomenon explained in Fig. 2. After presenting the model we will outline the differences between the model and the behavior of real ants. The considered model consists of a graph $G = (V, E)$, where V consists of two nodes, namely v_s (representing the nest of the ants) and v_d (representing the food source). Furthermore, E consists of two links, namely e_1 and e_2, between v_s and v_d. To e_1 we assign a length of l_1, and to e_2 a length of l_2 such that $l_2 > l_1$. In other words, e_1 represents the short path between v_s and v_d, and e_2 represents the long path. Real ants deposit pheromone on the paths on which they move. Thus, the chemical pheromone trails are modeled as follows. We introduce an artificial pheromone value τ_i for each of the two links e_i, $i = 1, 2$. Such a value indicates the strength of the pheromone trail on the corresponding path. Finally, we introduce n_a artificial ants. Each ant behaves as follows: Starting from v_s (i.e., the nest), an ant chooses with probability

$$\mathbf{p}_i = \frac{\tau_i}{\tau_1 + \tau_2} \quad , i = 1, 2, \tag{1}$$

between path e_1 and path e_2 for reaching the food source v_d. Obviously, if $\tau_1 > \tau_2$, the probability of choosing e_1 is higher, and vice versa. For returning from v_d to v_s, an ant uses the same path as it chose to reach v_d,[4] and it changes the artificial pheromone value associated with the used edge. In more detail, having chosen edge e_i an ant changes the artificial pheromone value τ_i as follows:

$$\tau_i \leftarrow \tau_i + \frac{Q}{l_i}, \tag{2}$$

where the positive constant Q is a parameter of the model. In other words, the amount of artificial pheromone that is added depends on the length of the chosen path: the shorter the path, the higher the amount of added pheromone.

The foraging of an ant colony is in this model iteratively simulated as follows: At each step (or iteration) all the ants are initially placed in node v_s. Then, each ant moves from v_s to v_d as outlined above. As mentioned in the caption of Fig. 2(d), in nature the deposited pheromone is subject to an evaporation over time. We simulate this pheromone evaporation in the artificial model as follows:

$$\tau_i \leftarrow (1 - \rho) \cdot \tau_i \quad , i = 1, 2 \tag{3}$$

The parameter $\rho \in (0, 1]$ is a parameter that regulates the pheromone evaporation. Finally, all ants conduct their return trip and reinforce their chosen path as outlined above.

We implemented this system and conducted simulations with the following settings: $l_1 = 1$, $l_2 = 2$, $Q = 1$. The two pheromone values were initialized to 0.5 each. Note that in our artificial system we cannot start with artificial pheromone values of 0. This would lead to a division by 0 in Eq. 1. The results

---

[4] Note that this can be enforced because the setting is symmetric, i.e., the choice of a path for moving from v_s to v_d is equivalent to the choice of a path for moving from v_d to v_s.

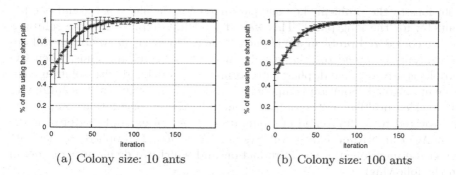

(a) Colony size: 10 ants   (b) Colony size: 100 ants

Fig. 3. Results of 100 independent runs (error bars show the standard deviation for each 5th iteration). The x-axis shows the iterations, and the y-axis the percentage of the ants using the short path

of our simulations are shown in Fig. 3. They clearly show that over time the artificial colony of ants converges to the short path, i.e., after some time all ants use the short path. In the case of 10 ants (i.e., $n_a = 10$, Fig. 3(a)) the random fluctuations are bigger than in the case of 100 ants (Fig. 3(b)). This indicates that the shortest path finding capability of ant colonies results from a cooperation between the ants.

The main differences between the behavior of the real ants and the behavior of the artificial ants in our model are as follows:

1. While real ants move in their environment in an asynchronous way, the artificial ants are synchronized, i.e., at each iteration of the simulated system, each of the artificial ants moves from the nest to the food source and follows the same path back.
2. While real ants leave pheromone on the ground whenever they move, artificial ants only deposit artificial pheromone on their way back to the nest.
3. The foraging behavior of real ants is based on an implicit evaluation of a solution (i.e., a path from the nest to the food source). By implicit solution evaluation we mean the fact that shorter paths will be completed earlier than longer ones, and therefore they will receive pheromone reinforcement more quickly. In contrast, the artificial ants evaluate a solution with respect to some quality measure which is used to determine the strength of the pheromone reinforcement that the ants perform during their return trip to the nest.

Ant System for the TSP: The First ACO Algorithm

The model that we used in the previous section to simulate the foraging behavior of real ants in the setting of Fig. 2 cannot directly be applied to CO problems. This is because we associated pheromone values directly with solutions to the problem (i.e., one parameter for the short path, and one parameter for the long path). This way of modeling implies that the solutions to the considered problem are already known. However, in combinatorial optimization we intend to *find* an unknown optimal solution. Thus, when CO problems are considered, pheromone values are associated with solution components instead. Solution components are the units from which solutions to the tackled problem are assembled. Generally, the set of solution components is expected to be finite and of moderate size. As an example we present the first ACO algorithm, called Ant System (AS) [46, 51], applied to the TSP, which we mentioned in the introduction and which we define in more detail in the following:

Definition 1. *In the TSP is given a completely connected, undirected graph $G = (V, E)$ with edge weights. The nodes V of this graph represent the cities, and the edge weights represent the distances between the cities. The goal is to find a closed path in G that contains each node exactly once (henceforth called a tour) and whose length is minimal. Thus, the search space S consists of all tours in G. The objective function value $f(s)$ of a tour $s \in S$ is defined as the sum of the edge weights of the edges that are in s.*

Concerning the AS approach, the edges of the given TSP graph can be considered solution components, i.e., for each $e_{i,j}$ is introduced a pheromone value $\tau_{i,j}$. The task of each ant consists in the construction of a feasible TSP solution, i.e., a feasible tour. In other words, the notion of *task of an ant* changes from *"choosing a path from the nest to the food source"* to *"constructing a feasible solution to the tackled optimization problem"*. Note that with this change of task, the notions of nest and food source lose their meaning.

Each ant constructs a solution as follows. First, one of the nodes of the TSP graph is randomly chosen as start node. Then, the ant builds a tour in the TSP graph by moving in each construction step from its current node (i.e., the city in which it is located) to another node which it has not visited yet. At each step the traversed edge is added to the solution under construction. When no unvisited nodes are left the ant closes the tour by moving from her current node to the node in which it started the solution construction. This way of constructing a solution implies that an ant has a memory T to store the already-visited nodes. Each solution construction step is performed as follows. Assuming the ant to be in node v_i, the subsequent construction step is done with probability

$$\mathbf{p}(e_{i,j}) = \frac{\tau_{i,j}}{\sum_{\{k \in \{1,\ldots,|V|\}|v_k \notin T\}} \tau_{i,k}} \quad , \forall\, j \in \{1,\ldots,|V|\}, v_j \notin T \ . \qquad (4)$$

Once all ants of the colony have completed the construction of their solution, pheromone evaporation is performed as follows:

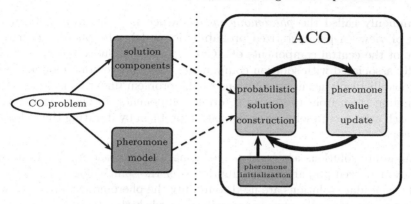

Fig. 4. The ACO framework

$$\tau_{i,j} \leftarrow (1 - \rho) \cdot \tau_{i,j} \quad , \vee\ \tau_{i,j} \in \mathcal{T} \tag{5}$$

Then the ants perform their return trip. Hereby, an ant—having constructed a solution s—performs for each $e_{i,j} \in s$ the following pheromone deposit:

$$\tau_{i,j} \leftarrow \tau_{i,j} + \frac{Q}{f(s)}, \tag{6}$$

where Q is again a positive constant and $f(s)$ is the objective function value of the solution s. As explained in the previous section, the system is iterated—applying n_a ants per iteration—until a stopping condition (e.g., a time limit) is satisfied.

Even though the AS algorithm has proved that the ants' foraging behavior can be transferred into an algorithm for discrete optimization, it gas generally been found to be inferior to state-of-the-art algorithms. Therefore, over the years several extensions and improvements of the original AS algorithm were introduced. They are all covered by the definition of the ACO framework, which we will outline in the following.

2.2 Ant Colony Optimization: A General Description

The ACO framework, as we know it today, was first defined by Dorigo and colleagues in 1999 [48]. The recent book by Dorigo and Stützle gives a more comprehensive description [52]. The definition of the ACO framework covers most—if not all—existing ACO variants for discrete optimization problems. In the following, we give a general description of this framework.

The basic way of working of an ACO algorithm is graphically shown in Fig. 4. Given a CO problem to be solved, one first has to derive a finite set $\mathcal{C}$ of solution components which are used to assemble solutions to the CO problem. Second, one has to define a set of *pheromone values* $\mathcal{T}$. This set of values is commonly called the *pheromone model*, which is—seen from a technical point of view—a parameterized probabilistic model. The pheromone model is one of the central components of ACO. The pheromone values $\tau_i \in \mathcal{T}$ are usually associated with solution components.[5] The pheromone model is used to probabilistically generate solutions to the problem under consideration by assembling them from the set of solution components. In general, the ACO approach attempts to solve an optimization problem by iterating the following two steps:

- candidate solutions are constructed using a pheromone model, that is, a parameterized probability distribution over the solution space;
- the candidate solutions are used to modify the pheromone values in a way that is deemed to bias future sampling towards high-quality solutions. The pheromone update aims to concentrate the search in regions of the search space containing high-quality solutions. It implicitly assumes that good solutions consist of good solution components.

In the following we give a more detailed description of solution construction and pheromone update.

Solution Construction

Artificial ants can be regarded as probabilistic constructive heuristics that assemble solutions as sequences of solution components. The finite set of solution components $\mathcal{C} = \{c_1, \ldots, c_n\}$ is hereby derived from the discrete optimization problem under consideration. For example, in the case of AS applied to the TSP (see previous section) each edge of the TSP graph was considered a solution component. Each solution construction starts with an empty sequence $s = \langle\rangle$. Then, the current sequence s is at each construction step extended by adding a feasible solution component from the set $\mathcal{N}(s) \subseteq \mathcal{C} \setminus s$.[6] The specification of $\mathcal{N}(s)$ depends on the solution construction mechanism. In the example of AS applied to the TSP (see previous section) the solution construction mechanism restricted the set of traversable edges to the ones that connected the ants' current node to unvisited nodes. The choice of a solution component from $\mathcal{N}(s)$ is at each construction step performed probabilistically with respect to the pheromone model. In most ACO algorithms the respective probabilities—also called the *transition probabilities*—are defined as follows:

$$\mathbf{p}(c_i \mid s) = \frac{[\tau_i]^\alpha \cdot [\eta(c_i)]^\beta}{\sum\limits_{c_j \in \mathcal{N}(s)} [\tau_j]^\alpha \cdot [\eta(c_j)]^\beta} \ , \quad \forall \, c_i \in \mathcal{N}(s), \tag{7}$$

---

[5] Note that the description of ACO as given for example in [48] allows pheromone values also to be associated with links between solution components. However, for the purpose of this introduction it is sufficient to assume pheromone values associated with components.

[6] Note that for this set operation the sequence s is regarded as an ordered set.

where η is an optional weighting function, that is, a function that, sometimes depending on the current sequence, assigns at each construction step a heuristic value $\eta(c_j)$ to each feasible solution component $c_j \in \mathcal{N}(s)$. The values that are given by the weighting function are commonly called the *heuristic information*. Furthermore, the exponents α and β are positive parameters whose values determine the relation between pheromone information and heuristic information. In the previous section's TSP example, we chose not to use any weighting function η, and we set α to 1.

Pheromone Update

Different ACO variants mainly differ in the update of the pheromone values they apply. In the following, we outline a general pheromone update rule in order to provide the basic idea. This pheromone update rule consists of two parts. First, a *pheromone evaporation*, which uniformly decreases all the pheromone values, is performed. From a practical point of view, pheromone evaporation is needed to avoid a too-rapid convergence of the algorithm towards a suboptimal region. It implements a useful form of *forgetting*, favoring the exploration of new areas in the search space. Second, one or more solutions from the current and/or from earlier iterations are used to increase the values of pheromone trail parameters on solution components that are part of these solutions:

$$\tau_i \leftarrow (1 - \rho) \cdot \tau_i + \rho \cdot \sum_{\{s \in \mathcal{S}_{upd}|c_i \in s\}} w_s \cdot F(s), \tag{8}$$

for $i = 1, \ldots, n$. $\mathcal{S}_{upd}$ denotes the set of solutions that are used for the update. Furthermore, $\rho \in (0, 1]$ is a parameter called evaporation rate, and $F : S \mapsto I\!R^+$ is a so-called quality function such that $f(s) < f(s') \Rightarrow F(s) \geq F(s')$, $\forall s \neq s' \in S$. In other words, if the objective function value of a solution s is better than the objective function value of a solution s', the quality of solution s will be at least as high as the quality of solution s'. Equation (8) also allows an additional weighting of the quality function, i.e., $w_s \in I\!R^+$ denotes the weight of a solution s.

Instantiations of this update rule are obtained by different specifications of $\mathcal{S}_{upd}$ and by different weight settings. In most cases, $\mathcal{S}_{upd}$ is composed of some of the solutions generated in the respective iteration (henceforth denoted by $\mathcal{S}_{iter}$) and the best solution found since the start of the algorithm (henceforth denoted by s_{bs}). Solution s_{bs} is often called the best-so-far solution. A well-known example is the *AS-update* rule, that is, the update rule of AS (see also Sect. 2.1). The AS-update rule, which is well known due to the fact that AS was the first ACO algorithm to be proposed in the literature, is obtained from update rule (8) by setting $\mathcal{S}_{upd} \leftarrow \mathcal{S}_{iter}$ and $w_s = 1$, $\forall s \in \mathcal{S}_{upd}$. An example of a pheromone update rule that is more used in practice is the *IB-update* rule (where IB stands for *iteration-best*). The IB-update rule is

Table 1. A selection of ACO variants

ACO variant	Authors	Main reference
Elitist AS (EAS)	Dorigo	[46]
	Dorigo, Maniezzo, and Colorni	[51]
Rank-based AS (RAS)	Bullnheimer, Hartl, and Strauss	[26]
$\mathcal{MAX}$–$\mathcal{MIN}$ Ant System ($\mathcal{MMAS}$)	Stützle and Hoos	[164]
Ant Colony System (ACS)	Dorigo and Gambardella	[49]
Hyper-Cube Framework (HCF)	Blum and Dorigo	[16]

given by $\mathcal{S}_{upd} \leftarrow \{s_{ib} = \text{argmax}\{F(s) \mid s \in \mathcal{S}_{iter}\}\}$ with $w_{s_{ib}} = 1$, that is, by choosing only the best solution generated in the respective iteration for updating the pheromone values. This solution, denoted by s_{ib}, is weighted by 1. The IB-update rule introduces a much stronger bias towards the good solutions found than the AS-update rule. However, this increases the danger of premature convergence. An even stronger bias is introduced by the *BS-update* rule, where BS refers to the use of the best-so-far solution s_{bs}. In this case, $\mathcal{S}_{upd}$ is set to $\{s_{bs}\}$ and s_{bs} is weighted by 1, that is, $w_{s_{bs}} = 1$. In practice, ACO algorithms that use variations of the IB-update or the BS-update rule and that additionally include mechanisms to avoid premature convergence achieve better results than algorithms that use the AS-update rule. Examples are given in the following section.

Well-Performing ACO Variants

Even though the original AS algorithm achieved encouraging results for the TSP problem, it was found to be inferior to state-of-the-art algorithms for the TSP as well as for other CO problems. Therefore, several extensions and improvements of the original AS algorithm were introduced over the years. An overview is provided in Table 1. These ACO variants mostly differ in the pheromone update rule that is applied.

In addition to these ACO variants, the ACO community has developed additional algorithmic features for improving the search process performed by ACO algorithms. A prominent example is the so-called candidate list strategy, which is a mechanism to restrict the number of available choices at each solution construction step. Usually, this restriction applies to a number of the best choices with respect to their transition probabilities (see Eq. 7). For example, in the case of the application of ACS (see Table 1) to the TSP, the restriction to the closest cities at each construction step both improved the final solution quality and led to a significant speedup of the algorithm (see [61]). The reasons for this are as follows: First, in order to construct high-quality solutions it is often enough to consider only the "promising" choices at each construction step. Second, to consider fewer choices at each construction step speeds up the solution construction process, because the reduced number of choices reduces the computation time needed to make a choice.

Applications of ACO Algorithms

As mentioned before, ACO was introduced by means of the proof-of-concept application to the TSP. Since then, ACO algorithms have been applied to many optimization problems. First, classical problems other than the TSP, such as assignment problems, scheduling problems, graph coloring, the maximum clique problem, or vehicle routing problems were tackled. More recent applications include, for example, cell placement problems arising in circuit design, the design of communication networks, bioinformatics problems, and problems arising in continuous optimization. In recent years some researchers have also focused on the application of ACO algorithms to multiobjective problems and to dynamic or stochastic problems.

The bioinformatics and biomedical fields in particular show an increasing interest in ACO. Recent applications of ACO to problems arising in these areas include the applications to protein folding [153, 154], to multiple sequence alignment [127], to DNA sequencing by hybridization [20], and to the prediction of major histocompatibility complex (MHC) class II binders [86]. ACO algorithms are currently among the state-of-the-art methods for solving, for example, the sequential ordering problem [62], the resource constraint project scheduling problem [120], the open shop scheduling problem [14], assembly line balancing [15], and the 2D and 3D hydrophobic polar protein folding problem [154]. In Table 2 we provide a list of representative ACO applications. For a more comprehensive overview that also covers the application of ant-based algorithms to routing in telecommunication networks we refer the interested reader to [52].

2.3 Recent Trends

Theoretical Work on ACO

The first theoretical works on ACO algorithms appeared in 2002. They deal with the question of algorithm convergence [75, 76, 163]. In other words: will a given ACO algorithm find an optimal solution when given enough resources? This is an interesting question, because ACO algorithms are stochastic search procedures in which the pheromone update could prevent them from ever reaching an optimum.

Recently, researchers have been dealing with the relation of ACO algorithms to other methods for learning and optimization. The work presented in [7] relates ACO to the fields of optimal control and reinforcement learning, whereas [183] describes the common aspects of ACO algorithms and probabilistic learning algorithms such as stochastic gradient ascent (SGA) and the

Table 2. A representative selection of ACO applications

Problem	Authors	Reference
Traveling salesman problem	Dorigo, Maniezzo, and Colorni	[46, 50, 51]
	Dorigo and Gambardella	[49]
	Stützle and Hoos	[164]
Quadratic assignment problem	Maniezzo	[109]
	Maniezzo and Colorni	[111]
	Stützle and Hoos	[164]
Scheduling problems	Stützle	[162]
	den Besten, Stützle, and Dorigo	[41]
	Gagné, Price, and Gravel	[59]
	Merkle, Middendorf, and Schenk	[120]
	Blum (resp., Blum and Sampels)	[14, 19]
Vehicle routing problems	Gambardella, Taillard, and Agazzi	[63]
	Reimann, Doerner, and Hartl	[145]
Timetabling	Socha, Sampels, and Manfrin	[160]
Set packing	Gandibleux, Delorme, and T'Kindt	[64]
Graph coloring	Costa and Hertz	[38]
Shortest supersequence problem	Michel and Middendorf	[123]
Sequential ordering	Gambardella and Dorigo	[62]
Constraint satisfaction problems	Solnon	[161]
Data mining	Parpinelli, Lopes, and Freitas	[134]
Maximum clique problem	Bui and Rizzo Jr	[25]
Edge-disjoint paths problem	Blesa and Blum	[13]
Cell placement in circuit design	Alupoaei and Katkoori	[2]
Communication network design	Maniezzo, Boschetti, and Jelasity	[110]
Bioinformatics problems	Shmygelska, Aguirre-Hernández, and Hoos	[153]
	Moss and Johnson	[127]
	Karpenko, Shi, and Dai	[86]
	Shmygelska and Hoos	[154]
	Korb, Stützle, and Exner	[93]
	Blum and Yábar Vallès	[20]
Industrial problems	Bautista and Pereira	[3]
	Blum, Bautista, and Pereira	[15]
	Silva, Runkler, Sousa, and Palm	[156]
	Gottlieb, Puchta, and Solnon	[69]
	Corry and Kozan	[37]
Continuous optimization	Bilchev and Parmee	[6]
	Monmarché, Venturini, and Slimane	[125]
	Dréo and Siarry	[54]
	Socha and Dorigo	[159]
	Socha and Blum	[158]
Multiobjective problems	Guntsch and Middendorf	[74]
	Lopéz-Ibáñez, Paquete, and Stützle	[106]
	Doerner, Gutjahr, Hartl, Strauss, and Stummer	[45]
Dynamic (or stochastic) problems	Guntsch and Middendorf	[73]
	Bianchi, Gambardella, and Dorigo	[5]
Music	Guéret, Monmarché, and Slimane	[72]

cross-entropy (CE) method. Meuleau and Dorigo have shown in [121] that ACO's pheromone update is very similar to stochastic gradient ascent in the space of pheromone values.

While convergence proofs can provide insight into the working of an algorithm, they are usually not very useful to the practitioner who wants to implement efficient algorithms. More relevant for practical applications might be the research efforts that were aimed at a better understanding of the behavior of ACO algorithms. Representative works are the ones on negative search bias [17] and the study of models of ACO algorithms [117, 118]. For a recent survey on theoretical work on ACO see [47].

Applying ACO to Continuous Optimization Problems

Many practical optimization problems can be formulated as continuous optimization problems, that is, problems in which the decision variables have continuous domains. While ACO algorithms were originally introduced to solve discrete problems, their adaptation to solve continuous optimization problems enjoys increasing attention. Early applications of ant-based algorithms to continuous optimization include algorithms such as Continuous ACO (CACO) [6], API [125], and Continuous Interacting Ant Colony (CIAC) [54]. However, all these approaches are conceptually quite different from ACO for discrete problems. The latest approach called $ACO_\mathbb{R}$, which was proposed by Socha in [157, 159], is closest to the spirit of ACO for discrete problems. While ACO algorithms for discrete optimization problems construct solutions by sampling at each construction step a discrete probability distribution that is derived from the pheromone information, $ACO_\mathbb{R}$ utilizes a continuous probability density function (PDF) for generating solutions. This density function is produced, for each solution construction, from an archive of solutions that the algorithm keeps and updates at all times. The archive update corresponds to the pheromone update in ACO algorithms for discrete optimization problems. Recently, $ACO_\mathbb{R}$ was applied to neural network training [158].

Hybridizing ACO with Branch & Bound Derivatives

Beam search (BS) is a classical tree search method that was introduced in the context of scheduling [131], but has since then been successfully applied to many other CO problems (e.g., see [40]). BS algorithms are incomplete derivatives of branch & bound algorithms, and are therefore approximate methods. The central idea behind BS is to construct a number of k_{bw} (the so-called beam width) solutions in parallel and non-independently. At each construction step the algorithm selects at most k_{bw} partial solutions by utilizing bounding information. Even though both ACO and BS have the common feature that they are based on the idea of constructing candidate solutions step-by-step, the ways by which the two methods explore the search space are quite different. While BS is a deterministic algorithm that uses a lower bound for guiding the search process, ACO algorithms are adaptive and probabilistic procedures. Furthermore, BS algorithms reduce the search space in the hope of not excluding all optimal solutions, while ACO algorithms consider the whole search space. Based on these observations Blum introduced a hybrid between ACO and BS which was labelled *Beam-ACO* [14, 15]. Beam-ACO is an ACO algorithm in which the standard ACO solution construction mechanism is replaced by a probabilistic beam search procedure. Work that is in a similar vein can be found in [109, 112].

ACO and Constraint Programming

Another interesting hybridization example concerns the use of constraint programming (CP) techniques (see [114]) for restricting the search performed by an ACO algorithm to promising regions of the search space. The motivation for this type of hybridization is as follows: Generally, ACO algorithms are competitive with other optimization techniques when applied to problems that are not overly constrained. However, when highly constrained problems such as scheduling or timetabling are concerned, the performance of ACO algorithms generally degrades. Note that this is also the case for other metaheuristics. The reason is to be found in the structure of the search space: When a problem is not overly constrained, it is usually not difficult to find feasible solutions. The difficulty rather lies in the optimization part, namely the search for good feasible solutions. On the other hand, when a problem is highly constrained the difficulty is rather in finding any feasible solution. This is where CP comes into play, because these problems are the target problems for CP applications. The idea of hybridizing ACO with CP is simple [122]. At each iteration, first constraint propagation is applied in order to reduce the remaining search tree. Then, solutions are constructed in the standard ACO way with respect to the reduced search tree. After the pheromone update, additional constraints might be added to the system.

Applying ACO in a Multilevel Framework

Multilevel techniques have been employed for quite a long time, especially in the area of multigrid methods (see [23] for an overview). More recently, they have been brought into focus by Walshaw for the application to CO. Walshaw and coworkers applied multilevel techniques to graph-based problems such as mesh partitioning [177]. The basic idea of a multilevel scheme is simple. Starting from the original problem instance, smaller and smaller problem instances are obtained by successive coarsening until some stopping criteria are satisfied. This creates a hierarchy of problem instances in which the problem instance of a given level is always smaller (or of equal size) to the problem instance of the next lower level. Then, a solution is computed to the smallest problem instance and successively transformed into a solution of the next higher level until a solution for the original problem instance is obtained. At each level, the obtained solution might be subject to a refinement process, for example, an ACO algorithm. Applications of ACO in multilevel frameworks include [94, 95, 20].

3 Particle Swarm Optimization

Particle swarm optimization (PSO) is a population-based stochastic optimization technique modelled on the social behaviors observed in animals or insects, e.g., bird flocking, fish schooling, and animal herding [92]. It was originally proposed by James Kennedy and Russell Eberhart in 1995 [91]. Since its inception, PSO has gained increasing popularity among researchers and practitioners as a robust and efficient technique for solving difficult optimization problems. In PSO, individual *particles* of a swarm represent potential solutions, which move through the problem search space seeking an optimal, or good enough, solution. The particles broadcast their current positions to neighboring particles. The position of each particle is adjusted according to its *velocity* (i.e., rate of change) and the difference between its current position, respectively the best position found by its neighbors, and the best position it has found so far. As the model is iterated, the swarm focuses more and more on an area of the search space containing high-quality solutions.

PSO has close ties to *artificial life* models. Early works by Reynolds on a flocking model *Boids* [146], and Heppner's studies on rules governing large numbers of birds flocking synchronously [78], indicated that the emergent group dynamics such as the bird flocking behavior are based on local interactions. These studies were the foundation for the subsequent development of PSO for the application to optimization. PSO is in some way similar to *cellular automata* (CA), which are often used for generating interesting self-replicating patterns based on very simple rules, e.g., John Conway's *Game of Life*. CAs have three main attributes: (1) individual cells are updated in parallel; (2) the value of each new cell depends only on the old values of the cell and its neighbors; and (3) all cells are updated using the same rules [149]. Particles in a swarm are analogous to CA cells, whose states are updated in many dimensions simultaneously.

The term *particle swarm* was coined by James Kennedy and Russell Eberhart, who were responsible for inventing the original PSO. Initially they intended to model the movements of flocks of birds and schools of fish. As their model further evolved to handle optimization, the visual plots they used started to display something more like *swarms* of mosquitoes. The term *particle* was used simply because the notion of velocity was adopted in PSO and *particle* seemed to be the most appropriate term in this context.

This section on PSO is organized as follows. In Sect. 3.1 we first present the original PSO developed by Kennedy and Eberhart. This is followed by descriptions of a number of key improvements and generalizations to the basic PSO algorithm. We then give an overview of several PSO variants that represent important progress made in this area, and a list of representative examples of PSO applications. In Sect. 3.2 we outline some recent trends in PSO research, including its theoretical works and its application in the areas of multiobjective optimization, dynamic optimization, and constraint handling.

3.1 Particle Swarm Optimization: An Introduction

In PSO, the velocity of each particle is modified iteratively by its *personal best* position (i.e., the best position found by the particle so far), and the best position found by particles in its neighborhood. As a result, each particle searches around a region defined by its personal best position and the best position from its neighborhood. Henceforth we use $\mathbf{v}_i$ to denote the velocity of the ith particle in the swarm, $\mathbf{x}_i$ to denote its position, $\mathbf{p}_i$ to denote the *personal best* position and $\mathbf{p}_g$ the best position found by particles in its neighborhood. In the original PSO algorithm, $\mathbf{v}_i$ and $\mathbf{x}_i$, for $i = 1, \ldots, n$, are updated according to the following two equations [91]:

$$\mathbf{v}_i \leftarrow \mathbf{v}_i + \boldsymbol{\varphi}_1 \otimes (\mathbf{p}_i - \mathbf{x}_i) + \boldsymbol{\varphi}_2 \otimes (\mathbf{p}_g - \mathbf{x}_i), \tag{9}$$

$$\mathbf{x}_i \leftarrow \mathbf{x}_i + \mathbf{v}_i, \tag{10}$$

where $\boldsymbol{\varphi}_1 = c_1 \mathbf{R}_1$ and $\boldsymbol{\varphi}_2 = c_2 \mathbf{R}_2$. $\mathbf{R}_1$ and $\mathbf{R}_2$ are two separate functions each returning a vector comprising random values uniformly generated in the range [0,1]. c_1 and c_2 are acceleration coefficients. The symbol $\otimes$ denotes pointwise vector multiplication. Equation (9) shows that the velocity term $\mathbf{v}_i$ of a particle is determined by three parts, the "momentum", the "cognitive", and the "social" part. The "momentum" term $\mathbf{v}_i$ represents the previous velocity term which is used to carry the particle in the direction it has travelled so far; the "cognitive" part, $\boldsymbol{\varphi}_1 \otimes (\mathbf{p}_i - \mathbf{x}_i)$, represents the tendency of the particle to return to the best position it has visited so far; the "social" part, $\boldsymbol{\varphi}_2 \otimes (\mathbf{p}_g - \mathbf{x}_i)$, represents the tendency of the particle to be attracted towards the position of the best position found by the entire swarm.

Position $\mathbf{p}_g$ in the "social" part is the best position found by particles in the neighborhood of the ith particle. Different neighborhood topologies can be used to control information propagation between particles. Examples of neighborhood topologies include ring, star, and von Neumann. Constricted information propagation as a result of using small neighborhood topologies such as von Neumann has been shown to perform better on complex problems, whereas larger neighborhoods generally perform better on simpler problems [116]. Generally speaking, a PSO implementation that chooses $\mathbf{p}_g$ from within a restricted local neighborhood is referred to as *lbest* PSO, whereas choosing $\mathbf{p}_g$ without any restriction (hence from the entire swarm) results in a *gbest* PSO. Algorithm 1 summarizes a basic PSO algorithm.

Figure 3.1 shows each component of the velocity term $\mathbf{v}_i$ in vector form, and the resulting position, $\mathbf{x}_i$ (updated), for the ith particle. Note that the inertia coefficient w is used to scale the previous velocity term, normally to reduce the "momentum" of the particle. More discussion on w will be provided in the next section.

Earlier studies showed that the velocity as defined in Eq. (9) has a tendency to explode to a large value, resulting in particles exceeding the boundaries of the search space. This is more likely to happen especially when a particle is far from $\mathbf{p}_g$ or $\mathbf{p}_i$. To overcome this problem, a velocity clamping method can be adopted where the maximum allowed velocity value is set to V_{max} in each dimension of $\mathbf{v}_i$. This method does not necessarily prevent particles

Algorithm 1 The PSO algorithm, assuming maximization

Randomly generate an initial swarm
repeat
    for each particle i **do**
        if $f(\mathbf{x}_i) > f(\mathbf{p}_i)$ **then** $\mathbf{p}_i \leftarrow \mathbf{x}_i$
        $\mathbf{p}_g = \max(\mathbf{p}_{neighbours})$
        Update velocity (see Eq. (9))
        Update position (see Eq. (10))
    end for
until termination criterion is met

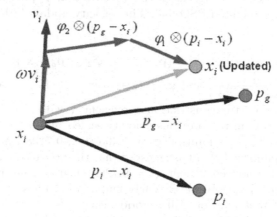

Fig. 5. Visualizing PSO components as vectors

from leaving the search space nor from converging. However, it does limit the particle step size, thereby preventing further divergence of particles.

Inertia Weight

Observe that the positions $\mathbf{p}_i$ and $\mathbf{p}_g$ in Eq. (9) can be collapsed into a single term $\mathbf{p}$ without losing any information:

$$\mathbf{v}_i \leftarrow \mathbf{v}_i + \boldsymbol{\varphi} \otimes (\mathbf{p} - \mathbf{x}_i), \tag{11}$$

$$\mathbf{x}_i \leftarrow \mathbf{x}_i + \mathbf{v}_i, \tag{12}$$

where $\mathbf{p} = \frac{\varphi_1 \mathbf{p}_i + \varphi_2 \mathbf{p}_g}{\varphi_1 + \varphi_2}$, and $\boldsymbol{\varphi} = \varphi_1 + \varphi_2$. Note that $\mathbf{p}$ represents the weighted average of the $\mathbf{p}_i$ and $\mathbf{p}_g$. It can be seen that the previous velocity term in Eq. (11) tends to keep the particle moving in the current direction. A coefficient *inertia weight*, w, can be used to control this influence on the new velocity. The velocity update (see Eq. (9)) can be now revised as:

$$\mathbf{v}_i \leftarrow w\mathbf{v}_i + \varphi_1 \otimes (\mathbf{p}_i - \mathbf{x}_i) + \varphi_2 \otimes (\mathbf{p}_g - \mathbf{x}_i) \tag{13}$$

The inertia-weighted PSO can converge under certain conditions even without using V_{max} [33]. For $w > 1$, velocities increase over time, causing particles to diverge eventually beyond the boundaries of the search space. For $w < 0$, velocities decrease over time, eventually reaching 0, resulting in convergence behavior. Eberhart and Shi suggested the use of a time-varying inertia weight, gradually decreasing its value typically from 0.9 to 0.4 (with $\varphi = 4.0$) [55].

Clerc described a general PSO algorithm that uses a *constriction coefficient*. Among the models suggested, the Constriction Type 1 PSO is equivalent to the inertia-weighted PSO [33]. The velocity update in Eq. (13) can be rewritten as:

$$\mathbf{v}_i \leftarrow \chi(\mathbf{v}_i + \varphi_1 \otimes (\mathbf{p}_i - \mathbf{x}_i) + \varphi_2 \otimes (\mathbf{p}_g - \mathbf{x}_i)), \tag{14}$$

where $\chi = \frac{2}{\left|2-\varphi-\sqrt{\varphi^2-4\varphi}\right|}$, and $\varphi = c_1 + c_2, \varphi > 4$. If φ is set to 4.1, and $c_1 = c_2 = 2.05$, then the constriction coefficient χ will be 0.7298. Applying χ in Eq. (14) results in the previous velocity scaled by 0.7298, and the "cognitive" and "social" parts multiplied by 1.496 (i.e., 0.7298 times 2.05). Both theoretical and empirical results suggested that the above configuration using a constant constriction coefficient $\chi = 0.7298$ ensures convergent behavior [55] without using V_{max}. However, early empirical studies by Eberhart and Shi suggested that it may be still a good idea to use velocity clamping together with the constriction coefficient, which showed improved performance on certain problems.

Fully Informed Particle Swarm

Equation (11) indicates that a particle tends to converge towards a point determined by $\mathbf{p} = \frac{\varphi_1\mathbf{p}_i+\varphi_2\mathbf{p}_g}{\varphi_1+\varphi_2}$, where $\varphi = \varphi_1+\varphi_2$. In the fully informed particle swarm (FIPS) as proposed by Mendes [116], $\mathbf{p}$ can be further generalized to any number of terms:

$$\mathbf{p} = \frac{\sum_{k\in\mathcal{N}} \mathbf{r}[0, \frac{c_{max}}{|\mathcal{N}|}] \otimes \mathbf{p}_k}{\sum_{k\in\mathcal{N}} \varphi_k}, \tag{15}$$

where $\mathbf{p}_k$ denotes the best previous position found by the kth particle in $\mathcal{N}$, which is a set of neighbors including the current particle itself. Note again that the division is a point wise operator here. If we set $k = 2$, $\mathbf{p}_1 = \mathbf{p}_i$, and $\mathbf{p}_2 = \mathbf{p}_g$, with both $\mathbf{p}_i, \mathbf{p}_g \in \mathcal{N}$, then the Constriction Type 1 PSO is just a special case of the more general PSO defined in Eq. (11). A significant implication of Eq. (15) is that it allows us to think more freely about employing terms of influence other than just $\mathbf{p}_i$ and $\mathbf{p}_g$ [116] [90].

PSO Variants

Although the canonical PSO was designed for continuous optimization, it can be extended to operate on binary search spaces. Kennedy and Eberhart developed a simple binary PSO by altering the velocity term in the canonical PSO into a probability threshold to determine if x_i is 0 or 1 [92]. PSO can be also extended to solve discrete or mixed (continuous and discrete) optimization problems [180, 32]. PSO can be adapted to work with discrete variables by simply discretizing the values after using them in the velocity and position update equations. Clerc provided several examples of PSO applied to combinatorial problems such as the knapsack, the traveling salesman, and the quadratic assignment problems [32].

An adaptive PSO version, *tribes*, was developed by Clerc [32], where the swarm size is determined by strategies for generating new particles as well as for removing poorly performing particles. The concept of a tribe is used to group particles that inform each other. Clerc's goal was to develop a PSO which can find the parameters on its own (e.g., swarm size), and still maintain a relatively good performance.

Kennedy proposed a PSO variant, bare-bones PSO, which does not use the velocity term [89]. In the bare-bones PSO each dimension of the new position of a particle is randomly selected from a Gaussian distribution with the mean being the average of p_i and p_g and the standard deviation being the distance between p_i and p_g:

$$x_i \leftarrow \mathcal{N}\left(\frac{p_i + p_g}{2}, \|p_i - p_g\|\right) \tag{16}$$

Note that there is no velocity term used in Eq. (16). The new particle position is simply generated via the Gaussian distribution. Sampling distributions other than Gaussian can also be employed [32, 147].

It has been observed that the canonical PSO tends to prematurely converge to local optima. To combat this problem, several PSO variants have incorporated a diversity maintenance mechanism. For example, ARPSO (attractive and repulsive PSO) was proposed to use a diversity measure to trigger an alternation between phases of attraction and repulsion [148]. A dissipative PSO was described in [179] to increase randomness. Similarly, a PSO with self-organized criticality was introduced in [107]. A PSO variant based on fitness-distance-ratio (FDR-PSO) was proposed in [173], to encourage interactions among particles that are both fit and close to each other. FDR-PSO was shown to give superior performance to the canonical PSO. FDR-PSO can be seen as using a dynamically defined neighborhood topology. Various neighborhood topologies have been adopted to restrict particle interactions [165, 116]. In particular, the von Neumann neighborhood topology has been shown to provide good performance across a range of test functions [116]. In [83], an H-PSO (Hierarchical PSO) was proposed, where a hierarchical tree structure is adopted to restrict the interactions among particles. Each particle is influenced only by its own personal best position and by the best position of the particle that is directly above it in the hierarchy. Another recently proposed PSO, CLPSO [105], which is in some way similar to FDR-PSO, allows incorporation of learning from more previous personal best positions. Gaussian distribution was employed in a PSO variant as a mutation operator in [79]. A cooperative PSO, similar to those previously developed coevolutionary algorithms, was also proposed in [171].

PSO variants have also been developed for solving multimodal optimization problems, where multiple equally good optima are sought. Niching methods such as crowding and fitness sharing that have been developed for evolutionary algorithms can be easily incorporated into PSO algorithms. Some representative PSO niching variants include NichePSO [142], SPSO (Speciation-based PSO) [101, 135, 8], and a PSO algorithm using a stretching function [138].

Applications of PSO Algorithms

PSO algorithms have been applied to optimization problems ranging from classical problems such as scheduling, the traveling salesman problem, neural network training, and task assignment, to highly specialized applications such as reactive power and voltage control [180], biomedical image registration [176], and even music composition [10]. In recent years, PSO is also a popular choice of many researchers for handling multiobjective optimization [155] and dynamic optimization problems [102].

One of the earliest applications of PSO was the evolution of neural network structures. Eberhart et al. used PSO to replace the traditional backpropagation learning algorithm in a multilayer perceptron [57]. Because of its fast convergence behavior, using PSO for neural network training can sometimes save a considerable amount of computation time compared with other optimization methods.

Table 3 shows a list of examples of PSO applications that can be found in the literature. For more information on PSO applications we refer the interested reader to [32].

3.2 Recent Trends

Theoretical Work on PSO

Since PSO was first introduced by Kennedy and Eberhart in 1995 [91], several studies have been carried out on understanding the convergence properties of PSO [33, 132, 170, 168]. Since an analysis of the convergence behavior of a swarm of multiple interactive particles is difficult, many of these works focus on studying the convergence behaviors of a simplified PSO system.

Kennedy provided the first analysis of a simplified particle behavior in [88], where particle trajectories for a range of variable choices were given. In [132],

Table 3. A representative selection of PSO applications

Problem	Authors	Reference
Traveling salesman problem	Onwubolu and Clerc	[130]
Flowshop scheduling	Rameshkumar, Suresh and Mohanasundaram	[143]
Task assignment	Salman, Imtiaz and Al-Madani	[150]
Neural networks	Kennedy, Eberhart, and Shi	[92]
	Mendes, Cortez, Rocha, and Neves	[115]
	Conradie, Miikkulaninen and Aldrich	[35]
	Gudisz and Venayagamoorthy	[71]
	Settles, Rodebaugh and Soule	[152]
Bioinformatics	Correa, Freitas and Johnson	[36]
	Georgiou, Pavlidis, Parsopoulos and Vrahatis	[66]
Industrial applications	Katare, Kalos and West	[87]
	Marinke, Matiko, Araujo and Coelho	[113]
Reactive power and voltage control	Yoshida, Kawata, et. al	[180]
PID controller	Gaing	[60]
Biomedical image registration	Wachowiak et. al	[176]
Floor planning	Sun, Hsieh, Wang and Lin	[166]
Quantizer design	Zha and Venayagamoorthy	[182]
Power systems	Venayagamoorthy	[174]
Clustering analysis	Chen and Ye	[30]
Constraint handling	Pulido and Coello	[140]
	Liang and Suganthan	[104]
Electromagnetic applications	Mikki and Kishk	[124]
Multiobjective problems	Moore and Chapman	[126]
	Coello and Lechuga	[34]
	Fieldsend and Singh	[58]
	Hu and Eberhart	[81]
	Parsopoulos and Vrahatis	[137]
	Li	[100]
Dynamic problems	Carlisle and Dozier	[28]
	Hu and Eberhart	[82]
	Eberhart and Shi	[56]
	Carlisle and Dozier	[29]
	Blackwell and Branke	[11, 12]
	Jason and Middendorf	[84]
	Parrott and Li	[135]
	Li, Blackwell, and Branke	[102]
Music	Blackwell and Bentley	[10]

Ozcan and Mohan showed that a particle in a one-dimensional PSO system, with its $\mathbf{p_i}$, $\mathbf{p_g}$, φ_1, and φ_2 kept constant, follows the path of a sinusoidal wave, where the amplitude and frequency of the wave are randomly decided.

A formal theoretical analysis of the convergence properties of a simplified PSO was provided by Clerc [33]. Clerc [33] represented the PSO system as defined in equations (11) and (12) as a dynamic system in state-space form. By simplifying the PSO to a deterministic dynamic system, its convergence can be shown based on the eigenvalues of the state matrix. A similar work was also carried out by Bergh and Engelbrecht [170], where regions of parameter space that guarantee convergence are identified. The conditions for convergence derived from both studies [33, 170] are: $w < 1$ and $w > \frac{1}{2}(c_1 + c_2) - 1$.

In a more recent work [172], Bergh and Engelbrecht generalized the above analysis by including the inertia weight w, and also provided a formal convergence proof of particles in this representation. Furthermore, they studied the particle trajectory with a relaxed assumption to allow stochastic values for φ_1 and φ_2. They demonstrated that a particle can exhibit a combination of divergent and convergent behaviors with certain probabilities when different values of φ_1 and φ_2 are used.

In [85], Kadirkamanathan et al. recently provided a new approach to the convergence analysis of PSO without the assumption of non-random PSO. The analysis of stochastic particle dynamics was made feasible by representing particle dynamics as a nonlinear feedback controlled system as formulated by Lure [53]. The convergence analysis was carried out using the concept of passive systems and Lyapunov stability [175]. Some conservative conditions for convergence were derived in this study.

PSO for Multiobjective Optimization

Multiobjective optimization (MO) problems represent an important class of real-world problems. Typically such problems involve trade-offs. For example, a car manufacturer may wish to maximize its profit, but meanwhile also to minimize its production cost. These objectives are typically conflicting to each other. For example, a higher profit could increase the production cost. Generally, there is no single optimal solution. Often the manufacturer needs to consider many possible "trade-off" solutions before choosing the one that suits its need. The curve or surface (for more than two objectives) describing the optimal trade-off solutions between objectives is known as the *Pareto front*. A multiobjective optimization algorithm is required to find solutions as close as possible to the Pareto front, while maintaining a good solution diversity along the Pareto front.

To apply PSO to multiobjective optimization problems, several issues have to be taken into consideration:

1. How to choose $\mathbf{p_g}$ (i.e., a leader) for each particle? The PSO needs to favor non-dominated particles over dominated ones, and drive the population towards different parts of the Pareto front, not just towards a single point. This requires that particles be allocated to different leaders.
2. How to identify non-dominated particles with respect to all particles' current positions and personal best positions? And how to retain these solutions during the search process? One strategy is to combine all particles' personal best positions and current positions, and then extract the non-dominated solutions from the combined population.
3. How to maintain particle diversity so that a set of well-distributed solutions can be found along the Pareto front? Some classic niching methods (e.g., crowding or sharing) can be adopted for this purpose.

The first PSO for solving multiobjective optimization was proposed by Moore and Chapman in 1999 [126]. An *lbest* PSO was used, and $\mathbf{p_g}$ was chosen from a local neighborhood using a ring topology. All personal best positions were kept in an archive. At each particle update, the current position is compared with solutions in this archive to see if the current position represents a non-dominated solution. The archive is updated at each iteration to ensure it contains only non-dominated solutions.

Interestingly it was not until 2002 that the next publication on PSO for multiobjective optimization appeared. In [34], Coello and Lechuga proposed MOPSO (Multiobjective PSO) which uses an external archive to store non-dominated solutions. The diversity of solutions is maintained by keeping only one solution within each hypercube which is predefined by a user in the objective space. In [137], Parsopoulos and Vrahatis adopted a more traditional weighted-sum approach. However, by using gradually changing weights, their approach was able to find a diverse set of solutions along the Pareto front. In [58], Fieldsend and Singh proposed a PSO using a *dominated tree* structure to store non-dominated solutions found. The selection of leaders was also based on this structure. To maintain a better diversity, a *turbulence* operator was adopted to function as a 'mutation' operator in order to perturb the velocity value of a particle.

With the aim of increasing the efficiency of extracting non-dominated solutions from a swarm, Li proposed NSPSO (Non-dominated Sorting PSO) [100], which follows the principal idea of the well-known NSGA II algorithm [39]. In NSPSO, instead of comparing solely a particle's personal best with its potential offspring, all particles' personal best positions and offspring are first combined to form a temporary population. After this, domination comparisons for all individuals in this temporary population are carried out. This approach will ensure more non-dominated solutions can be discovered through the domination comparison operations than the above-mentioned multiobjective PSO algorithms.

Many more multiobjective PSO variants have been proposed in recent years. A survey conducted by Sierra and Coello in 2006 shows that there are currently 25 different PSO algorithms for handling multiobjective optimization problems. Interested readers should refer to [155] for more information on these different approaches.

PSO for Dynamic Optimization

Many real-world optimization problems are dynamic and require optimization algorithms capable of adapting to the changing optima over time. For example, traffic conditions in a city change dynamically and continuously. What might be regarded as an optimal route at one time might not be optimal the next minute. In contrast to optimization towards a static optimum, in a dynamic environment the goal is to track as closely as possible the dynamically changing optima. Figure 6 shows an example of a three-peak dynamic environment.

A defining characteristic of PSO is its fast convergent behavior and inherent adaptability [92]. Particles can adaptively adjust their positions based

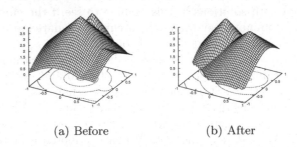

(a) Before   (b) After

Fig. 6. Three-peak dynamic environment, before and after movement of optima. Note that the small peak to the right of the figure becomes hidden and that the highest point switches optimum

on their dynamic interactions with other particles in the population. This makes PSO especially appealing as a potential solution to dynamic optimization problems. Several studies have suggested various approaches to applying PSO to solve dynamic optimization problems [28, 29, 56, 82, 103, 11, 12, 135]. These studies showed that the original PSO must be adapted to meet the additional challenges presented by dynamic optimization problems. In particular, the following questions need to be addressed:

1. How do we detect a change that has actually occurred?
2. Which response strategies are appropriate to use once a change is detected?
3. How do we handle the issue of 'out-of-date' memory as particles' personal best positions become invalid once the environment has changed?
4. How do we handle the trade-off issue between convergence (in order to locate optima) and diversity (in order to relocate changed optima)?

One of the early works on using PSO for dynamic optimization was by Eberhart and Shi in [56], where they used an inertia-weighted PSO to track the optimum of a three-dimensional unimodal parabolic function which changes its maxima every 100 iterations. It was found under certain circumstances that the PSO's performance was comparable to or better than that of previously published evolutionary algorithms.

For detection, Carlisle and Dozier used a *sentry* particle which is randomly chosen at each iteration [28]. The sentry particle gets evaluated before each iteration and compares its fitness with its previous fitness value. If the two values are different, indicating the environment has changed, then the whole population gets alerted and several possible responses can then be triggered. A simple strategy was also proposed by Hu and Eberhart to re-evaluate p_g and a second-best particle to detect if a change has occurred [82].

Various response strategies have been proposed. To deal with the issue of 'out-of-date' memory as the environment changes, Carlisle and Dozier proposed to periodically replace all personal best positions by their corresponding current positions when a change has been detected [29]. This allows particles to forget their past experience and use only up-to-date knowledge about the new environment. Hu and Eberhart studied the effects of re-randomizing various proportions of the swarm to maintain some degree of diversity in order to better track the optima after a change [82]. However, this approach suffers from possible information loss since the re-randomized portion of the population does not retain any information that might be useful from the past iterations.

In order to maintain better particle diversity throughout a run, Blackwell and Bentley introduced charged swarms where mutually repelling *charged* particles orbit a nucleus of neutral particles (conventional PSO particles) [9]. Whereas the charged particles allow the swarm to better adapt to changes in the environment, the neutral particles play the role of continuing to converge towards the optimum.

Inspired by multi-population EA approaches such as the self-organizing scouts [24], Blackwell and Branke proposed an interacting multi-swarm PSO as a further improvement to the charged swarms [11]. The multi-swarm PSO aims at maintaining multiple swarm populations on different peaks. Multiple swarms are prevented from converging to the same optimum by randomizing the worse of two swarms that come too close. The multi-swarm PSO also replaces the charged particles with quantum particles whose position is solely based on a probability function centered around the swarm attractor. The resulting multi-quantum swarms outperform charged and standard PSOs on the moving peaks function. This multi-swarm approach is particularly attractive because of its improved adaptability in a more complex multimodal dynamic environment where multiple peaks exist and need to be tracked.

With a similar aim to locate and track multiple peaks in a dynamic environment, Parrott and Li in [101, 135] proposed a species-based PSO (SPSO) incorporating a speciation algorithm first proposed by Pétrowski [139]. The SPSO uses a local "species seed" which provides the local p_g to particles whose positions are within a user-specified radius of the seed. This encourages swarms to converge onto multiple local optima instead of a single global optimum, hence developing multiple sub-populations in parallel. In addition, the dynamic SPSO uses a parameter p_{max} to limit the number of particles allowed in a species (or swarm), with the excess particles reinitialized at random positions in the total search space. In [102], Li et al. also demonstrated that the quantum particle model in [11] can be incorporated into SPSO to improve its optima-tracking performance for the moving peaks problem [24].

In another work [84], Janson and Middendorf proposed a PSO using a dynamic and hierarchical neighborhood structure to handle dynamic optimization problems. They demonstrated that such a structure is useful for maintaining some particle diversity in a dynamic environment.

PSO for Constraint Handling

Many real-world problems require an optimization algorithm to find solutions that satisfy a certain number of constraints. The most common approach for solving constrained problems is the use of a penalty function, where the constrained problem is transformed into an unconstrained one, by penalizing the constraints and creating a single objective function. Parsopoulos and Vrahatis proposed a PSO where non-stationary penalty functions are used [136]. The penalty value is dynamically modified during a run. This method is problem dependent; however its results are generally superior to those obtained through stationary functions. In Toscano and Coello's PSO algorithm [141], if both particles compared are infeasible, then the particle that has the lowest value in its total violation of constraints wins. One major disadvantage of using penalty functions, in which case all constraints must be combined into a single objective function (this is also called the weighted-sum approach), is that a user must specify a weight coefficient for each constraint. However, finding optimal weight coefficients is no easy task. A preferred approach is a multiobjective one where the concept of "dominance" can be used to identify better solutions which are non-dominated solutions with respect to the current population. The merit of this multiobjective approach is that the user is no longer required to specify any weight coefficient.

Another useful technique as described by Clerc is "confinement by dichotomy" [32], which makes use of an iterative procedure to find points that are close to the boundaries defined by constraints. Both the dichotomy and multiobjective methods are general enough that they are applicable to most constrained optimization problems.

4 Further Examples of Swarm Intelligence in Optimization

ACO and PSO are two very successful examples of swarm intelligence, yet there are many more applications based on SI principles. Some representative examples are given in the following.

4.1 Applications Inspired by the Division of Labour

Algorithms based on the division of labour in ant colonies and wasp colonies are an important example of the use of swarm intelligence principles in technical applications. Much of the relevant works go back to the study of Wilson [178], who showed that the concept of division of labour in colonies of ants from the *Pheidole* genus allows the colony to adapt to changing demands. Workers in these colonies are generally divided into two groups: Small minors and larger majors. The minors are mostly doing quotidian tasks, whereas the

majors do seed milling, storing of abdominal food, or defense tasks. By experimentally reducing the number of minors, Wilson observed that some of the
majors switched to tasks usually fulfilled by minors. The division of labour
was later modelled by Theraulaz et al. [167] and Bonabeau et al. [22] by means
of response threshold models. The model permits a set of threshold values for
each individual, one threshold value for each type of task. A threshold value,
which may be fixed or dynamically changing over time, can be interpreted as
the degree of specialization for the respective task. Furthermore, each task
emits a stimulus in order to attract the attention of the individuals, which—
depending on the stimulus and the corresponding threshold value—decide
whether to accept or to decline the task.

The above-mentioned response threshold models inspired several technical applications. Campos et al. [27], Cicirello and Smith [31], and Nouyan et
al. [129] deal with a static, or a dynamic task allocation problem where trucks
have to be painted in a number of painting booths. Another application concerns media streaming in peer-to-peer networks. Here, a peer must adapt to
changes in the supply and demand of media streams. For this purpose, Sasabe
et al. [151] propose a novel caching algorithm based on a response threshold
model. In [181], Yu and Ram propose a multi-agent system for the scheduling of dynamic job shops with flexible routing and sequence-dependent setups
based on the division of labour in social insects. Finally, Merkle et al. [119]
use a response threshold model for the self-organized task allocation for computing systems with reconfigurable components.

4.2 Ant-Based Clustering and Sorting

In 1991 Deneubourg et al. [43] proposed a model to describe the clustering as
well as the sorting behavior of ants. Here, clustering refers to the gathering
of items in order to form heaps. This happens, for example, when ants of the
species *Pheidole pallidula* cluster the bodies of dead nest mates (also known
as cemetery formation). Sorting, on the other hand, refers to the spatial ar-
rangement of different objects according to their properties, a behavior which
can be observed, for example, in nests of the species *Leptothorax unifasciatus*.
Ants of this species compactly cluster eggs and microlarvae at the center of
the brood area, whereas the largest larvae are placed at the periphery of the
brood area. In computer simulations in [43] ants were modelled as agents randomly moving in their environment in which items were initially scattered.
Agents were able to pick up items, to transport them, and to drop them. The
probabilities for these actions were derived from the distribution of items in
the agents' local neighborhood. For example, items that are isolated had a
higher probability of being picked up. As a result, a clustering and sorting of
items in the agents environment was obtained.

Mostly based on the above mentioned model by Deneubourg et al., several algorithms for clustering and sorting appeared in the literature. The first one was an algorithm proposed in [108] that extended the original model in order to be able to handle numerical data. Later papers deal with the models' use for the two-dimensional visualization of document collections such as Web data (see, for example, [77]) and for graph partitioning (see, for example, [97]).

4.3 Other Applications

Recently, research on swarm robotics has taken much inspiration from swarm intelligence. For example, the path finding and orientation skills of the desert ant *Cataglyphis* were used as an archetype for building a robot orientation unit [98]. Models for the division of labor between members of an ant colony were used to regulate the joint work of robots (see, for example, [1]). In [96] the collective transport of ants inspired the design of controllers of robots for doing coordinated work. More detailed and up-to-date information can be found in Chap. 3 of this book.

Acknowledgements

This work was supported by grants TIN2005-08818 (OPLINK) and TIN2007-66523 (FORMALISM) of the Spanish government, and by the EU project FRONTS (FP7-ICT-2007-1) funded by the European Comission under the FET Proactive Initiative *Pervasive Adaptation*. In addition, Christian Blum acknowledges support from the *Ramón y Cajal* program of the Spanish Ministry of Science and Technology of which he is a research fellow.

References

1. W. Agassounoun, A. Martinoli, and R. Goodman. A scalable, distributed algorithm for allocating workers in embedded systems. In Terry Bahill, editor, *Proceedings of the 2001 IEEE Systems, Man and Cybernetics Conference*, pages 3367–3373. IEEE Press, 2001.
2. S. Alupoaei and S. Katkoori. Ant colony system application to macrocell overlap removal. *IEEE Transactions on Very Large Scale Integration (VLSI) Systems*, 12(10):1118–1122, 2004.
3. J. Bautista and J. Pereira. Ant algorithms for a time and space constrained assembly line balancing problem. *European Journal of Operational Research*, 177(3), 2007.
4. G. Beni. The concept of cellular robotic systems. In *Proceedings of the IEEE International Symposium on Intelligent Systems*, pages 57–62. IEEE Press, Piscataway, NJ, 1988.

5. L. Bianchi, L. M. Gambardella, and M. Dorigo. An ant colony optimization approach to the probabilistic traveling salesman problem. In J. J. Merelo, P. Adamidis, H.-G. Beyer, J.-L. Fernández-Villacanas, and H.-P. Schwefel, editors, *Proceedings of PPSN VII, Seventh International Conference on Parallel Problem Solving from Nature*, volume 2439 of *Lecture Notes in Computer Science*, pages 883–892. Springer, Berlin, Germany, 2002.

6. B. Bilchev and I. C. Parmee. The ant colony metaphor for searching continuous design spaces. In T. C. Fogarty, editor, *Proceedings of the AISB Workshop on Evolutionary Computation*, volume 993 of *Lecture Notes in Computer Science*, pages 25–39. Springer, Berlin, Germany, 1995.

7. M. Birattari, G. Di Caro, and M. Dorigo. Toward the formal foundation of ant programming. In M. Dorigo, G. Di Caro, and M. Sampels, editors, *Ant Algorithms – Proceedings of ANTS 2002 – Third International Workshop*, volume 2463 of *Lecture Notes in Computer Science*, pages 188–201. Springer, Berlin, Germany, 2002.

8. S. Bird and X. Li. Adaptively choosing niching parameters in a PSO. In Mike Cattolico, editor, *Genetic and Evolutionary Computation Conference, GECCO 2006, Proceedings, Seattle, Washington, USA, July 8-12, 2006*, pages 3–10. ACM, 2006.

9. T. Blackwell and P. J. Bentley. Dynamic search with charged swarms. In *Proc. the Workshop on Evolutionary Algorithms Dynamic Optimization Problems (EvoDOP 2003)*, pages 19–26, 2002.

10. T. Blackwell and P. J. Bentley. Improvised music with swarms. In David B. Fogel, Mohamed A. El-Sharkawi, Xin Yao, Garry Greenwood, Hitoshi Iba, Paul Marrow, and Mark Shackleton, editors, *Proceedings of the 2002 Congress on Evolutionary Computation CEC 2002*, pages 1462–1467. IEEE Press, 2002.

11. T. Blackwell and J. Branke. Multi-swarm optimization in dynamic environments. In *EvoWorkshops*, volume 3005 of *Lecture Notes in Computer Science*, pages 489–500. Springer, 2004.

12. T. Blackwell and J. Branke. Multi-swarms, exclusion and anti-convergence in dynamic environments. *IEEE Transactions on Evolutionary Computation*, 10(4):459–472, 2006.

13. M. Blesa and C. Blum. Ant colony optimization for the maximum edge-disjoint paths problem. In G. R. Raidl, S. Cagnoni, J. Branke, D. W. Corne, R. Drechsler, Y. Jin, C. G. Johnson, P. Machado, E. Marchiori, R. Rothlauf, G. D. Smith, and G. Squillero, editors, *Applications of Evolutionary Computing, Proceedings of EvoWorkshops 2004*, volume 3005 of *Lecture Notes in Computer Science*, pages 160–169. Springer, Berlin, Germany, 2004.

14. C. Blum. Beam-ACO—Hybridizing ant colony optimization with beam search: An application to open shop scheduling. *Computers & Operations Research*, 32(6):1565–1591, 2005.

15. C. Blum, J. Bautista, and J. Pereira. Beam-ACO applied to assembly line balancing. In M. Dorigo, L. M. Gambardella, A. Martinoli, R. Poli, and T. Stützle, editors, *Ant Colony Optimization and Swarm Intelligence – Proceedings of ANTS 2006 – Fifth International Workshop*, volume 4150 of *Lecture Notes in Computer Science*, pages 96–107. Springer, Berlin, Germany, 2006.

16. C. Blum and M. Dorigo. The hyper-cube framework for ant colony optimization. *IEEE Transactions on Systems, Man, and Cybernetics – Part B*, 34(2):1161–1172, 2004.

17. C. Blum and M. Dorigo. Search bias in ant colony optimization: On the role of competition-balanced systems. *IEEE Transactions on Evolutionary Computation*, 9(2):159–174, 2005.
18. C. Blum and A. Roli. Metaheuristics in combinatorial optimization: Overview and conceptual comparison. *ACM Computing Surveys*, 35(3):268–308, 2003.
19. C. Blum and M. Sampels. An ant colony optimization algorithm for shop scheduling problems. *Journal of Mathematical Modelling and Algorithms*, 3(3):285–308, 2004.
20. C. Blum and M. Yábar Vallès. Multi-level ant colony optimization for DNA sequencing by hybridization. In F. Almeida, M. Blesa, C. Blum, J. M. Moreno, M. Pérez, A. Roli, and M. Sampels, editors, *Proceedings of HM 2006 – 3rd International Workshop on Hybrid Metaheuristics*, volume 4030 of *Lecture Notes in Computer Science*, pages 94–109. Springer-Verlag, Berlin, Germany, 2006.
21. E. Bonabeau, M. Dorigo, and G. Theraulaz. *Swarm Intelligence: From Natural to Artificial Systems*. Oxford University Press, New York, NY, 1999.
22. E. Bonabeau, G. Theraulaz, and J.-L. Deneubourg. Fixed response thresholds and the regulation of division of labor in social societies. *Bulletin of Mathematical Biology*, 60:753–807, 1998.
23. A. Brandt. Multilevel computations: Review and recent developments. In S. F. McCormick, editor, *Multigrid Methods: Theory, Applications, and Supercomputing, Proceedings of the 3rd Copper Mountain Conference on Multigrid Methods*, volume 110 of *Lecture Notes in Pure and Applied Mathematics*, pages 35–62. Marcel Dekker, New York, 1988.
24. J. Branke. *Evolutionary Optimization in Dynamic Environments*. Kluwer Academic Publishers, Norwell, MA, 2002.
25. T. N. Bui and J. R. Rizzo Jr. Finding maximum cliques with distributed ants. In K. Deb et al., editor, *Proceedings of the Genetic and Evolutionary Computation Conference (GECCO 2004)*, volume 3102 of *Lecture Notes in Computer Science*, pages 24–35. Springer, Berlin, Germany, 2004.
26. B. Bullnheimer, R. Hartl, and C. Strauss. A new rank-based version of the Ant System: A computational study. *Central European Journal for Operations Research and Economics*, 7(1):25–38, 1999.
27. M. Campos, E. Bonabeau, G. Theraulaz, and J.-L. Deneubourg. Dynamic scheduling and division of labor in social insects. *Adaptive Behavior*, 8(3):83–96, 2000.
28. A. Carlisle and G. Dozier. Adapting particle swarm optimization to dynamic environments. In the *Proceedings of the International Conference on Artificial Intelligence (ICAI 2000)*, pages 429–434, Las Vegas, Nevada, USA, 2000.
29. A. Carlisle and G. Dozier. Tracking changing extrema with adaptive particle swarm optimizer. In *Proceedings of the 5th Biannual World Automation Congress*, pages 265–270, Orlando FL, USA, 2002.
30. C.-Y. Chen and F. Ye. Particle swarm optimization algorithm and its application to clustering analysis. In *IEEE International Conference on Networking, Sensing and Control*, volume 2, pages 789–794, 2004.
31. V. A. Cicirello and S. S. Smith. Wasp-like agents for distributed factory coordination. *Journal of Autonomous Agents and Multi-Agent Systems*, 8:237–266, 2004.
32. M. Clerc. *Particle Swarm Optimization*. ISTE Ltd, UK, 2006.

33. M. Clerc and J. Kennedy. The particle swarm—explosion, stability, and convergence in a multidimensional complex space. *IEEE Transactions on Evolutionary Computation*, 6:58–73, 2002.

34. C. Coello Coello and M. Salazar Lechuga. MOPSO: A Proposal for Multiple Objective Particle Swarm Optimization. In *Congress on Evolutionary Computation (CEC 2002)*, volume 2, pages 1051–1056, Piscataway, New Jersey, May 2002. IEEE Service Center.

35. A.V.E. Conradie, R. Miikkulaninen, and C. Aldrich. Adaptive control utilizing neural swarming. In *Proc. of Genetic and Evolutionary Computation Conference (GECCO)*, New York, USA, 2002.

36. E.S. Correa, A. Freitas, and C.G. Johnson. A new discrete particle swarm algorithm applied to attribute selection in a bioinformatics data set. In *GECCO 2006: Proceedings of the 8th Annual Conference on Genetic and Evolutionary Computation*, volume 1, pages 35–42, Seattle, Washington, USA, 2006. ACM Press.

37. P. Corry and E. Kozan. Ant colony optimization for machine layout problems. *Computational Optimization and Applications*, 28(3):287–310, 2004.

38. D. Costa and A. Hertz. Ants can color graphs. *Journal of the Operational Research Society*, 48:295–305, 1997.

39. K. Deb, A. Pratap, S. Agrawal, and T. Meyarivan. A fast and elitist multiobjective genetic algorithm: NSGA II. *IEEE Transactions on Evolutionary Computation*, 6(2):182–197, 2002.

40. F. Della Croce, M. Ghirardi, and R. Tadei. Recovering beam search: enhancing the beam search approach for combinatorial optimisation problems. In *Proceedings of PLANSIG 2002 – 21st workshop of the UK Planning and Scheduling Special Interest Group*, pages 149–169, 2002.

41. M. L. den Besten, T. Stützle, and M. Dorigo. Ant colony optimization for the total weighted tardiness problem. In M. Schoenauer, K. Deb, G. Rudolph, X. Yao, E. Lutton, J. J. Merelo, and H.-P. Schwefel, editors, *Proceedings of PPSN VI, Sixth International Conference on Parallel Problem Solving from Nature*, volume 1917 of *Lecture Notes in Computer Science*, pages 611–620. Springer, Berlin, Germany, 2000.

42. J.-L. Deneubourg, S. Aron, S. Goss, and J.-M. Pasteels. The self-organizing exploratory pattern of the *Argentine* ant. *Journal of Insect Behaviour*, 3:159–168, 1990.

43. J.-L. Deneubourg, S. Goss, N. Franks, A. Sendova-Franks, C. Detrain, and L. Chrétien. The dynamics of collective sorting: Robot-like ants and ant-like robots. In *Proceedings of the First International Conference on Simulation of Adaptive Behaviour: From Animals to Animats 1*, pages 356–365. MIT Press, Cambridge, MA, 1991.

44. G. Di Caro and M. Dorigo. AntNet: Distributed stigmergetic control for communications networks. *Journal of Artificial Intelligence Research*, 9:317–365, 1998.

45. K. Doerner, W. J. Gutjahr, R. F. Hartl, C. Strauss, and C. Stummer. Pareto ant colony optimization: A metaheuristic approach to multiobjective portfolio selection. *Annals of Operations Research*, 131:79–99, 2004.

46. M. Dorigo. *Optimization, Learning and Natural Algorithms* (in Italian). PhD thesis, Dipartimento di Elettronica, Politecnico di Milano, Italy, 1992.

47. M. Dorigo and C. Blum. Ant colony optimization theory: A survey. *Theoretical Computer Science*, 344(2-3):243–278, 2005.

48. M. Dorigo, G. Di Caro, and L. M. Gambardella. Ant algorithms for discrete optimization. *Artificial Life*, 5(2):137–172, 1999.
49. M. Dorigo and L. M. Gambardella. Ant Colony System: A cooperative learning approach to the traveling salesman problem. *IEEE Transactions on Evolutionary Computation*, 1(1):53–66, 1997.
50. M. Dorigo, V. Maniezzo, and A. Colorni. Positive feedback as a search strategy. Technical Report 91-016, Dipartimento di Elettronica, Politecnico di Milano, Italy, 1991.
51. M. Dorigo, V. Maniezzo, and A. Colorni. Ant System: Optimization by a colony of cooperating agents. *IEEE Transactions on Systems, Man, and Cybernetics – Part B*, 26(1):29–41, 1996.
52. M. Dorigo and T. Stützle. *Ant Colony Optimization*. MIT Press, Cambridge, MA, 2004.
53. C. A. Dosoer and M. Vidyasagar. *Feedback Systems: Input–Ouput Properties*. Academics, New York, 1975.
54. J. Dréo and P. Siarry. A new ant colony algorithm using the heterarchical concept aimed at optimization of multiminima continuous functions. In M. Dorigo, G. Di Caro, and M. Sampels, editors, *Ant Algorithms – Proceedings of ANTS 2002 – Third International Workshop*, volume 2463 of *Lecture Notes in Computer Science*, pages 216–221. Springer, Berlin, Germany, 2002.
55. R. Eberhart and Y. Shi. Comparing inertia weights and constriction factors in particle swarm optimization. In *Proc. of IEEE Int. Conf. Evolutionary Computation*, pages 84–88, 2000.
56. R. C. Eberhart and Y. Shi. Tracking and optimizing dynamic systems with particle swarms. In *Proc. the 2001 Congress on Evolutionary Computation (CEC 2001)*, pages 94–100. IEEE Press, 2001.
57. R. C. Eberhart, P. K. Simpson, and R. W. Dobbins. *Computational Intelligence PC Tools*. Academic Press, Boston, 1996.
58. J. E. Fieldsend and S. Singh. A multiobjective algorithm based upon particle swarm optimisation, an efficient data structure and turbulence. In *Proceedings of the 2002 UK Workshop on Computational Intelligence*, pages 37–44, Birmingham, UK, September 2002.
59. C. Gagné, W. L. Price, and M. Gravel. Comparing an ACO algorithm with other heuristics for the single machine scheduling problem with sequence-dependent setup times. *Journal of the Operational Research Society*, 53:895–906, 2002.
60. Z. L. Gaing. A particle swarm optimization approach for optimum design of PID controller in AVR system. *IEEE Transactions on Energy Conversion*, 19(2):384–391, June 2004.
61. L. M. Gambardella and M. Dorigo. Solving symmetric and asymmetric TSPs by ant colonies. In T. Baeck, T. Fukuda, and Z. Michalewicz, editors, *Proceedings of the 1996 IEEE International Conference on Evolutionary Computation (ICEC'96)*, pages 622–627. IEEE Press, Piscataway, NJ, 1996.
62. L. M. Gambardella and M. Dorigo. Ant Colony System hybridized with a new local search for the sequential ordering problem. *INFORMS Journal on Computing*, 12(3):237–255, 2000.
63. L. M. Gambardella, É. D. Taillard, and G. Agazzi. MACS-VRPTW: A multiple ant colony system for vehicle routing problems with time windows. In D. Corne, M. Dorigo, and F. Glover, editors, *New Ideas in Optimization*, pages 63–76. McGraw-Hill, London, UK, 1999.

64. X. Gandibleux, X. Delorme, and V. T'Kindt. An ant colony optimisation algorithm for the set packing problem. In M. Dorigo, M. Birattari, C. Blum, L. M. Gambardella, F. Mondada, and T. Stützle, editors, *Proceedings of ANTS 2004 – Fourth International Workshop on Ant Colony Optimization and Swarm Intelligence*, volume 3172 of *Lecture Notes in Computer Science*, pages 49–60. Springer, Berlin, Germany, 2004.

65. M. R. Garey and D. S. Johnson. *Computers and Intractability: A Guide to the Theory of NP-Completeness*. W. H. Freeman, 1979.

66. V. Georgiou, N. Pavlidis, K. Parsopoulos, and M. Vrahatis. Optimizing the performance of probabilistic neural networks in a bioinformatics task. In *Proceedings of the EUNITE 2004 Conference*, pages 34–40, 2004.

67. F. Glover. Future paths for integer programming and links to artificial intelligence. *Computers & Operations Research*, 13:533–549, 1986.

68. F. Glover and G. Kochenberger, editors. *Handbook of Metaheuristics*. Kluwer Academic Publishers, Norwell, MA, 2002.

69. J. Gottlieb, M. Puchta, and C. Solnon. A study of greedy, local search, and ant colony optimization approaches for car sequencing problems. In S. Cagnoni, J. J. Romero Cardalda, D. W. Corne, J. Gottlieb, A. Guillot, E. Hart, C. G. Johnson, E. Marchiori, J.-A. Meyer, M. Middendorf, and G. R. Raidl, editors, *Applications of Evolutionary Computing, Proceedings of EvoWorkshops 2003*, volume 2611 of *Lecture Notes in Computer Science*, pages 246–257. Springer, Berlin, Germany, 2003.

70. P.-P. Grassé. La reconstruction du nid et les coordinations inter-individuelles chez *bellicositermes natalensis et cubitermes sp*. La théorie de la stigmergie: Essai d'interprétation des termites constructeurs. *Insectes Sociaux*, 6:41–81, 1959.

71. V. G. Gudise and G. K. Venayagamoorthy. Comparison of particle swarm optimization and backpropagation as training algorithms for neural networks. In *IEEE Swarm Intelligence Symposium 2003 (SIS 2003)*, pages 110–117, Indianapolis, Indiana, USA, 2003.

72. C. Guéret, N. Monmarché, and M. Slimane. Ants can play music. In M. Dorigo, M. Birattari, C. Blum, L. M. Gambardella, F. Mondada, and T. Stützle, editors, *Proceedings of ANTS 2004 – Fourth International Workshop on Ant Colony Optimization and Swarm Intelligence*, volume 3172 of *Lecture Notes in Computer Science*, pages 310–317. Springer, Berlin, Germany, 2004.

73. M. Guntsch and M. Middendorf. Pheromone modification strategies for ant algorithms applied to dynamic TSP. In E. J. W. Boers, J. Gottlieb, P. L. Lanzi, R. E. Smith, S. Cagnoni, E. Hart, G. R. Raidl, and H. Tijink, editors, *Applications of Evolutionary Computing: Proceedings of EvoWorkshops 2001*, volume 2037 of *Lecture Notes in Computer Science*, pages 213–222. Springer, Berlin, Germany, 2001.

74. M. Guntsch and M. Middendorf. Solving multiobjective permutation problems with population based ACO. In C. M. Fonseca, P. J. Fleming, E. Zitzler, K. Deb, and L. Thiele, editors, *Proceedings of the Second International Conference on Evolutionary Multi-Criterion Optimization (EMO 2003)*, volume 2636 of *Lecture Notes in Computer Science*, pages 464–478. Springer, Berlin, Germany, 2003.

75. W. J. Gutjahr. A graph-based ant system and its convergence. *Future Generation Computer Systems*, 16(9):873–888, 2000.

76. W. J. Gutjahr. ACO algorithms with guaranteed convergence to the optimal solution. *Information Processing Letters*, 82(3):145–153, 2002.
77. J. Handl and B. Meyer. Improved ant-based clustering and sorting in a document retrieval interface. In J. J. Merelo, P. Adamidis, H.-G. Beyer, J.-L. Fernández-Villacanas, and H.-P. Schwefel, editors, *Proceedings of PPSN VII, Seventh International Conference on Parallel Problem Solving from Nature*, volume 2439 of *Lecture Notes in Computer Science*, pages 913–923. Springer, Berlin, Germany, 2002.
78. F. Heppner and U. Grenander. A stochastic nonlinear model for coordinated bird flocks. In S. Krasner, editor, *The Ubiquity of Chaos*, Washington, DC, 1990. AAAS Publications.
79. N. Higashi and H. Iba. Particle swarm optimization with Gaussian mutation. In *Proc. of the 2003 IEEE Swarm Intelligence Symposium (SIS'03)*, pages 72–79, 2003.
80. H. H. Hoos and T. Stützle. *Stochastic Local Search: Foundations and Applications*. Elsevier, Amsterdam, The Netherlands, 2004.
81. X. Hu and R. Eberhart. Multiobjective optimization using dynamic neighborhood particle swarm optimization. In *Proceedings of the 2002 Congress on Evolutionary Computation CEC 2002*, pages 1677–1681. IEEE Press, 2002.
82. X. Hu and R. C. Eberhart. Adaptive particle swarm optimisation: detection and response to dynamic systems. In *Proc. Congress on Evolutionary Computation*, pages 1666–1670, 2002.
83. S. Janson and M. Middendorf. A hierarchical particle swarm optimizer and its adaptive variant. *IEEE Transactions on Systems, Man, and Cybernetics, Part B*, 35(6):1272–1282, 2005.
84. S. Janson and M. Middendorf. A hierarchical particle swarm optimizer for noisy and dynamic environments. *Genetic Programming and Evolvable Machines*, 7(4):329–354, 2006.
85. V. Kadirkamanathan, K. Selvarajah, and P. Fleming. Stability analysis of the particle dynamics in particle swarm optimizer. *IEEE Transactions on Evolutionary Computation*, 10(3):245–255, June 2006.
86. O. Karpenko, J. Shi, and Y. Dai. Prediction of MHC class II binders using the ant colony search strategy. *Artificial Intelligence in Medicine*, 35(1-2):147–156, 2005.
87. S. Katare, A. Kalos, and D. West. A hybrid swarm optimizer for efficient parameter estimation. In *Proceedings of the 2004 Congress on Evolutionary Computation CEC 2004)*, pages 309–315. IEEE Press, 2004.
88. J. Kennedy. The behaviour of particles. In *Evolutionary Programming VII: Proceedings of the 7th Annual Conference*, volume 1447 of *Lecture Notes in Computer Science*, pages 581–589, San Diego, CA, 1998. Springer, Berlin, Germany.
89. J. Kennedy. Bare bones particle swarms. In *Proceedings of the IEEE Swarm Intelligence Symposium 2003 (SIS 2003)*, pages 80–87, Indianapolis, Indiana, USA, 2003.
90. J. Kennedy. In search of the essential particle swarm. In *Proc. of the 2006 IEEE Congress on Evolutionary Computation*, pages 6158–6165. IEEE Press, 2006.
91. J. Kennedy and R. C. Eberhart. Particle swarm optimization. In *Proceedings of the 1995 IEEE International Conference on Neural Networks*, volume 4, pages 1942–1948. IEEE Press, Piscataway, NJ, 1995.

92. J. Kennedy, R. C. Eberhart, and Y. Shi. *Swarm Intelligence*. Morgan Kaufmann Publishers, San Francisco, CA, 2004.
93. O. Korb, T. Stützle, and T. E. Exner. PLANTS: Application of ant colony optimization to structure-based drug design. In M. Dorigo, L. M. Gambardella, A. Martinoli, R. Poli, and T. Stützle, editors, *Ant Colony Optimization and Swarm Intelligence – Proceedings of ANTS 2006 – Fifth International Workshop*, volume 4150 of *Lecture Notes in Computer Science*, pages 247–258. Springer, Berlin, Germany, 2006.
94. P. Korošec, J. Šilc, and B. Robič. Mesh-partitioning with the multiple ant-colony algorithm. In M. Dorigo, M. Birattari, C. Blum, L. M. Gambardella, F. Mondada, and T. Stützle, editors, *Proceedings of ANTS 2004 – Fourth International Workshop on Ant Colony Optimization and Swarm Intelligence*, volume 3172 of *Lecture Notes in Computer Science*, pages 430–431. Springer, Berlin, Germany, 2004.
95. P. Korošec, J. Šilc, and B. Robič. Solving the mesh-partitioning problem with an ant-colony algorithm. *Parallel Computing*, 30:785–801, 2004.
96. C. R. Kube and E. Bonabeau. Cooperative transport by ants and robots. *Robotics and Autonomous Systems*, 30:85–101, 2000.
97. P. Kuntz, D. Snyers, and P. Layzell. A stochastic heuristic for visualizing graph clusters in a bi-dimensional space prior to partitioning. *Journal of Heuristics*, 5(3):327–351, 1998.
98. D. Lambrinos, R. Möller, T. Labhart, R. Pfeifer, and R. Wehner. A mobile robot employing insect strategies for navigation. *Robotics and Autonomous Systems*, 30:39–64, 2000.
99. E. Lawler, J. K. Lenstra, A. H. G. Rinnooy Kan, and D. B. Shmoys. *The Travelling Salesman Problem*. John Wiley & Sons, New York, NY, 1985.
100. X. Li. A Non-dominated Sorting Particle Swarm Optimizer for Multiobjective Optimization. In E. Cantú-Paz et al., editor, *Genetic and Evolutionary Computation—GECCO 2003. Proceedings, Part I*, pages 37–48. Springer. Lecture Notes in Computer Science Vol. 2723, July 2003.
101. X. Li. Adaptively choosing neighbourhood bests using species in a particle swarm optimizer for multimodal function optimization. In K. Deb, editor, *Proceedings of Genetic and Evolutionary Computation Conference 2004 (GECCO'04) (LNCS 3102)*, pages 105–116, 2004.
102. X. Li, J. Branke, and T. Blackwell. Particle swarm with speciation and adaptation in a dynamic environment. In Mike Cattolico, editor, *Genetic and Evolutionary Computation Conference, GECCO 2006, Proceedings, Seattle, Washington, USA, July 8-12, 2006*, pages 51–58. ACM, 2006.
103. X. Li and K.H. Dam. Comparing particle swarms for tracking extrema in dynamic environments. In *Proc. of the 2003 IEEE Congress on Evolutionary Computation*, pages 1772–1779, 2003.
104. J. J. Liang and P. N. Suganthan. Dynamic multi-swarm particle swarm optimizer with a novel constraint-handling mechanism. In *Proc. of the 2006 IEEE Congress on Evolutionary Computation*, pages 9–16. IEEE Press, 2006.
105. J. J. Liang, A. K. Qin, P. N. Suganthan, and S. Baskar. Comprehensive learning particle swarm optimizer for global optimization of multimodal functions. *IEEE Trans. Evol. Comput.*, 10(3):281–295, June 2006.
106. M. López-Ibáñez, L. Paquete, and T. Stützle. On the design of ACO for the biobjective quadratic assignment problem. In M. Dorigo, M. Birattari,

C. Blum, L. M. Gambardella, F. Mondada, and T. Stützle, editors, *Proceedings of ANTS 2004 – Fourth International Workshop on Ant Colony Optimization and Swarm Intelligence*, volume 3172 of *Lecture Notes in Computer Science*, pages 214–225. Springer, Berlin, Germany, 2004.

107. M. Lovbjerg and T. Krink. Extending particle swarm optimizers with self-organized criticality. In *Proc. of the 2002 IEEE Congr. Evol. Comput.*, pages 1588–1593. IEEE Press, 2002.

108. E. D. Lumer and B. Faieta. Diversity and adaptation in populations of clustering ants. In D. Cliff, P. Husbands, J.-A. Meyer, and S. W. Wilson, editors, *Proceedings of the 3rd International Conference on Simulation of Adaptive Behaviour: From Animals to Animats 3 (SAB 94)*, pages 501–508. MIT Press, 1994.

109. V. Maniezzo. Exact and approximate nondeterministic tree-search procedures for the quadratic assignment problem. *INFORMS Journal on Computing*, 11(4):358–369, 1999.

110. V. Maniezzo, M. Boschetti, and M. Jelasity. An ant approach to membership overlay design. In M. Dorigo, M. Birattari, C. Blum, L. M. Gambardella, F. Mondada, and T. Stützle, editors, *Proceedings of ANTS 2004 – Fourth International Workshop on Ant Colony Optimization and Swarm Intelligence*, volume 3172 of *Lecture Notes in Computer Science*, pages 37–48. Springer, Berlin, Germany, 2004.

111. V. Maniezzo and A. Colorni. The Ant System applied to the quadratic assignment problem. *IEEE Transactions on Data and Knowledge Engineering*, 11(5):769–778, 1999.

112. V. Maniezzo and M. Milandri. An ant-based framework for very strongly constrained problems. In M. Dorigo, G. Di Caro, and M. Sampels, editors, *Ant Algorithms – Proceedings of ANTS 2002 – Third International Workshop*, volume 2463 of *Lecture Notes in Computer Science*, pages 222–227. Springer, Berlin, Germany, 2002.

113. R. Marinke, I. Matiko, E. Araujo, and L. Coelho. Particle swarm optimization (PSO) applied to fuzzy modeling in a thermal-vacuum system. In *Fifth International Conference on Hybrid Intelligent Systems (HIS'05)*, pages 67–72. IEEE Computer Society, 2005.

114. K. Marriott and P. Stuckey. *Programming With Constraints*. MIT Press, Cambridge, MA, 1998.

115. R. Mendes, P. Cortez, M. Rocha, and J. Neves. Particle swarms for feedforward neural networks training. In *International Joint Conference on Neural Networks*, pages 1895–1889. Honolulu (Hawaii), USA, 2002.

116. R. Mendes, J. Kennedy, and J. Neves. The fully informed particle swarm: simpler, maybe better. *IEEE Transactions on Evolutionary Computation*, 8(3):204–210, June 2004.

117. D. Merkle and M. Middendorf. Modelling ACO: Composed permutation problems. In M. Dorigo, G. Di Caro, and M. Sampels, editors, *Ant Algorithms – Proceedings of ANTS 2002 – Third International Workshop*, volume 2463 of *Lecture Notes in Computer Science*, pages 149–162. Springer, Berlin, Germany, 2002.

118. D. Merkle and M. Middendorf. Modelling the dynamics of ant colony optimization algorithms. *Evolutionary Computation*, 10(3):235–262, 2002.

119. D. Merkle, M. Middendorf, and A. Scheidler. Self-organized task allocation for computing systems with reconfigurable components. In *Proceedings of the 20th*

International Parallel and Distributed Processing Symposium (IPDPS 2006), 8 pages, IEEE press, 2006.

120. D. Merkle, M. Middendorf, and H. Schmeck. Ant colony optimization for resource-constrained project scheduling. *IEEE Transactions on Evolutionary Computation*, 6(4):333–346, 2002.

121. N. Meuleau and M. Dorigo. Ant colony optimization and stochastic gradient descent. *Artificial Life*, 8(2):103–121, 2002.

122. B. Meyer and A. Ernst. Integrating ACO and constraint propagation. In M. Dorigo, M. Birattari, C. Blum, L. M. Gambardella, F. Mondada, and T. Stützle, editors, *Proceedings of ANTS 2004 – Fourth International Workshop on Ant Colony Optimization and Swarm Intelligence*, volume 3172 of *Lecture Notes in Computer Science*, pages 166–177. Springer, Berlin, Germany, 2004.

123. R. Michel and M. Middendorf. An island model based ant system with lookahead for the shortest supersequence problem. In A. E. Eiben, T. Bäck, M. Schoenauer, and H.-P. Schwefel, editors, *Proceedings of PPSN-V, Fifth International Conference on Parallel Problem Solving from Nature*, volume 1498 of *Lecture Notes in Computer Science*, pages 692–701. Springer, Berlin, Germany, 1998.

124. S. Mikki and A. Kishk. Investigation of the quantum particle swarm optimization technique for electromagnetic applications. In *2005 IEEE Antennas and Propagation Society International Symposium*, volume 2A, pages 45–48, 2005.

125. N. Monmarché, G. Venturini, and M. Slimane. On how *Pachycondyla apicalis* ants suggest a new search algorithm. *Future Generation Computer Systems*, 16:937–946, 2000.

126. J. Moore and R. Chapman. Application of particle swarm to multiobjective optimization. Department of Computer Science and Software Engineering, Auburn University, 1999.

127. J. D. Moss and C. G. Johnson. An ant colony algorithm for multiple sequence alignment in bioinformatics. In D. W. Pearson, N. C. Steele, and R. F. Albrecht, editors, *Artificial Neural Networks and Genetic Algorithms*, pages 182–186. Springer, Berlin, Germany, 2003.

128. G. L. Nemhauser and A. L. Wolsey. *Integer and Combinatorial Optimization*. John Wiley & Sons, New York, 1988.

129. S. Nouyan, R. Ghizzioli, M. Birattari, and M. Dorigo. An insect-based algorithm for the dynamic task allocation problem. *Künstliche Intelligenz*, 4:25–31, 2005.

130. G. Onwubolu and M. Clerc. Optimal path for automated drilling operations by a new heuristic approach using particle swarm optimization. *International Journal of Production Research*, 42(3/01):473–491, February 2004.

131. P. S. Ow and T. E. Morton. Filtered beam search in scheduling. *International Journal of Production Research*, 26:297–307, 1988.

132. E. Ozcan and C.K. Mohan. Analysis of a simple particle swarm optimization system. In *Intelligent Engineering Systems Through Artificial Neural Networks*, pages 253–258, 1998.

133. C. H. Papadimitriou and K. Steiglitz. *Combinatorial Optimization— Algorithms and Complexity*. Dover Publications, Inc., New York, NY, 1982.

134. R. S. Parpinelli, H. S. Lopes, and A. A. Freitas. Data mining with an ant colony optimization algorithm. *IEEE Transactions on Evolutionary Computation*, 6(4):321–332, 2002.

135. D. Parrott and X. Li. Locating and tracking multiple dynamic optima by a particle swarm model using speciation. *IEEE Transactions on Evolutionary Computation*, 10(4):440–458, August 2006.

136. K. Parsopoulos and M. Vrahatis. Particle swarm optimization method for constrained optimization problems. *Intelligent Technologies—Theory and Applications: New Trends in Intelligent Technologies*, 76:214–220, 2002.

137. K. Parsopoulos and M. Vrahatis. Particle swarm optimization method in multiobjective problems. In *Proceedings of the 2002 ACM Symposium on Applied Computing (SAC 2002)*, pages 603–607. Madrid, Spain, ACM Press, 2002.

138. K. Parsopoulos and M. Vrahatis. On the computation of all global minimizers through particle swarm optimization. *IEEE Transactions on Evolutionary Computation*, 8(3):211–224, June 2004.

139. A. Pétrowski. A clearing procedure as a niching method for genetic algorithms. In *Proceedings of the 3rd IEEE International Conference on Evolutionary Computation*, pages 798–803, 1996.

140. G. Pulido and C. Coello Coello. A constraint-handling mechanism for particle swarm optimization. In *Proc. of the 2004 IEEE Congress on Evolutionary Computation*, pages 1396–1403. IEEE Press, 2004.

141. G. T. Pulido and C. Coello Coello. A constraint-handling mechanism for particle swarm optimization. In *Proceedings of the 2004 IEEE Congress on Evolutionary Computation*, pages 1396–1403, Portland, Oregon, 20-23 June 2004. IEEE Press.

142. A. P. Engelbrecht, R. Brits and F. van den Bergh. A niching particle swarm optimizer. In *Proceedings of the 4th Asia-Pacific Conference on Simulated Evolution and Learning 2002 (SEAL 2002)*, pages 692–696, 2002.

143. K. Rameshkumar, R. Suresh, and K. Mohanasundaram. Discrete particle swarm optimization (DPSO) algorithm for permutation flowshop scheduling to minimize makespan. In *First International Conference of Advances in Natural Computation*, pages 572–581, 2005.

144. C. R. Reeves, editor. *Modern Heuristic Techniques for Combinatorial Problems*. John Wiley & Sons, Inc., New York, NY, 1993.

145. M. Reimann, K. Doerner, and R. F. Hartl. D-ants: Savings based ants divide and conquer the vehicle routing problems. *Computers & Operations Research*, 31(4):563–591, 2004.

146. C.W. Reynolds. Flocks, herds and schools: a distributed behavioral model. *Computer Graphics*, 21(4):25–34, 1987.

147. T. Richer and T. Blackwell. The *Lévy* particle swarm. In *Congress on Evolutionary Computation (CEC 2006)*, pages 808– 815. IEEE press, 2006.

148. J. Riget and J. Vesterstroem. A diversity-guided particle swarm optimizer— the ARPSO. *Technical Report 2002-02, Department of Computer Science, University of Aarhus*, 2002.

149. R. Rucker. *Seek!* Four Walls Eight Windows, New York, 1999.

150. A. Salman, A. Imtiaz, and S. Al-Madani. Particle swarm optimization for task assignment problem. In *IASTED International Conference on Artificial Intelligence and Applications (AIA 2001)*, Marbella, Spain, 2001.

151. M. Sasabe, N. Wakamiya, M. Murata, and H. Miyahara. Effective methods for scalable and continuous media streaming on peer-to-peer networks. *European Transactions on Telecommunications*, 15:549–558, 2004.

152. M. Settles, B. Rodebaugh, and T. Soule. Comparison of genetic algorithm and particle swarm optimizer when evolving a recurrent neural network. In *Genetic and Evolutionary Computation Conference 2003 (GECCO 2003)*, pages 151–152, Chicago, USA, 2003.
153. A. Shmygelska, R. Aguirre-Hernández, and H. H. Hoos. An ant colony optimization algorithm for the 2D HP protein folding problem. In M. Dorigo, G. Di Caro, and M. Sampels, editors, *Ant Algorithms – Proceedings of ANTS 2002 – Third International Workshop*, volume 2463 of *Lecture Notes in Computer Science*, pages 40–52. Springer, Berlin, Germany, 2002.
154. A. Shmygelska and H. H. Hoos. An ant colony optimisation algorithm for the 2D and 3D hydrophobic polar protein folding problem. *BMC Bioinformatics*, 6(30):1–22, 2005.
155. M. Reyes Sierra and C. Coello Coello. Multi-objective particle swarm optimizers: A survey of the state-of-the-art. *International Journal of Computational Intelligence Research*, 2(3):287–308, 2006.
156. C. A. Silva, T. A. Runkler, J. M. Sousa, and R. Palm. Ant colonies as logistic processes optimizers. In M. Dorigo, G. Di Caro, and M. Sampels, editors, *Ant Algorithms – Proceedings of ANTS 2002 – Third International Workshop*, volume 2463 of *Lecture Notes in Computer Science*, pages 76–87. Springer, Berlin, Germany, 2002.
157. K. Socha. ACO for continuous and mixed-variable optimization. In M. Dorigo, M. Birattari, C. Blum, L. M. Gambardella, F. Mondada, and T. Stützle, editors, *Proceedings of ANTS 2004 – Fourth International Workshop on Ant Colony Optimization and Swarm Intelligence*, volume 3172 of *Lecture Notes in Computer Science*, pages 25–36. Springer, Berlin, Germany, 2004.
158. K. Socha and C. Blum. An ant colony optimization algorithm for continuous optimization: An application to feed-forward neural network training. *Neural Computing & Applications*, 2007. In press.
159. K. Socha and M. Dorigo. Ant colony optimization for continuous domains. *European Journal of Operational Research*, 2007. In press.
160. K. Socha, M. Sampels, and M. Manfrin. Ant algorithms for the university course timetabling problem with regard to the state-of-the-art. In S. Cagnoni, J. J. Romero Cardalda, D. W. Corne, J. Gottlieb, A. Guillot, E. Hart, C. G. Johnson, E. Marchiori, J.-A. Meyer, M. Middendorf, and G. R. Raidl, editors, *Applications of Evolutionary Computing, Proceedings of EvoWorkshops 2003*, volume 2611 of *Lecture Notes in Computer Science*, pages 334–345. Springer, Berlin, Germany, 2003.
161. C. Solnon. Ant can solve constraint satisfaction problems. *IEEE Transactions on Evolutionary Computation*, 6(4):347–357, 2002.
162. T. Stützle. An ant approach to the flow shop problem. In *Proceedings of the 6th European Congress on Intelligent Techniques & Soft Computing (EUFIT'98)*, pages 1560–1564. Verlag Mainz, Aachen, Germany, 1998.
163. T. Stützle and M. Dorigo. A short convergence proof for a class of ACO algorithms. *IEEE Transactions on Evolutionary Computation*, 6(4):358–365, 2002.
164. T. Stützle and H. H. Hoos. $\mathcal{MAX}$-$\mathcal{MIN}$ Ant System. *Future Generation Computer Systems*, 16(8):889–914, 2000.
165. P.N. Suganthan. Particle swarm optimiser with neighbourhood operator. In *Congress on Evolutionary Computation (CEC 1999)*, pages 1958–1962, Washington, USA, 1999.

166. T.-Y. Sun, S.-T. Hsieh, H.-M. Wang, and C.-W. Lin. Floorplanning based on particle swarm optimization. In *IEEE Computer Society Annual Symposium on Emerging VLSI Technologies and Architectures 2006*, pages 5–10. IEEE Press, 2006.

167. G. Theraulaz, E. Bonabeau, and J.-L. Deneubourg. Response threshold reinforcement and division of labour in insect societies. *Proceedings: Biological Sciences*, 265(1393):327–332, 1998.

168. I. C. Trelea. The particle swarm optimization algorithm: convergence analysis and parameter selection, 2003.

169. R. Unger and J. Moult. Finding the lowest free-energy conformation of a protein is an NP-hard problem: Proofs and implications. *Bulletin of Mathematical Biology*, 55(6):1183–1198, 1993.

170. F. van den Bergh. *Analysis of Particle Swarm Optimizers*. PhD thesis, Department of Computer Science, University of Pretoria, Pretoria, South Africa, 2002.

171. F. van den Bergh and A.P. Engelbrecht. A cooperative approach to particle swarm optimization. *IEEE Trans. Evol. Compu.*, 8:225–239, Jun. 2004.

172. F. van den Bergh and A.P. Engelbrecht. A study of particle swarm optimization particle trajectories. *Information Sciences*, 176:937–971, 2006.

173. K. Veeramachaneni, T. Peram, C. Mohan, and L. Osadciw. Optimization using particle swarm with near neighbor interactions. In *Proc. of Genetic and Evolutionary Computation Conference*, pages 110 – 121, Chicago, Illinois, 2003.

174. G. K. Venayagamoorthy. Optimal control parameters for a UPFC in a multimachine using PSO. In *Proceedings of the 13th International Intelligent Systems Application to Power Systems 2005*, pages 488–493, 2005.

175. M. Vidyasagar. *Nonlinear Systems Analysis*. Prentice Hall, Englewood Cliffs, NJ, 1993.

176. M. Wachowiak, R. Smolikova, Y. Zheng, J. Zurada, and A. Elmaghraby. An approach to multimodal biomedical image registration utilizing particle swarm optimization. *IEEE Transactions on Evolutionary Computation*, 8(3):289–301, June 2004.

177. C. Walshaw and M. Cross. Mesh partitioning: A multilevel balancing and refinement algorithm. *SIAM Journal on Scientific Computing*, 22(1):63–80, 2000.

178. E. O. Wilson. The relation between caste ratios and division of labour in the ant genus *phedoile*. *Behavioral Ecology and Sociobiology*, 16(1):89–98, 1984.

179. X. Xie, W. Zhang, and Z. Yang. A dissipative particle swarm optimization. In *Proc. Congr. Evol. Comput. 2002 (CEC 2002)*, pages 1456–1461. IEEE Press, 2002.

180. H. Yoshida, K. Kawata, Y. Fukuyama, S. Takayama, and Y. Nakanishi. A particle swarm optimization for reactive power and voltage control considering voltage security assessment. *IEEE Transactions on Power Systems*, 15(4):1232–1239, November 2001.

181. X. Yu and B. Ram. Bio-inspired scheduling for dynamic job shops with flexible routing and sequence-dependent setups. *International Journal of Production Research*, 44(22):4793–4813, 2006.

182. W. Zha and G. K. Venayagamoorthy. Neural networks based non-uniform scalar quantizer design with particle swarm optimization. In *Proceedings 2005 IEEE Swarm Intelligence Symposium (SIS 2005)*, pages 143–148. IEEE Press, 2005.

183. M. Zlochin, M. Birattari, N. Meuleau, and M. Dorigo. Model-based search for combinatorial optimization: A critical survey. *Annals of Operations Research*, 131(1–4):373–395, 2004.

Swarm Robotics

Erol Şahin[1], Sertan Girgin[2], Levent Bayındır[1], and Ali Emre Turgut[1]

[1] KOVAN Research Laboratory
   Middle East Technical University, Ankara, Turkey
   {erol,levent,aturgut}@ceng.metu.edu.tr
[2] Team SequeL
   INRIA Futurs Lille, Villeneuve d'Ascq, France
   sertan.girgin@inria.fr

Summary. Swarm robotics is a novel approach to the coordination of large numbers of robots and has emerged as the application of swarm intelligence to multi-robot systems. Different from other swarm intelligence studies, swarm robotics puts emphases on the physical embodiment of individuals and realistic interactions among the individuals and between the individuals and the environment. In this chapter, we present a brief review of this new approach. We first present its definition, discuss the main motivations behind the approach, as well as its distinguishing characteristics and major coordination mechanisms. Then we present a brief review of swarm robotics research along four axes; namely design, modelling and analysis, robots and problems.

1 Introduction

Swarm robotics represents a novel approach to the coordination of large numbers of robots whose main inspiration stems from the observation of social insects [10, 9]. These insects, such as ants, wasps and termites, are known to coordinate their behaviors to accomplish tasks that are beyond the capabilities of a single individual; ants can carry large preys to their nest; termites can build large mounds from mud within which a desired level of temperature and moisture is maintained [5]. The emergence of such synchronized behavior at the system level is rather impressive for researchers working on multi-robot systems, since it emerges despite the individuals being relatively incapable, despite the lack of centralized coordination and despite the simplicity of interactions.

The term swarm intelligence [4] was originally conceived as a "buzz word" by Beni in the 1980s [3] to denote a class of cellular robotic systems [2]. Later, the term moved on to cover a wide range of studies from optimization to social insect studies, losing its robotics context in the meantime. Recently the term swarm robotics has started to be used as the application of swarm intelligence to physically embodied systems.

2 What Is Swarm Robotics?

Given the plethora of terms being used for describing different approaches used in multi-robot systems, such as "distributed robotics" or "collective robotics," the distinguishing characteristics of swarm robotics from the rest need to be clarified. This concern was first explicitly stated in [10] and a definition was provided as follows.

Definition 1. *"Swarm robotics is the study of how a large number of relatively simple physically embodied agents can be designed such that a desired collective behavior emerges from the local interactions among the agents and between the agents and the environment."[9]*

2.1 System-Level Properties

The system-level operation of a swarm robotic system should exhibit three functional properties that are observed in natural swarms and remain as desirable properties of multi-robot systems.

- **Robustness.** The swarm robotic system should be able to operate despite disturbances from the environment or the malfunction of its individuals. A number of factors can be observed in social insects behind the robustness of their operation. First, swarms are inherently redundant systems; the loss of an individual can be immediately compensated by another one. Second, coordination is decentralized and therefore the destruction of a particular part of the swarm is unlikely to stop its operation. Third, the individuals that make up the swarm are relatively simple, making them less prone to failure. Fourth, sensing is distributed; hence the system is robust against the local perturbances in the environment.
- **Flexibility.** The individuals of a swarm should be able to coordinate their behaviors to tackle tasks of different nature. For instance, the individuals in an ant colony can collectively find the shortest path to a food source or carry a large prey through the utilization of different coordination strategies.
- **Scalability.** The swarm should be able to operate under a wide range of group sizes and support a large number of individuals without impacting performance considerably. That is, the coordination mechanisms and strategies to be developed for swarm robotic systems should ensure the operation of the swarm under varying swarm sizes.

2.2 Distinguishing Characteristics

We will now summarize the main distinguishing characteristics of swarm robotics research (see [9] for a full discussion). First, the research should be relevant to the coordination of a swarm of robots. That is, the individuals should have a physical embodiment, be situated, and be able to physically interact with their environment. Moreover, the coordination mechanisms being studied should promise to be scalable for a wide range of swarm sizes.

Second, the robotic system being studied should be rather homogeneous. That is, the individuals that makes up the swarm should be rather identical, at least at the level of interactions. Coordination strategies developed for heterogeneous multi-robot systems, which consist of individuals that differ in their interactions due to their physical embodiment or their behavioral control, fall outside of the swarm robotics approach.

Third, the individuals should be relatively simple. The simplicity criterion in the definition does not directly refer to the hardware and software complexity of the robots, but is rather meant to emphasize the limitations in their individual capabilities relative to the task. The members of the swarm system should be relatively incapable or inefficient on their own with respect to the task at hand. That is, either (i) the task should be hard or impossible to be carried out by a single robot, and the cooperation of a group of robots should be essential, or (ii) the deployment of a group of robots should improve the performance and robustness of the handling of the task.

Fourth, the individuals should have local interaction abilities. This constraint ensures that the coordination between the robots is distributed, and that it is more likely to scale with the size of the swarm. Mechanisms that rely on global interaction capabilities are likely to be bounded by the bandwidth and the range of communication channel and may create unscalable coordination mechanisms.

2.3 Coordination Mechanisms

Studies in physical and biological systems have revealed that there are a number of coordination mechanisms that are at work in natural systems which can act as sources of inspiration for coordinating swarm robotic systems. Two of the main coordination mechanisms are: self-organization and stigmergy.

Self-organization, defined as "a process in which patterns at the global level of a system emerge solely from numerous interactions among the lower-level components of the system" [5], is common in natural systems. Studies of self-organization in natural systems show that an interplay of positive and negative feedback of local interactions among the individuals is essential [4]. In these systems, positive feedback is typically generated through autocatalytic behaviors; that is the change inflicted in the swarm-environment system by the execution of a behavior increases the triggering of the very same behavior. Such a positive feedback cycle is then counterbalanced by a negative feedback mechanism, which typically stems from a "depletion of physical resources" [4]. In addition to these mechanisms, self-organization also depends on the existence of randomness and multiple interactions within the system.

Studies of self-organization in natural systems often develop models that are built with simplified interactions in the environment and abstract behavioral mechanisms in individuals. The self-organization models of social insects and animals have already been used as inspiration sources since, in a sense, swarm robotics can be considered as the engineering and utilization of self-organization in physically embodied swarms.

Stigmergy, defined as indirect communication of individuals through environment, was first proposed by Grasse [13] to explain the coordination mechanisms behind the building of nests in termites. Stigmergic communication is common in many social insects; ants are known to lay pheromones on the ground to mark the paths to food sources and these pheromones act as attractants to be followed by ants. Stigmergy is of interest to swarm robotics since it provides a communication mechanism that is local, distributed and scalable.

3 Research Directions

During the last 4-5 years, interest in swarm robotics has been on the rise. The growing interest in this new approach is being fueled by the advances in mechatronics and other technologies, such as MEMS, which have started to shrink the size and the cost of robots for mass production and opened the way towards the deployment of large-scale swarm robotic systems in real-world applications. In the discussion below, we will provide a brief review of the swarm robotics studies in four categories; namely design, modelling and analysis, robots, and problems.

3.1 Design

The main problem of a swarm robotic system can be stated as follows: How should one design individuals, both in terms of their physical embodiment as well as their behavioral control, such that a desired swarm-level behavior emerges from the interactions among the individuals as well as between the individuals and the environment? This goal, which can also be considered as the "engineering of self-organization" in multi-robot systems, is a challenging task that is difficult, if not impossible, to solve in general terms. The studies within this category can be grouped into two: *ad-hoc* and *principled* approaches.

In *ad-hoc approaches*, behaviors of individual robots are designed manually to achieve a desired swarm-level behavior. In this approach, usually, though not always, behaviors of social insects are usually adapted to the robots at hand. This process implicitly assumes that the behaviors used as inspiration are observed at a certain abstraction level that captures essential parameters that need to be adapted to robots and should yet reproduce similar swarm-level behaviors.

In *principled approaches*, instead of designing a specific swarm-level behavior, a general methodology through which desired swarm-level behaviors can be used to build necessary individual behaviors is proposed or utilized. One such approach is the use of artificial evolution. Evolutionary methods have been successfully used to develop behaviors within the Swarm-bots project [12]. In particular, the SwarmBot3D [22], a physics-based simulation environment for simulating the Swarm-bots robotic system at different levels of complexity, was used. In most of these studies, simple feedforward or recurrent multi-layer perceptrons were used to encode the behaviors. The evolved behaviors in the simulation environment were later successfully transferred to the physical robot system.

3.2 Modelling and Analysis

The behavior of a swarm robotic system at the system level emerges from the interactions of its individuals. These interactions, determined by the behaviors of the individuals and the environment, are inherently probabilistic. As a consequence of this, the behavioral outcome of swarm robotic systems is not straightforward and modelling and analysis of the swarm is desirable for at least two purposes. First, for a desired task to be accomplished, and for a proposed behavioral design at the individual level, one needs to obtain guarantees for system-level performance. Second, in most ad-hoc approaches, although the overall composition of individual behaviors may be known, the optimal values of the parameters may remain unknown. Systematic experiments with physical robots are often difficult to perform. Moreover, they can provide only limited guarantees and little insight into the operation of the system. The models that can be used towards this end can be reviewed in three groups. In *sensor-based modelling*, the sensing and actuation of the individual robots as well as robot-robot and robot-environment interactions are modelled. This kind of modelling, mostly used for building realistic simulators of robotic systems, allows us to conduct experiments in simulation and yet to obtain results that are in agreement with the ones obtained from physical robots. Although this type of modelling is common in building robotics simulators [21], models to be used in swarm robotics require more fidelity at the level of inter-robot interactions. However, the building of these models is subject to the trade-off between realism and simplicity – models and interactions need to be realistic to be useful, and, yet, at the same time they must be as simple as possible for speed.

One simulation platform built with all these issues in mind is Swarmbot3D [22], a physics-based simulator specifically developed for the swarm-bot robotic system. The simulator contained models of the s-bot robot at different levels of complexity and was verified against the physical robot. The simulator was used both to generate behaviors for different problems using evolutionary methods (see the previous subsection) as well as to systematically analyze the resulting system-level behaviors. These simulations, even at the lowest level of complexity, proved to be computationally intensive, and a system that can parallelize the simulations over a cluster of computers was developed in [29].

This type of modelling can be used as a constructive means to design behaviors, and provide insight into the behavior of the swarm through systematic experimentation.

In *microscopic* modelling, similar to the sensor-based approach, modelling is carried out at the individual level. The states of the individuals and the transitions among these states are modelled analytically. Such a modelling takes into account the characteristics of the environment, the physical embodiment and the behavioral control of the robots. Through such modelling, instead of simulating the individual interactions within the system, the model evolves the states of the individuals in time.

An excellent example of this type of modelling in swarm robotics can be found in [16]. The authors studied the stick pulling problem, in which two robots have to collaborate to pull sticks. They proposed a probabilistic model to represent the changes in the states of the robots. The model, which is essentially a set of rate equations, was built using the physical characteristics of the robots, such as the body shape and size of the robot as well as the placement and characteristics of the sensors, and the environment. It also took into account the behavioral design of the individual robots and used it as a basis for the transitions among the different states. Microscopic modelling was reported to be much faster than the sensor-based modelling and yet provided means to link the behavioral parameters to the system-level outcomes.

In *macroscopic modelling*, unlike in the previous two approaches, modelling is done at the swarm level. This type of modelling, in which the behavior of some average quantities that represent the state of the system is represented, is common in physics and chemistry. Contrary to sensor-based and microscopic models, macroscopic models need to be solved only once to obtain the steady state of the model. This allows one to find the optimum behavioral parameters without conducting any systematic experiments with the robots and provides a theoretical guarantee over the system-level behavior of the swarm. One example of such modelling can be found in [19]. In this study, an analytical macroscopic model of the stick-pulling problem, mentioned above, is proposed. In this model, the number of robots in a certain state as well as the number of unextracted sticks are used to represent the state of the system and the rate equations describing the change in them are derived. Using such a model, the authors are able to determine optimal parameters for the behaviors of the individual robots without making any systematic experiments.

3.3 Robots

One major research direction has been the development of physical swarm robotic systems since the building of a swarm robotic system takes more than gathering a number of copies of a generic robot platform. All the studies towards this end, have focused on developing mobile robots that are aimed to provide a research platform and are not intended for real-world operation.

Below we will discuss the extra requirements (or wish list from the researchers' viewpoint) expected from robots that would be used in swarm robotic systems.

- **Sensing and Signalling.** The main emphasis in swarm robotics is the interaction among the robots as well as the interaction of the robots with their environment, resulting in extra constraints for the robots to be used. In particular, (i) the interference among the sensing systems of the robots and the effect of environmental factors on them should be minimal, (ii) the robots should be able to distinguish other kin-robots (preferably as easily as proximity sensing), and (iii) the robots should be able to leave "marks" in the environment and be able to sense them (i.e. stigmergy). Furthermore, it is preferable that the robots are equipped with (or extendable to) some form of generic sensing capability to allow the researcher to test novel sensing strategies.

- **Communication.** Unlike in stand-alone robotic systems, communication by plugging cables into the robots is no longer a feasible option. Therefore the robots have to support wireless communication (i) between a console and the robots, to allow easier monitoring and debugging of algorithms on individual robots, (ii) among robots such as in the form of ad-hoc networks. The robots should also be programmable in parallel through a wireless communication channel since control algorithms are mostly the same for all the robots and programming the swarm as a whole would be a big time saver.

- **Physical Interaction.** The robots should be able to physically interact with each other and the environment since this is required by possible tasks such as self-assembly and self-organized construction.

- **Power.** The robots should have a long battery life. In most studies, the swarm may need to operate for a period that is long enough for the collective behavior to emerge, and the goal to be reached.

- **Cost.** The robots should be as cheap as possible, since, unlike stand-alone robots, they will be sold at least in groups of tens.

- **Size.** Size does matter in swarm robotic systems. The robots should be small enough not to make it necessary to increase the size of the test arena when experimenting with the system, and yet big enough not to limit the expandability of the robot or increase the cost of the swarm robots due to miniaturization in components.

- **Simulation.** Swarm robotic systems require realistic simulators. They are essential for speeding up the development of new control algorithms. Such simulators need to model the interactions between the robots as well as the interactions of the robots with their environment in a realistic way that is also verified against the physical robots.

Developing a single robot platform that would realize the whole wish list is a difficult, if not impossible, challenge. The design choices made regarding one requirement, such as size, pose additional constraints towards the reaching of other requirements, such as power and communication. In the rest of this section, we review some of the existing mobile robot platforms that are developed (or can be used) for conducting swarm robotics research and evaluate them based on the wish list stated above.

- Alice [6] is a small rectangular mobile robot with dimensions 22×21 mm. The robot, driven by two high efficiency SWATCH motors for locomotion, hosts a PIC16F877 microcontroller with 8K words flash EPROM program memory. Alice has four IR proximity sensors for obstacle detection and a short-range robot-to-robot communication module as well as an IR receiver for remote control. There are also a wide variety of modules, such as a linear camera, RF, or gripper modules, for extending its capabilities. Ten hours of autonomy are reported with two button batteries and 20 hours of autonomy is achieved with an additional LiPoly battery. The robot model is available in the Webots simulator.

- e-Puck [8] is a circular robot with a diameter of 70 mm. The robot, driven by two stepper motors for locomotion, hosts a dsPIC 30F6014A microcontroller with 144 KB program memory and 8 KB of RAM. ePuck has eight IR sensors used for measuring proximity to objects as well as ambient light. It has a speaker for audible feedback and three directional microphones which can be used for sound localization and a three-axis accelerometer. The robot has a color camera, a number of LEDs to signal or show its state and Bluetooth as the main wireless communication channel. The robots can be programmed via this Bluetooth module. e-Puck also provides an expansion bus and has optional ZigBee communication modules. Three hours of autonomy are reported using a 5 Wh Li-Ion battery. The robot model is available in the Webots simulator.

- Jasmine [23] is a small rectangular robot with dimensions 23×23 mm. The robot, driven by two small gear head motors for locomotion, has six IR sensors for proximity sensing and proximal communication. There is one powerful IR LED for detailed analysis of an object of interest and an IR communication module with host. Jasmine III has a modular design in which different sensing modules such as an ambient light sensor, a color sensor and different locomotion modules can be utilized. Two hours of autonomy are reported with LiPoly batteries. A simulator called LaRoSim was built for conducting experiments in simulation.

- s-bot [22] has a circular shape having a diameter of 116 mm. The robots have a locomotion sub-system consisting of both wheels and tracks which are driven by two DC gear head motors. s-bots are equipped with two grippers for studying problems such as self-assembly and coordinated movement. The robots have sensors of different modalities, including 15 IR proximity sensors for obstacle detection, four IR sensors below the robot facing the ground, torque sensors on the wheels, a force sensor between the base and the wheels, a three-axis accelerometer, an omni-directional camera and eight RGB LEDs for messaging between the s-bots. The robot is equipped with a 400 MHz custom XScale CPU board, 32 MB of flash memory and 64 MB of RAM. A Wi-Fi module is used for wireless communication. The robots have one hour battery life (Li-Ion). A custom-made simulator called SwarmBot3D is developed to simulate s-bots at different levels of complexity.

- Swarmbot [28] is a square-shaped robot with dimensions 130×130 mm. It has four wheels on each side driven by two DC gear head motors. The robot is equipped with an ARM Thumb CPU, an FPGA, eight bump sensors, four light sensors and a camera. The Swarmbot uses ISIS, an IR system that can sense the range, bearing and orientation of other neighboring robots. Additional modules are linear CCD, magnetic food and swarm-cam emitters which can be utilized on demand. There is an RF communication unit for debugging and programming purposes. The battery life of the robots is not reported.

- Centibots [25] are modified versions of Pioneer 2-AT and Amigobots. An inertial navigation system to estimate coordinates of the robots, a SICK laser range finder for map building and a CCD camera used to extend the sensing capabilities of the robots. An on-board computer, USB web cam for intruder and object-of-interest detection are added to Amigobots. There is a Wi-Fi wireless ad-hoc network between robots. An autonomy of three to six hours is reported for Pioneers and two hours for Amigobots.

- Kobot [34] is a circular mobile robot platform having a diameter of 120 mm. Two high-quality gear head DC motors are used for locomotion. Kobot has a modulated IR system that can provide proximity readings from objects and distinguish robots from obstacles. The sensing system, which uses modulated IR signals, is robust against environmental lighting conditions and minimizes interference among robots. Kobots use the IEEE 802.15.4/ZigBee protocol as wireless communication channel. Through this channel, the robots in a swarm can be programmed in parallel. Ten hours of battery life is reported with LiPoly batteries. A custom-made physics-based simulator is available.

3.4 Problems

So far, swarm robotics research is mostly confined to the development of proof-of-concept studies in simulators or robotic systems operating in laboratory environments. Below we will describe some of the problems that have been addressed in swarm robotics research and describe some of the exemplary studies addressing them.

- **Aggregation.** Self-organized aggregation, the grouping of individuals of a swarm into a cluster without using any environmental clues, is a common behavior observed in organisms ranging from bacteria to social insects and mammals. In swarm robotic systems it can be considered as one of the fundamental behaviors that can act as a precursor to other behaviors such as flocking and self-assembly. In [11, 30], aggregation behaviors were developed for myopic robots, robots that can perceive only a small part of the whole environment, confined to a large arena, using evolutionary approaches as well as a probabilistic controller inspired by social insects.

- **Dispersion.** Self-organized dispersion can be considered as the opposite of aggregation and is of interest in surveillance scenarios. In this problem (see, for example, [27, 26, 28]) the challenge is to obtain uniform spreading of a swarm of robots in a space, maximizing the area covered yet remaining connected through some form of communication channel.

- **Foraging.** This problem is inspired by the behavior of ants which search for food sources distributed around their nest. In this problem, the challenge is to find the optimum search strategies that maximize the ratio of returned food to the resources committed (such as the number of individuals performing foraging or signalling strategies) in an environment. Different foraging strategies have been explored and analyzed [31, 17, 20], and models of foraging have been developed [18, 15].

- **Self-assembly.** This behavior is observed in ants, where they form chains through connecting to each other to build bridges or float-like structures to stay above water. The problem of self-assembly can be defined as the self-organized creation of structures through the formation of physical connections among a swarm of individual robots. Self-assembly has been studied in physical robots [7, 24] such that a desired self-assembled structure is formed.

- **Connected Movement.** This problem can be described as follows: How can a swarm of mobile robots, physically connected to each other, coordinate their movement such that the group moves smoothly in an environment and avoids environmental obstacles, such as holes, in a coordinated way. This problem has been studied in [32, 33] as part of the Swarm-bots project. In these studies, evolutionary approaches were used to evolve behaviors that can control a number of connected robots to avoid holes within the environment. The robots, which are physically connected to each other through their grippers, were able to sense the forces acting on their bodies through traction sensors and were able to detect holes underneath them.

- **Cooperative Transport.** Ants are known to transport large preys to their nest through coordinating their pushing and pulling actions. Such a coordination ability is obviously valuable for swarm robotic systems since it allows individuals to join forces, generating a combined force large enough to pull a heavy object. This problem is partially related to the connected movement, with the difference that it includes a passive object that needs to be transported. In [14] a recurrent neural network controller is evolved to obtain solitary and group transport behaviors in a physics-based simulator. The angular position of the goal (marked with a light source) and the distance as well as the angular position of the prey and a connection sensor indicating whether the robot is connected to other robots or not were used to control the motors of the robot.

- **Pattern Formation.** This is a rather generic term for the problem of how a desired geometrical pattern can be obtained and maintained by a swarm of robots without any centralized coordination. Pattern formation may refer either to geometric or to functional pattern formation. In geometric pattern formation, the challenge is to develop behaviors such that individuals of a swarm form a desired geometrical pattern, similar to the formation of crystals. In this task, the environment is assumed to be uniform and the focus is on the use of inter-robot interactions to create such patterns. In functional pattern formation, the pattern to be formed is dictated by the environment. In natural swarms, the surrounding of a prey by a group of predators or the formation of pulling chains by weaver ants can be considered as examples of functional pattern formation, where the geometrical shape or size of the patterns formed are partially determined by the task at hand.

- **Self-organized Construction.** This problem can be formulated as follows: How can a number of passive objects, randomly distributed in an environment, be clustered together by a swarm of robots. This problem, sometimes also referred to as "aggregation," has been one of first problems studied. Beckers et al. [1] studied how a swarm of physical robots can cluster frisbees spread in an environment, and showed that despite the lack of communication and signalling among robots, frisbee clusters can be obtained.

4 Conclusion

In this chapter we provided a brief review of swarm robotics as a new approach to the control and coordination of multi-robot systems. We stated the inspiration behind this approach, the desirable properties, and the requirements to clarify the defining characteristics of this approach in relation to other existing studies. Then we reviewed the studies in this new field, grouping them into four categories. Due to the lack of a good review article in this rather new field, we have opted to present the reader with an overall picture of the field in rather general terms and pointed out some of the most interesting studies.

Acknowledgements

This work was partially funded by the KARİYER: Kontrol Edilebilir Robot Oğulları Career Project (Project no: 104E066) awarded to Erol Şahin by TÜBİ-TAK (Turkish Scientific and Technical Council).

References

1. R. Beckers, O. E. Holland, and J.-L. Deneubourg. From local actions to global tasks: Stigmergy and collective robotics. In R.A. Brooks and P. Maes, editors, *Proceedings of the 4th International Workshop on the Synthesis and Simulation of Living Systems (Artificial Life IV)*, pages 181–189, Cambridge, MA, USA, July 1994. MIT Press.
2. G. Beni. From swarm intelligence to swarm robotics. In E. Şahin and W. Spears, editors, *Proceedings of the First International Workshop on Swarm Robotics (at SAB 2004)*, volume 3342 of *Lecture Notes in Computer Science*, pages 1–9. Springer, Berlin, Germany, 2005.
3. G. Beni and J. Wang. Swarm intelligence. In *Proc. of the Seventh Annual Meeting of the Robotics Society of Japan*, pages 425–428, Tokyo, Japan, 1989.
4. E. Bonabeau, M. Dorigo, and G. Theraulaz. *Swarm Intelligence: From Natural to Artificial Systems*. Santa Fe Institute Studies on the Sciences of Complexity. Oxford University Press, 1999.
5. S. Camazine, J.-L. Deneubourg, N.R. Franks, J. Sneyd, G. Theraulaz, and E. Bonabeau. *Self-Organisation in Biological Systems*. Princeton University Press, NJ, USA, 2001.
6. G. Caprari and R. Siegwart. Mobile micro-robots ready to use: Alice. In *Proceedings of IEEE/RSJ International Conference on Intelligent Robots and Systems*, pages 3295–3300. IEEE press, 2005.
7. A. Christensen, R. O'Grady, and M. Dorigo. A mechanism to self-assemble patterns with autonomous robots. Technical report, TR/IRIDIA/2007-009, 2007.
8. C. M. Cianci, X. Raemy, J. Pugh, and A. Martinoli. Communication in a swarm of miniature robots: The e-Puck as an educational tool for swarm robotics. In E. Şahin, W. M. Spears, and A. F. T. Winfield, editors, *Proceedings of the Second International Workshop on Swarm Robotics (at SAB 2006)*, volume 4433 of *Lecture Notes in Computer Science*, pages 103–115. Springer, Berlin, Germany, 2007.
9. E. Şahin. Swarm robotics: From sources of inspiration to domains of application. In E. Şahin and W. Spears, editors, *Proceedings of the First International Workshop on Swarm Robotics (at SAB 2004)*, volume 3342 of *Lecture Notes in Computer Science*, pages 10–20. Springer, Berlin, Germany, 2005.
10. M. Dorigo and E. Şahin. Special issue: Swarm robotics. *Autonomous Robots*, 17:111–113, 2004.
11. M. Dorigo, V. Trianni, E. Sahin, R. Gross, T.H. Labella, G. Baldassarre, S. Nolfi, J.-L. Deneubourg, F. Mondada, D. Floreano, and L.M. Gambardella. Evolving self-organizing behaviors for a swarm-bot. *Autonomous Robots*, 17(2-3):223–245, 2004.
12. M. Dorigo, E. Tuci, R. Gross, V. Trianni, T.H. Labella, S. Nouyan, and C. Ampatzis. The swarm-bots project. In E. Şahin and W. Spears, editors, *Swarm Robotics Workshop: State-of-the-art Survey*, volume 3342 of *Lecture Notes in Computer Science*, pages 31–44. Springer, Berlin, Germany, 2005.
13. P. Grasse. La reconstruction du nid et les coordinations inter-individuelles chez *Bellicositermes natalensis* et *Cubitermes sp.* la theorie de la stigmergie: Essai. *Insectes Sociaux*, 6:41–83, 1959.
14. R. Gross, M. Bonani, F. Mondada, and M. Dorigo. Autonomous self-assembly in mobile robotics. *IEEE Transactions on Robotics*, 22(6), 2006.

15. H. Hamann and H. Worn. An analytical and spatial model of foraging in a swarm of robots. In E. Şahin, W. M. Spears, and A. F. T. Winfield, editors, *Proceedings of the Second International Workshop on Swarm Robotics (at SAB 2006)*, volume 4433 of *Lecture Notes in Computer Science*, pages 43–55. Springer, Berlin, Germany, 2007.

16. A.J. IIjspeert, A. Martinoli, A. Billard, and L.M. Gambardella. Collaboration through the exploitation of local interactions in autonomous collective robotics: The stick pulling experiment. *Autonomous Robots*, 11:149?171, 2001.

17. M. J. B. Krieger, J.-B. Billeter, and L. Keller. Ant-like task allocation and recruitment in cooperative robots. *Nature*, 406:992–995, 2000.

18. K. Lerman and A. Galstyan. Mathematical model of foraging in a group of robots: Effect of interference. *Autonomous Robots*, 13:127–141, 2002.

19. K. Lerman, A. Galstyan, A. Martinoli, and A.J. Ijspeert. A macroscopic analytical model of collaboration in distributed robotic systems. *Artificial Life*, 7(4):375 – 393, 2001.

20. W. Liu, A. Winfield, J. Sa, J. Chen, and L. Dou. Strategies for energy optimisation in a swarm of foraging robots. In E. Şahin, W. M. Spears, and A. F. T. Winfield, editors, *Proceedings of the Second International Workshop on Swarm Robotics (at SAB 2006)*, volume 4433 of *Lecture Notes in Computer Science*, pages 14–26. Springer, Berlin, Germany, 2007.

21. O. Michel. Webots: Professional mobile robot simulation. *Journal of Advanced Robotics Systems*, 1(1):39–42, 2004.

22. F. Mondada, G. C. Pettinaro, A. Guignard, I. Kwee, D. Floreano, J.-L. Deneubourg, S. Nolfi, L. M. Gambardella, and M. Dorigo. SWARM-BOT: a New Distributed Robotic Concept. *Autonomous Robots, Special Issue on Swarm Robotics*, 17(2-3):193–221, 2004.

23. University of Stuttgart. Open-source microrobotic project, 2007.

24. R. O'Grady, R. Groß, A. L. Christensen, F. Mondada, M. Bonani, and M. Dorigo. Performance benefits of self-assembly in a swarm-bot. Technical report, TR/IRIDIA/2007-008, 2007.

25. C. Ortiz, K. Konolige, R. Vincent, B. Morisset, A. Agno, M. Eriksen, D. Fox, B. Limketkai, J. Ko, B. Stewart, and D. Schulz. Centibots: Very large scale distributed robotic teams. In D. L. McGuinness and G. Ferguson, editors, *Proceedings of the Nineteenth National Conference on Artificial Intelligence (AAAI 2004)*, pages 1022–1023. AAAI Press/The MIT Press, 2004.

26. D. Payton, M. Daily, R. Estowski, M. Howard, and C. Lee. Pheromone robotics. *Autonomous Robots*, 11(3):319–324, 2001.

27. D. Payton, R. Estkowski, and M. Howard. Pheromone robotics and the logic of virtual pheromones. In E. Şahin and W. Spears, editors, *Proceedings of the First International Workshop on Swarm Robotics (at SAB 2004)*, number 3342 in Lecture Notes in Computer Science, pages 45–57. Springer, Berlin, Germany, 2005.

28. J. Smith and J. McLurkin. In R. Alami, R. Chatila, and H. Asama, editors, *Distributed Autonomous Robotic Systems*, chapter Distributed Algorithms for Dispersion in Indoor Environments using a Swarm of Autonomous Mobile Robots. Springer, Berlin, Germany, 2007.

29. O. Soysal, E. Bahceci, and E. Şahin. PES: A system for parallelized fitness evaluation of evolutionary methods. In A. Yazici and C. Sener, editors, *Proceedings of the Eighteenth International Symposium on Computer and Information*

Sciences (ISCIS), volume 2869 of *Lecture Notes in Computer Science*, pages 889–896. Springer, Berlin, Germany, 2003.

30. O. Soysal and E. Şahin. A macroscopic model for self-organized aggregation in swarm robotic systems. In E. Şahin, W. M. Spears, and A. F. T. Winfield, editors, *Proceedings of the Second International Workshop on Swarm Robotics (at SAB 2006)*, volume 4433 of *Lecture Notes in Computer Science*, pages 27–42, Springer, Berlin, Germany, 2007.

31. K. Sugawara and M. Sano. Cooperative acceleration of task performance: Foraging behavior of interacting multi-robots system. *Physica D*, 100:343–354, 1997.

32. V. Trianni and M. Dorigo. Emergent collective decisions in a swarm of robots. In *Proceedings of the IEEE Swarm Intelligence Symposium*, pages 241–248. IEEE press, 2005.

33. V. Trianni, S. Nolfi, and M. Dorigo. Cooperative hole avoidance in a *Swarm-bot*. *Robotics and Autonomous Systems*, 54(2):97–103, 2006.

34. A. E. Turgut, F. Gökçe, H. Çelikkanat, L. Bayındır, and E Şahin. Kobot: A mobile robot designed specifically for swarm robotics research. Technical Report METU-CENG-TR-2007-05, Dept. of Computer Engineering, Middle East Technical University, 2007.

Routing Protocols for Next-Generation Networks Inspired by Collective Behaviors of Insect Societies: An Overview

Muddassar Farooq[1] and Gianni A. Di Caro[2]

[1] Next Generation Intelligent Networks Research Center
National University of Computer and Emerging Sciences (NUCES)
Islamabad, Pakistan
muddassar.farooq@udo.edu

[2] "Dalle Molle" Institute for Artificial Intelligence (IDSIA)
Lugano, Switzerland
gianni@idsia.ch

Summary. In this chapter we discuss the properties and review the main instances of network routing algorithms whose bottom-up design has been inspired by *collective behaviors of social insects* such as *ants* and *bees*. This class of bio-inspired routing algorithms includes a relatively large number of algorithms mostly developed during the last ten years. The characteristics inherited by the biological systems of inspiration almost naturally empower these algorithms with characteristics such as *autonomy, self-organization, adaptivity, robustness*, and *scalability*, which are all desirable if not necessary properties to deal with the challenges of current and next-generation networks. In the chapter we consider different classes of wired and wireless networks, and for each class we briefly discuss the characteristics of the main ant- and bee-colony-inspired algorithms which can be found in literature. We point out their distinctive features and discuss their general pros and cons in relationship to the state of the art.

1 Introduction

The constant improvement in communication technologies and the related dramatic increase in user demand to be connected anytime and anywhere to both the wealth of information accessible through the Internet and other users and communities have boosted the pervasive deployment of wireless and wired networked systems. These systems are characterized by their being *large* or very large, highly *heterogeneous* in terms of communication technologies, protocols, and services, and very *dynamic*, due to continual changes in topology, traffic patterns, and number of active users and services. *Intelligent* [10] and *autonomic* [72] management, control, and service provisioning in these complex networks, and in the future networks resulting from their integration and evolution, require the definition of novel protocols and techniques for all the architectural components of the network.

In this chapter we focus on the *routing* component, which is at the very core of the functioning of every network since it implements the strategies used by network nodes to discover and use paths to forward data or information from sources to destinations. An effective design of the routing protocol can provide the basic support to unleash the intrinsic power of the highly pervasive, heterogeneous, and dynamic complex networks of the next generation. In this perspective, the routing path selection must be realized in a fully *automatic* and *distributed* way, and it must be *dynamic*, to take into account the constant evolution of the network state, which is defined by multiple concurrent factors such as topology, traffic flows and available services.

The literature in the domain of routing is very extensive. Routing research has fully accompanied the evolution of networking to constantly adapt the routing protocols to the different novel communication technologies and to the changes in user demand. In this chapter we review routing protocols and algorithms which have been specifically designed taking inspiration from, and reverse engineering the characteristics of, processes observed in *insect societies*. This class of routing protocols is indeed relatively large. The first notable examples date back to the beginning of the second half of the 1990s [27, 113, 123, 151], and a number of further implementations rapidly followed the first ones and gained the attention of the scientific community. In this chapter we will limit the discussion to the most popular and effective instances of this specific class of routing protocols.

The fact that insect societies have, and, more in general, nature has, served as a major source of inspiration for the design of novel routing algorithms can be understood by noticing that these biological systems are characterized by the presence of a set of distributed, autonomous, minimalist units that through local interactions self-organize to produce system-level behaviors which show life long adaptivity to changes and perturbations in the external environment. Moreover, these systems are usually resilient to minor internal failures and losses of units, and scale quite well by virtue of their modular and fully distributed design. All these characteristics, both in terms of system organization and resulting properties, meet most of the necessary and desired properties of routing protocols for next-generation networks. This fact makes it potentially very attractive to look at insect societies to draw inspiration for the design of novel routing protocols featuring *autonomy, distributedness, adaptivity, robustness*, and *scalability*. These are desirable properties not only in the domain of network routing but also in a number of other domains. As a matter of fact, in the last 20 years, collective behaviors of insect societies related to operations such as *foraging, labor division, nest building and maintenance and cemetery formation* have provided the impetus for a growing body of scientific work, mostly in the fields of telecommunications, distributed systems, operations research, and robotics (e.g., see [7, 24, 43, 46, 48] for references and overviews). Behaviors observed in colonies of *ants* and of *termites* have fueled the large majority of this work. More recently, also *bee colonies* are attracting a growing interest. In the following we review network routing algorithms inspired by these three classes of social insects. The vast majority of the reviewed algorithms are derived from ant colonies, and in particular, from their ability to discover and follow *shortest paths* between their nest and sources of food [59].

All the algorithms that we will discuss later in the chapter are character- ized by the fact of their being composed of a potentially very large number of *autonomous* and *fully distributed* controllers, and of having been designed according to a *bottom-up* approach relying on basic *self-organizing* abilities of the system. These characteristics, together with the biological inspiration from behaviors of *insect societies*, are the very fingerprints of the *swarm intel- ligence* (SI) paradigm [7]. These peculiar design guidelines contrast with those of the more common *top-down* approach followed for the design of the major- ity of "classical" routing protocols. In typical top-down design a centralized algorithm with well-known properties is implemented in a distributed system. Clearly, this requires us to modify the original algorithm to cope with the intrinsic limitations of a distributed architecture in terms of full state observ- ability and delays in the propagation of the information. The main effect of these modifications consists in the fact that several properties of the original algorithm do not hold anymore if the network dynamics is non-stationary, which is the most common case. Still, it is relatively easy to assert some gen- eral formal properties of the system. On the other hand, with the bottom-up approach, the design starts with the definition of the behavior and interac- tion modalities of the individual node with the perspective of obtaining the desired global behavior as the result of the joint actions of all nodes inter- acting with one another and with the environment at the local level. It is in general "easier" to follow a bottom-up approach, and the resulting algorithm is usually more flexible, scalable, and capable of adapting to a variety of dif- ferent situations. This is precisely the case for the SI algorithms that we will review. The negative aspect of this way of proceeding is that it is usually hard to state the formal properties and the expected behavior of the system. One of the objectives of this chapter consists in showing the common traits and properties of SI routing algorithms derived from insect societies, compar- ing them to the characteristics and properties of established state-of-the-art routing algorithms not based on SI, and evaluating the relative merits.

For space reasons and without loss of generality, we will restrict the classes of networks that we will consider. More specifically, we will focus the discussion on routing algorithms for non-optical *connectionless* and *connection-oriented wired networks* offering *best-effort* and/or *guaranteed quality* services, and for *wireless mobile ad hoc networks* (MANETs) [106]. These are wide and general classes of networks that include a large number of network instances of both practical and theoretical interest. Concerning SI-based routing algorithms for other important classes of networks the interested reader can consult for in- stance [58, 91] for the case of *optical networks*, [119] for the case of *satellite networks*, and [15, 88, 97, 108, 109] for the case of *sensor networks*. In [145] the interested reader can find a general overview of nature-inspired routing algorithms, while in [2], a more general discussion on the design of algorithms for modern telecommunication networks using design patterns derived from the observation of biological systems can be found.

1.1 Organization of the Chapter

The remaining content of the chapter is organized for considering separately the ant- and the bee-colony-inspired frameworks and their applications to each one of the considered classes of telecommunication networks. For each algorithm we will point out the general design characteristics and performance.

- Section 2 briefly introduces *network routing*, and discusses the general characteristics of routing and the associated challenges for each one of the considered network classes.
- Section 3 provides a comprehensive set of *classification features* that we will use to characterize routing protocols and to which we will refer to throughout the chapter to highlight the main differences among the different protocols and, more specifically, between the SI protocols reviewed here and the more standard, established ones which are widely deployed in real-world networks.
- Section 4 and its two subsections describe respectively the ant and bee colony behaviors that have fueled the design of so many network routing algorithms. In particular, Subsect. 4.1 introduces the *Ant Colony Optimization (ACO)* metaheuristic, which is based on the reverse-engineering of the ant colony shortest-path behavior, and which has provided the main practical guidelines for the design of the ant-colony-inspired algorithms.
- Section 5 and all its subsections are devoted to the discussion of routing protocols derived from ACO. First, the general principles behind ACO and ACO for routing are discussed in Subsect. 5.1. In Subsects 5.2 and 5.3, we describe in some detail AntNet and ABC, which are the main reference algorithms that have guided the design of most of the other algorithms. In Subsects. 5.4 to 5.7 we discuss the characteristics of a number of ACO routing algorithms. The algorithms are grouped per network type and are considered in chronological order.
- Section 6 and its two subsections are devoted to the discussion of routing protocols derived from bee colonies. In practice, we discuss in some detail two main implementations, BeeHive for wired connectionless networks and BeeAdHoc for MANETs.
- Section 7 summarizes the presented results and draws some general conclusions about the efficacy and the future perspectives of the SI approach to the design of novel routing protocols for next-generation networks.

2 Generalities and Challenges of Network Routing

The behavior of the network routing protocol drives network dynamics and critically affects performance. In fact, it implements the strategies used by network nodes to determine and use paths to forward data or information from sources to destinations. Generally speaking, the *routing protocol* defines what information is going to be used to make routing decisions, how this information is communicated among the nodes, and how it is encoded in the node's *routing table*, which is the local database of routing information. A routing table maintains the necessary information to define for each end node of interest and for each locally available output interface the quality and cost associated with the selection of the interface as the next hop to forward data toward the end node. The *routing algorithm*, being part of the protocol, makes use of this information to actually select the paths and forward data along them. The challenges faced by a routing protocol and the measure of its efficacy depend on the characteristics of the network at hand. In the following we briefly review these aspects for the considered network types. The interested reader can find more accurate discussions in networking textbooks.

Transmission Mode: Connection-Oriented vs. Connectionless

One basic distinction among network types is based on the adopted point-to-point *switching* technique. The two main classes of networks can be singled out: *circuit-switched* and *store-and-forward*. In circuit-switched networks, prior to start sending end-to-end data, it is necessary to seek out and establish a physical dedicated path between the two end points. No buffers for data are needed. Once the connection is set up, the only delay is propagation time. A telephone network is a typical example of a circuit-switched network. In store-and-forward networks, an intermediate node along the path stores each incoming block of data, inspect it for errors, and retransmits it along the path to the destination. *Message*, *packet*, and *cell switching* refer respectively to the cases of a store-and-forward network in which the transferred block is a complete message, a variable-length block of data with a size upper bound, or a small, fixed-size block of data. The switching method most widely used in networks, such as the Internet, is the *packet-switching* one. It can support different transmission modes. The *connection-oriented* mode shares the same principles as the circuit-switching technique. Prior to packet sending, a path connection (*virtual circuit*) must be established between the two endpoints. The virtual circuit can be a dedicated physical connection or a logical one, shared among different data sessions. The task of the routing system is to find and use full end-to-end paths. Typical measures of performance in this case are the *session acceptance ratio*, the *delivered throughput*, and statistics of the packet latencies such as the *average end-to-end delay*. The latter two performance metrics are reference metrics for almost any type of network, since they summarize two basic aspects related to the quantity and the quality of the service a network can deliver. In connectionless (*datagram*) networks, a packet is injected into the network without requiring establishing any connection, physical or virtual, and without any guarantee that the packet will be delivered at the destination. Each relay node deals with the packet independently of the other nodes and makes use of packet header information to decide how to route the packet. In this case, routing tables and data forwarding across the nodes should be consistent to let the packet traveling over existing and loop-less routes

Delivered Service: Best-Effort vs. Guaranteed-Quality

One major distinction can be made between networks offering *best-effort* services and those offering *quality of service (QoS)*. In best-effort networks the user applications are served with no guarantees on the quality of the delivered service. On the other hand, in QoS networks the user can specify constraints on the quality of the obtained service (e.g., in terms of end-to-end delay, delay jitter, bandwidth, etc.) and the network is expected to either meet these requirements or reject the application. In QoS networks the general challenge of routing consists in the ability to rapidly and robustly identify one or more paths that meet the QoS requirements of current traffic sessions while providing at the same time an efficient utilization of the network resources in order to be ready to satisfy the QoS requests of future sessions. There are several network models that can allow provisioning of QoS. The most popular ones are: *IntServ, DiffServ*, and *Multi Protocol Label Switching (MPLS)* (e.g., see [139]). In IntServ the network must find and reserve resources for each single QoS flow. DiffServ is based on the organization of data traffic into multiple classes, with each class associated with different QoS requirements. Each packet is placed into a specific class and each router is configured to take different routing and scheduling actions depending on the class of the data packet. MPLS is a data-carrying mechanism which emulates some basic properties of a circuit-switched network over a packet-switched network. Once an end-to-end path has been found, it is uniquely identified at the nodes by means of labels and can be then efficiently used to forward data flows.

Topology and Connectivity: Wired vs. Wireless Mobile Ad Hoc Networks

In *wired networks* hosts and routers are connected through one-to-one cables creating a fixed network topology which undergoes only low-rate modifications due to addition or removal of resources and to temporary failures. Point-to-point communication links are usually reliable and have large bandwidth. Terminals are equipped with good computational resources and are not concerned with power supply issues. The challenges for a routing protocol are the changing traffic patterns, the heavy loads, the small topological modifications, and the usually large number of nodes which scale up over time.

Wireless networks with mobile users present radically different characteristics and challenges. In this chapter we are interested in one specific class of wireless mobile networks, the *mobile ad hoc networks (MANETs)* [106], which during the past few years have become a very active area of research due to their unique characteristics. In a MANET all nodes are mobile and can enter and leave the network at any time. They communicate with each other via medium-range wireless connections that can constantly be established and broken because of mobility. There is no ground infrastructure to rely on. All nodes are peers and can serve as routers to each other. Data packets are forwarded from node to node in a multi-hop fashion. The wireless channel is shared among the peer nodes and the access must be arbitrated according to some distributed Medium Access Control (MAC) protocol, which results in a rather low and irregular amount of effective available bandwidth. Terminals have usually less computational power than in the wired case and are powered by on-board batteries with limited lifetime. All these aspects, such as mobility, shared channel, low bandwidth, short battery lifetime, and distributed multi-hop forwarding, impose severe challenges and restrictions to the routing protocol. A good protocol is one that can effectively *adapt* to dramatic topological changes, needs relatively *low control overhead*, provides *high throughput* and *low packet delays*, and saves as much as possible of *battery power* to let the users and their mobile devices participate as long as possible in the network activities. It is clearly very hard to meet in a satisfactory way all these conflicting objectives; therefore, a rather large number of different routing algorithms have rapidly appeared in the literature (e.g., see [11, 106, 122]). A common feature of MANET routing algorithms is that they are all adaptive.

State-of-the-Art Routing Algorithms

Long-standing research on network routing has resulted in a rather large number of routing protocols and algorithms showing different characteristics according to the different types of networks and offered services they are meant for. Clearly, it is not possible to properly account for this large literature here. In this paragraph we limit ourselves to a brief discussion of a small number of state-of-the-art algorithms that are often mentioned to assess the relative performance of the reviewed swarm intelligence algorithms.

OSPF [87] and RIP [79] are among the most popular protocols for routing within Autonomous Systems (*interior protocols*) in use in the wired Internet, while BGP [158] is widely used to communicate among Autonomous Systems. OSPF belongs to the category of *link-state* algorithms. In these algorithms, each node periodically floods a comprehensive state description of all its communication links. This description is used at each receiving node to incrementally construct and update a complete weighted graph of the network. OSPF makes use of Dijkstra's shortest-path algorithm to calculate the routes based on this graph representation. While OSPF is mainly topology-adaptive, an earlier version of it, Shortest Path First (SPF) [73], was both topology- and traffic-adaptive. QOSPF [159] is an extension of OSPF to deal with QoS requests in conjunction with a resource reservation protocol such as *RSVP* [157]. In QOSPF, flooded link-state messages report on QoS information and resources used by active flows.

RIP and BGP are instances of *distance-vector* protocols. In this case, each node only knows the set of network destinations and maintains in the routing table the vector of the best distances (e.g., number of hops) to reach each destination. These distances are periodically sent to all the neighbors and are calculated incrementally from hop to hop using algorithms derived from the well-known Distributed Bellman-Ford algorithm [6], which is in turn based on *dynamic programming* [5]. In practice, when node i receives from its neighbor j a message saying that j's shortest distance estimate to destination d is n hops, i can safely set its best distance to d as $n+1$ hops if its current shortest distance estimate was $m > n+1$. This way of constructing distance estimates is prone to what is termed "counting to infinity": a very slow convergence to the right distance vectors after a destination becomes unreachable, with the concrete risk of loops and dangling routes. A notable recent distance-vector implementation which deals effectively with these problems and has also interesting additional properties is the Multi-path Distance-Vector Algorithm (MDVA) [138]. The algorithm is loop-free under stationary conditions and makes also use of multiple paths.

The Bellman-Ford's way of constructing estimates building on others' estimates is also termed *bootstrapping* and is widely used in the domain of *model-based reinforcement learning* [125]. More precisely, the notions of bootstrapping and reinforcement learning have guided the design of Q-routing [9] and of the derived PQ-routing [19], which are among the most notable contributions of artificial intelligence research to the domain of network routing.

Concerning MANETs, the reference algorithms are: Ad hoc On-demand Distance Vector routing (AODV) [99], Optimized Link State Routing (OLSR) [21], and Dynamic Source Routing (DSR) [69]. AODV is a *reactive distance-vector* algorithm, that is, routing information is only collected when necessary to route an active traffic session. OLSR is a *proactive link-state* algorithm directly derived from OSPF and adapted to deal with the dynamic aspects of MANETs. DSR is a *reactive source-routing* algorithm, that is, the header of each data packet carries the complete route to the destination in the form of ordered next hop nodes.

3 Classification Features of Network Routing Protocols

In principle, many different taxonomies can be adopted to effectively classify routing protocols (e.g., [53]). In the following, we identify a specific set of classification features which will serve to capture the distinctive characteristics of each considered SI algorithm and, at the same time, to point out the general differences existing between these algorithms and more standard, non-nature-inspired, protocols. The classification features we propose here are partly based on those considered by CISCO [20]:

Static vs. Dynamic. Static routing protocols are based on the use of routing tables which are defined *offline* by network administrators based on some prior knowledge of the network. Dynamic protocols update routing tables and routing decisions *online* to reflect changes in the network state. Most of the protocols currently in use on the Internet, such as the mentioned OSPF and RIP, mainly deal with *topological changes* deriving from run-time failures and/or addition or removal of network resources, and do not explicitly take into account *varying traffic patterns*. On the other hand, most of the SI algorithms are explicitly designed to be adaptive to both topological and traffic variations.

Single-Path vs. Alternate- and Multi-path. Single-path algorithms for routing make use of a single path at a time to forward traffic between two end-points. The path is determined to be the best one available according to the considered performance metrics. Alternate-path algorithms still make use of a single path but calculate and maintain also a backup path to be readily used in case of any problems or unavailability of the main reference path. Finally, multi-path algorithms discover, maintain, and use multiple paths to forward flows between the same source-destination pair. This allows it to multiplex the traffic, usually resulting in better failure resilience, utilization of network resources, and higher throughput with respect to the other two mentioned strategies.

Flat vs. Hierarchical Organization. Flat routing protocols consider all nodes in the (sub)network as peers and maintain an entry in the routing table for each of them. This allows peers to discover best individual routes at the cost of transmitting a relatively large amount of control packets and maintaining large routing tables. Routing algorithms based on hierarchical organization form logical groups of routers and organize them into areas, domains, and autonomous systems. This popular way of organizing the network requires two types of routers, *interior routers*, which route traffic within a domain, and *exterior routers*, which route traffic between domains. A hierarchical organization requires significantly smaller routing tables than a flat organization, requiring, in turn, smaller memory storage and less use of bandwidth to maintain routes.

Host vs. Router Intelligent. In host-intelligent protocols a host determines the entire route to a destination and appends it to each packet header.

This way of proceeding is also known as *source routing*. The other routers in the system simply forward packets to the next hop specified in the header and in principle do not need to maintain up-to-date routing information for destinations not addressed by local sessions. On the other hand, in *next hop routing protocols* routing decisions are taken by the single routers ("router-intelligent") that discover, maintain, and use paths on a per packet or flow basis.

Global vs. Local Representation. In routing protocols using a global representation, each node maintains a complete topological database of the network with the aim of constructing a network graph and applying (shortest) path finding algorithms to it. The popular class of link-state protocols exploits this strategy. On the other hand, protocols relying on local representations define the local routing policy on the sole basis of the local traffic and topology models. Distance-vector protocols make use of local representations. Link-state algorithms converge quicker, scale better, but require more CPU power and memory than distance-vector algorithms. Therefore, they are more expensive to implement and support. SI protocols, which tend to simplicity, are usually based on local representations.

Deterministic vs. Probabilistic Decisions. Deterministic algorithms use a deterministic *selection rule* applied to the information contained in the routing table to decide next hops. Usually this results always in the greedy selection of the best routing alternative. In contrast, probabilistic algorithms make use of a probabilistic selection rule. On the one hand this might result in locally sub-optimal choices; on the other hand, when multiple equivalent or comparable choices are available, the adoption of probabilistic routing selection will spread traffic across different concurrent paths implementing de facto a *multi-path* scheme and favoring *load balancing*. Clearly, a probabilistic scheme requires more computational and memory resources than a deterministic scheme to process each single packet and maintain all the necessary routing information [130]. A probabilistic decision scheme can be used also to forward control packets, not only data packets. In these cases, the probabilistic scheme can be exploited to provide a certain level of randomness in the way paths are discovered and set up. This is supposed to add robustness and flexibility to the routing system to better cope with the intrinsic network variability. As will be shown later, probabilistic schemes for both data and control packets are widely adopted in SI algorithms.

Constructive vs. Destructive Routing Table Making. Constructive protocols start with an empty set of routes and incrementally add routes till the final routing tables are constructed. In contrast, destructive algorithms begin by assuming that all possible paths in the network are valid. That is, they assume that the network is a fully connected graph. Starting from this initial assumption, destructive algorithms incrementally gather information to cut paths that do not actually exist in the physical network [133]. Protocols based on strong exploratory or random strategies are usually destructive, as is the case for many SI protocols for wired networks. On the other hand, when the network topology is highly dynamic, for example, routes constantly appear and disappear as in the case of MANETs, the usual approach is the constructive one.

Proactive vs. Reactive Behavior. Reactive protocols gather routing information only in response to an event, usually one which triggers the need for new routes, such as the start of a data session toward a new destination or the failure of an existing route in use. In proactive protocols, routing information is constantly gathered, so that it is readily available when is needed. In the literature, proactive behavior is often associated with the fact that the protocol proactively defines and maintains routes toward all the possible destinations in the network. *Hybrid protocols* result from any combination of reactive and proactive behaviors. Usually all the protocols for wired networks offering best-effort services are proactive. QoS protocols are hybrid, with the reactive component addressing the QoS requests and the proactive component serving for both the QoS and the best-effort routes. Protocols for MANETs are rather uniformly distributed among the three different types of behaviors.

Proactive gathering of routing information can in principle permit us to build sound statistical estimates of relevant aspects of the network dynamics that can be used in turn to *learn* and *adapt* the local routing policies with continuity. On the other hand, it is usually unfeasible to build sound statistical estimates when using a purely reactive strategy since there is no continuity of information gathering. Clearly, an adaptive learning approach will only be effective if the network dynamics show exploitable correlations over time at either the local or the global level, and do not hectically change with a high frequency.

Formal Guarantees vs. Emergent Behavior. Some algorithms come with formal guarantees concerning specific aspects of their behavior and performance. Properties that are particularly useful to be assessed regard: *failure resiliency*, establishment of *loop-less routes*, and *convergence* to an optimal route assignment. Fully deterministic algorithms designed according to top-down approaches have higher chances to enjoy verifiable properties than algorithms designed following a bottom-up approach and that make use of random components, which is often the case for SI algorithms. For this special class of algorithms, the resulting network behavior can be effectively categorized as "emergent", since it is usually hard to provide a precise formal description of the expected network response and performance. On the other hand, also in the case of top-down design, the above-mentioned properties can be usually asserted in special cases and only when steady stationary conditions are assumed, which is more the exception, rather than the rule, for network behavior.

4 From Insect Societies to Network Routing Protocols

Two specific classes of insect societies have inspired a relatively large volume of work in the specific domain of network routing: *ant* and *bee colonies*. More specifically, the ability of ant colonies to discover *shortest paths* between their nest and sources of food using a pheromone laying-following mechanism [59] has been reverse-engineered and put to work in the general optimization framework of the *Ant Colony Optimization (ACO)* metaheuristic [44, 45, 48]; see also Chap. 2 of this book. To date, ACO is a state-of-the-art metaheuristic for many problems in the domains of combinatorial optimization and network routing. More recently, the *communication* and *recruitment strategies* adopted for effective foraging within a beehive have inspired the development of some novel algorithms for routing problems.

In the following two subsections we discuss separately the general principles behind the ant- and bee-inspired approaches to network routing.

4.1 Shortest-Path Behavior in Ant Colonies and the Ant Colony Optimization Metaheuristic

It has been observed that foraging ants in a colony can converge on moving over the shortest among different paths connecting their nest to a food source [45, 59]. The main catalyst of this colony-level *shortest-path behavior* is a *volatile* chemical substance called *pheromone*. While moving, ants lay pheromone on the ground and, at each step, they preferentially decide, with a random component, to locally move toward the adjacent areas marked by higher pheromone intensity. Shorter paths between the nest and the food source can be completed quicker and more frequently by the ants moving back and forth, and will therefore be marked with higher pheromone intensity. These paths will then attract over time more and more foraging ants, which will in turn increase the pheromone level of these paths, until there is convergence of the majority of the ants onto the shortest path(s).

The local intensity of the pheromone field encodes a spatially distributed *measure of goodness* locally associated with each moving decision. It is the result of the repeated and concurrent *path sampling* experiences of the ants. In other words, it is the result of a *collective reinforcement learning process* [24, 125] happening at the colony level. This form of distributed learning and control based on indirect communication among *agents* (the ants) which locally modify the environment and react to these modifications leading to a phase of global coordination of the agent actions is called *stigmergy* [129]. In nature, ant colonies, as well as other social insects, make use of a variety of different pheromone signals for stigmergic communications. The different pheromones are secreted by different glands, and differ both in their chemical composition and their volatility. Recent studies have shown that this complex indirect signaling system based on *multiple pheromones* is efficiently exploited to react and coordinate in different ways to different stimuli in the environment [68]. For instance, the presence of a predator fuels the release of a danger type of pheromone, while the discovery of a prey to be carried into the nest stimulates the generation of an intense but short-lived type of pheromone which is different from the long-lived pheromone laid for the exploitation of an abundant source of food. Pheromones can be not only *attractive*, as the ones described so far, but also *repulsive*. For instance, a branching leading to a bad route can be marked with repulsive pheromone to avoid its future selection.

Stigmergic coordination is one of the keys to obtaining self-organized behaviors not only in ant colonies but more generally across social systems. When stigmergy is at work, a system's *protocols* (interfaces) play a prominent role compared to *modules* (agents) [22], which can be kept relatively *simple*. A good stigmergic model supplies robustness, scalability, and evolvability, and allows to fully exploit the potentialities of the modules and of modularity. Stigmergic systems are paradigmatic examples of the swarm intelligence approach.

The ability of ant colonies to "solve" distributed shortest-path problems using a number of minimalist agents and pheromone-mediated stigmergic communications has been exploited in the framework of the ACO metaheuristic, in which all the mechanisms at work in the ant colony shortest-path behavior have been reverse-engineered to define a nature-inspired metaheuristic for the (distributed) solution of *generalized shortest-path problems* in graph structures (notice that almost any network and combinatorial optimization problem can be formulated in terms of finding shortest paths in a graph [24]). The ACO metaheuristic features are *repeated path construction* by a distributed system of lightweight agents called *ants*, the use of a *stochastic decision policy* to incrementally construct each path by an ant that moves step-by-step from one node of the graph to an adjacent one, stigmergic communications among the ants through node-local stigmergic variables called *pheromone variables*, and collective *stigmergic learning* of the pheromone variables, which represent the parameters of the decision policy, that is, which encode the expected quality of each decision about the next node to include in the path under construction.

The application of the ACO metaheuristic to network routing is quite straightforward. This results both from the intrinsic distributed architecture of the metaheuristic and from the fact that the problem of defining optimized routing paths in a network environment can be configured as a particular instance of a shortest-path problem, with the weights of the edges being dynamic values depending on bandwidth, propagation delay, and input traffic (whose characteristics are usually unknown with precision in advance).

4.2 Useful Ideas from Honey Bee Colonies

More recently than ant colonies, honey bee colonies have attracted a strong interest as a potential source of inspiration for the design of optimization strategies for dynamic, time-varying, and multi-objective problems. Bee colonies show structural characteristics similar to those of ant colonies, such as the presence of a population of minimalist social individuals, and must face analogous problems such as distributed foraging and nest building and maintenance. Bees utilize a sophisticated communication protocol that enables them to communicate directly through bee-to-bee signals and, when required, similarly to ants, to use stigmergic feedback cues for bee-to-group or group-to-bee communication. In these two classes of insects, communication and cooperation are realized by radically different modalities due to the different nature of these insects (ants walk, while bees mainly fly). In particular, while in the case of ants communication is achieved via a pheromone trail that is laid on the ground while walking, in the case of bees it is a form of visual communication that plays an equivalent role. In the following we briefly point out and discuss the main mechanisms at work in a bee colony which have found their application in the design of routing algorithms.

Adaptive and Age-Related Division of Labor

A honey bee colony consists of morphologically uniform individuals with different temporary specializations [114]. The benefit of the organization is an increased flexibility to adapt to the changing environments. For instance, a nectar forager can become a water forager if the colony is running out on its water supplies. More specifically, in honey bees *division of labor* is mainly related to *age*: workers of different ages specialize in different tasks (this phenomenon is called *age polyethism* or behavioral development). Workers typically perform brood rearing for the first week, engage in other hive maintenance duties (wax secretion, guarding, undertaking, nectar processing) when they are "middle-aged" (two to three weeks old), and switch to foraging and colony defense when they are about three weeks old. These phases can be adaptively modified in response to the alteration of colony conditions.

Communication Inside the Colony and Worker Recruitment

As in the ant case, in a bee colony *foraging* is a critical aspect for the survival of the colony and is executed in a *fully distributed* and *competing* way. Foraging bees constantly leave the hive searching for new sources of nutrients, bring the nutrients back to the hive, and try to recruit other bees to exploit the food site found by competing with each other during the recruitment process. Foragers announce a food source of interest to their fellow foragers by doing a dance on the dance floor inside the hive [136, 137]. This dance is termed *waggle dance*. It is a particular figure-eight dance that encodes the direction of the food source in the angle from the sun, and the distance in the duration of each waggle run [114]. If the distance is very short the waggle dance resembles a *round dance*. Foragers respond to the waggle dance with a strong preference for choosing nearer food sites over distant ones in order to increase the net energetic efficiency of the colony. The waggle dance is a direct form of *agent-to-agent communication*.

Nectar foragers, upon return to the hive, sometimes also perform across the hive a quite strange dance termed *tremble dance*. The tremble dance means that the forager found a rich food source but upon return to the hive, after a certain threshold time, could not find a food storer bee to give her nectar. This suggests that the message of the tremble dance is to stimulate the bees inside the hive to increase and/or to switch to nectar-processing activities, and to inhibit the outside foragers from recruiting additional bees. Basically the tremble dance is intended to activate behaviors that keep a colony's nectar-processing rate matched with its nectar intake rate.

Stochastic Selection of Food Sites

The unemployed foragers refrain from extensively surveying the dance floor to identify the best food site. On the contrary, they observe maximally two or three dances on the dance floor and then decide to follow the indications of one of them according to a *stochastic rule*. As a result, a colony *distributes* its foraging force on multiple food sites such that when one rich food site has been almost fully exploited the colony is already exploiting other sites [114]. In this way an effective *balancing* between exploitation and exploration is automatically obtained. Sumpter [124] has developed a formal agent-based model using process algebra for the foraging behavior of honey bee colonies which provides some useful insights into the colony-level strategy for the distribution of the exploitation activities.

5 Routing Protocols Based on Ant Colony Optimization

5.1 General Structure and Properties of ACO Routing Protocols

The main characteristic of an ACO routing algorithm [24, 38] consists in the continual acquisition of routing information through *path sampling* and *discovery* using small control packets, the ants. The aim is to adaptively learn statistical estimates of the quality (e.g., expected end-to-end delay) of each local routing choice. The ants are generated concurrently and independently by the nodes, with the task to try out a path to an assigned destination. An ant going from source s to destination d collects information about the quality of the path, and, either on the way to d or while retracing its way back from d to s, it uses this information to update the routing tables at intermediate nodes, reinforcing the good paths. In other words, the repeated path sampling and consequent reinforcement of good routing decisions is a form of *distributed reinforcement learning based on stigmergy* (e.g., see [100, 24]). The routing table at node i is derived from the so-called *pheromone table* $\mathcal{T}^i$, which contains for each destination d of interest a vector $\bar{\tau}_d$ of real-valued entries τ_{nd}, one for each node n in the reachable neighbor of i, indicated hereafter with $\mathcal{N}^i$. These entries, which are the *pheromone variables*, are a *local* measure of the goodness of going over the neighbor n on the way to d.

They are continually updated according to the quality of the paths *sampled* by the ants. The repeated work of the ants results in the availability, at each node, of a *bundle of paths*, each with an estimated measure of quality based on pheromone. The information from the pheromone tables is usually combined with additional *heuristic information* η not depending on or derived from the sampling activities of the ants, to obtain the *selection probabilities* p which are used by the ants to find their way to the assigned destination d: at each node i they stochastically choose a next hop $n \in \mathcal{N}^i$ giving higher preference to those next hops associated with higher p_{nd} values, which are calculated as some function f of both pheromone and heuristic values, $p_{nd} = f(\tau_{nd}, \eta_{nd})$. The heuristic values have the same structure as the pheromone values and associate with each pair (next hop, destination) a heuristic measure of goodness. For instance, the number of packets waiting on the queue for link $i \to n$ can be used as a local measure of the goodness of using that link. However, not all the implementations make use of a heuristic correction to the pheromone values to derive the selection probability values.

In the case of connectionless networks, packets are usually routed more or less in the same way as the ants: packets are routed *stochastically*, choosing with a higher probability those links associated with higher pheromone and ant-routing values. This way data for a same destination are adaptively spread over *multiple paths* (but with a preference for the best paths), resulting in *load balancing*. In the case of connection-oriented networks, spreading can be done at the level of virtual or physical circuits. For both data packets and circuits, mechanisms are usually adopted to avoid low-quality paths, while ants are more explorative, so that also less good paths are occasionally sampled and maintained as *backup paths* for failures or sudden congestion. In this way *path exploration* is kept separate from the use of paths by data. If enough ants are sent to the different destinations, nodes can keep up-to-date information about the best paths, and automatically adapt their data load spreading.

Referring to the classification features of Sect. 3, ACO implementations for routing usually show the following characteristics: (i) they are all adaptive, with a special focus on traffic patterns, (ii) they usually provide and use multiple paths, (iii) they are mostly based on a flat organization, (iv) router-intelligent schemes are the most adopted ones, (v) global representations are barely used since the approach in a sense emphasizes simplicity and locality, (vi) probabilistic exploratory decisions are an integral part of all the implementations, (vii) they adopt either a constructive or a destructive approach depending on the network type, (viii) the majority of the implementations follow either a proactive or a hybrid scheme, and make use of some form of incremental learning to continuously adapt over time the routing tables to network changes, (ix) usually these algorithms come with no or little formal guarantees apart from some guarantees of probabilistic convergence to the optimal policy under stationary network conditions, and the probabilistic guarantee that a packet following a loop will be routed out of the loop in a short time.

The first ACO routing algorithms were developed at the beginning of the second half of the 1990s and were designed for wired networks: *AntNet* [29] for connectionless IP data networks and *ABC* [113] for circuit-switched telephone networks. A number of other ACO implementations for different routing problems have been developed since then. The majority of these subsequent implementations have based their design on the general features and architecture of either AntNet or ABC. Therefore, in the following we give a special attention to these two algorithms that can be considered as the main reference templates for ACO routing implementations and can help us understand the common architecture and characteristics of most of the other implementations.

In [24, 37, 38] Di Caro et al., starting from the observation of the existence of a core set of features common to most of the ACO-derived algorithms for routing, defined the *Ant Colony Routing* (ACR) framework. ACR includes basic ACO concepts but at the same time extends them with notions from the domains of reinforcement learning [125], multi-agent systems, and autonomic networking [72], and specializes them for the specific class of network routing problems. The ACR framework is intended to provide the basic guidelines for the design of novel adaptive protocols for routing in modern dynamic networks. Because of lack of space, we are not going to discuss here the ACR framework. However, it is worth pointing out that ACR explicates the two main mechanisms for *monitoring* and *learning* which are at the core of most of the routing algorithms derived from ACO: node-local monitoring of traffic dynamics for *inductive learning* of congestion and routing information, and non-local *sampling and probing* of full paths by using ant agents that implements a combination of *active learning* and *Monte Carlo learning* [125] strategies. The use of these techniques is in some sense not new to the field of networking. The use of inductive learning traces back to the work on learning automata of Narendra et al. [90, 92], while active probing has been widely used to estimate characteristics of network paths (e.g., [67]). However, the way their are combined, implemented, and used in ACO routing, and, more generally, in ACR, is innovative and highly effective.

The interested reader can find additional definitions, discussions, and analysis concerning the application of ACO to routing in [24, 29, 37, 38, 121].

5.2 AntNet: The Main Reference Algorithm for Connectionless Networks

AntNet (1997) [24, 25, 28, 29, 30] was proposed by Di Caro and Dorigo for dynamic best-effort routing in wired IP networks such as the Internet. The algorithm is explicitly designed to provide *traffic-adaptive routing*. Topological changes are not explicitly considered, so that route breaks due to link failures are only dealt with implicitly by reacting to the increase of the number of data packets waiting in the queue of the broken link. A flat network organization with router-intelligent hosts is assumed. Informally, the behavior of AntNet can be summarized as follows.

At the beginning of the operations routing tables are initialized with uniform equal values for all the neighbors, basically adopting a destructive approach. They are then adapted over time as a result of the ant-based activities. At regular fixed intervals and concurrently with data traffic, ant agents are *proactively* and independently launched from each network node s toward destination nodes d which are chosen by following a *random proportional* selection that favors the locally most requested destinations, or by implementing with a very small probability a *random uniform* selection. These ants are called *forward ants*. A forward ant is a sort of random experiment aimed at exploring the network searching for a *minimum delay path* connecting an ant's source and destination nodes, and gathering at the nodes information about the end-to-end delay for the followed path. Ants, once generated, act as *autonomous agents*. They communicate in an indirect, stigmergic way, through the information they locally read from and write to the nodes in three data structures: the *pheromone table* $\mathcal{T}$, the *parametric statistical model* $\mathcal{M}$, and the *data routing table* $\mathcal{R}$, that together define the routing information database locally available to issue routing decisions (see also Fig. 1).

The pheromone table is a stochastic matrix which is used by the ants as a routing table. Each pheromone estimate $\tau_{nd} \in \mathcal{T}^i$, $n \in \mathcal{N}^i$, is the result of the continual path sampling and learning activities of the ants, and is related to the inverse of the estimate of the expected minimum time to reach d. τ's values for the same destination d are normalized to 1 ($\sum_{n \in \mathcal{N}^i} \tau_{nd} = 1$). This allows to treat the pheromone values as probabilities and better evaluate the relative goodness of each neighbor. $\mathcal{M}^i$ is a parametric statistical model for the traffic and delay situation on the paths to the different destinations. $\mathcal{M}^i$ is a vector of N triples (μ_d, σ_d^2, W_d), with N being the number of destinations. μ_d is the sample exponential mean of the ants' traveling time to reach d, σ_d^2 is its variance, and W_d is the best end-to-end time observed during the last window of w ant samples. Finally, the data routing table $\mathcal{R}^i$ is the stochastic matrix used for routing data packets. It is derived from the pheromone table by an exponentiation and renormalization process that assigns to the best routes much higher selection probabilities than in the case of the pheromone table. This is because the ants are supposed to explore, while the data packets are supposed to exploit at best the paths found by the ants.

Forward ants *simulate data packets*. They move hop-by-hop toward their destination making use of the same priority queues used by data packets, experiencing in this way the same delays. During its journey to d, a forward ant stores in its memory the traveling time $t_{i \to j}$ between each hop $i \to j$ and the identifiers of the visited nodes along the followed path $\mathcal{P}_{s \to d}$. At each intermediate node i, a *stochastic decision policy* $\pi_\epsilon(\mathcal{T}^i, \mathcal{L}^i, \mathcal{P})$ is applied to select the next node $n \in \mathcal{N}^i$ to move to, where $\mathcal{N}^i$ is the set of neighbors of i.

The *selection probability* p_{nd} assigned to each neighbor $n \in \mathcal{N}^i$ is a measure of the goodness, relative to all the other $j \in \mathcal{N}^i, j \neq i$, of using the neighbor as next hop for final destination d. p values are calculated considering a combination of: (i) the pheromone value τ_{nd} which is the result of the continual, long-term path sampling and learning activities of the ant agents, (ii) the length in bytes to be sent of the link queue $l_n \in \mathcal{L}^i$ associated with n, which is a *heuristic* instantaneous measure of congestion of the path going through n, and (iii) the list of the visited nodes stored in the ants' memory, which is used to avoid loops. More precisely, each p_{nd} is defined as:

$$p_{nd} = \frac{\tau_{nd} + \alpha l_n}{1 + \alpha(|\mathcal{N}^i| - 1)} \tag{1}$$

if $n \notin \mathcal{P}_{s \to i}$, and 0 otherwise. In practice, with this formula, the selection probability of a next hop is calculated as the weighted sum of the estimate τ, which is the result of a continual process of incremental learning, and the instantaneous quality estimate l. Both τ and and l values are scaled between 0 and 1, in order to be summed consistently. $\alpha \in [0, 1]$ determines the relative importance of the long-term versus the instantaneous view of the goodness of each next hop decision. The denominator is just a normalization factor.

Once arrived at destination, the forward ant becomes a *backward ant*, which is *source-routed* to s: it goes back to its source node by moving along the same path $\mathcal{P}_{s \to d} = [s, v_1, v_2, \ldots, d]$ as before but in the opposite direction. For its return trip the ant makes use of high-priority queues to quickly retrace the path and update the routing information.

Arriving from neighbor j, at each visited node $i \in \mathcal{P}_{s \to d}$ the backward ant updates, for the choice of j as next hop, the routing information related to each node $\delta \in \mathcal{P}_{i \to d}$ visited by the forward ant when traveling from i to d. Basically, each node δ is considered as an intermediate destination. The backward ant first *evaluates* the goodness of the followed path and of its sub-paths, and then uses this evaluation to update the local routing information. Path evaluation is done by comparing the traveling times experienced along the path with the expected traveling times maintained in the statistical model $\mathcal{M}^i$. From the evaluation process, a path *reinforcement* value $r \in [0, 1]$ is defined as:

$$r = c_1 \frac{W_\delta}{T_{i\delta}} + c_2 F(T_{i\delta}, \mu_\delta, \sigma_\delta^2, W_\delta), \tag{2}$$

where c_1 and c_2 are weighting factors, $c_1 + c_2 = 1$, and F is a real function accounting for the statistical dispersion of the sampled values. In practice, the sampled path (and its sub-paths) gets a reinforcement proportional to how good the traveling time $T_{i\delta}$ just experienced by this ant is compared to what has been observed in the recent past. At the visited nodes i, r is used to update the pheromone entries as follows. The path to each "destination" δ going through the used neighbor j is reinforced, while, by normalization, the goodness of all the other alternatives is proportionally decreased:

$$\begin{aligned} \tau_{j\delta} &\leftarrow \tau_{j\delta} + r(1 - \tau_{j\delta}), \\ \tau_{k\delta} &\leftarrow \tau_{k\delta} - r\tau_{k\delta}, \quad \forall k \in \mathcal{N}^i, k \neq j. \end{aligned} \tag{3}$$

Fig. 1 shows the data structures used by the ants at the nodes, and illustrates the two core phases of operations in the AntNet: the decision step of the forward ant and the update process executed by the backward ant.

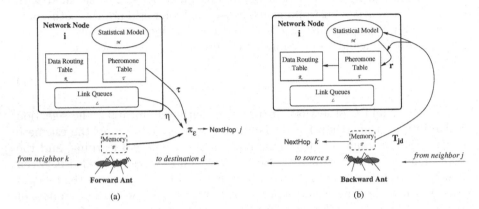

Fig. 1. The two core phases of AntNet shown at a node $i \in \mathcal{P}$ for an ant generated in s and targeted to d: (a) the decision step of the forward ant, and (b) the update and move step of the backward ant. The arrows help us to visualize from which data structures the ant gets the information to decide the next step during the two phases, and the logical sequence of updating steps happening during the backward phase

Once the ant has returned to its source node, it is removed from the network. *Data packets* are routed according to a stochastic decision policy similar to that of the ants but based on the information contained in the local data routing table $\mathcal{R}$, which is derived from the pheromone table used to route the ants preferring the best paths. In this way, data traffic is concurrently spread over the best available *multiple paths*, resulting in an optimized utilization of network resources and in automatic *load balancing*.

AntNet-FA [24, 32] (also known as *AntNet-CO*) is a minor but quite effective improvement of AntNet: forward ants also make use of high priority queues. In this way, forward ants quickly get to the destination, and do not need to carry traveling times; it is the backward ants that calculate incrementally the trip times while traveling backward. Coming from neighbor n, at node i the backward ant estimates the time necessary to cross the link $i \to n$ by looking at the number of bytes waiting in the l_{in} queue. The link crossing time T_{in} is obtained on the basis of a queue depletion model:

$$T_{in} = \frac{l_{in}}{b_{in}} + d_{in}, \tag{4}$$

where b is the link bandwidth and d is its propagation delay. The adopted model is simple but also quite reliable. AntNet-FA's strategy on the one hand permits to calculate source-destination trip times which are more up-to-date than those used by AntNet's backward ants, and on the other hand it allows a quicker gathering and spreading of routing information. This is a clear advantage in the case of large topologies and quickly changing input traffic.

AntNet's authors have evaluated their algorithm on the basis of a relatively large number of *simulation experiments* using a custom network simulator. The algorithm has been tested on a variety of different scenarios based on different topologies with number of nodes ranging from few units to 150, and considering UDP traffic patterns with different geographical and generation characteristics. *Throughput, 90th percentile of packet delays*, and *routing overhead* have been chosen as performance indices. The reported experiments show that AntNet robustly outperforms in terms of throughput and delay several different dynamic state-of-the-art algorithms: *Q-routing*[9], *PQ-routing* [19], *Shortest Path First (SPF)* [73], *Dynamic Bellman-Ford* [115], and *OSPF*. The improvement in performance is achieved without increasing the routing overhead. Moreover, AntNet-FA outperforms AntNet, with the difference becoming larger with increasing network size.

5.3 ABC: The Main Reference Algorithm for Connection-Oriented Networks

Schoonderwoerd et al. (1996) [112, 113] were the first to apply the ACO ideas to routing and load-balancing problems in networks. More precisely, they considered a telephone network in which the connection between sender and receiver is explicitly established by reserving a virtual circuit. In their network model, each node is a crossbar switch and can handle only a limited number of simultaneous calls. Connection links are seen as full-duplex channels with infinite capacity. Therefore, network bottlenecks are nodes' capacities. This means that the network is *cost-symmetric*: the congestion status over an end-to-end path is the same in both directions since it only depends on the spare connection capacity at the nodes (e.g., see [56]). The proposed routing algorithm, named `Ant-based control (ABC)`, aims at distributing the calls over multiple switches (i.e., *load balancing*) to minimize the number of calls that cannot be routed because of congestion.

ABC and AntNet share the same general organization and principles. The main differences between the two algorithms are due to the differences existing between the two network scenarios that have been addressed. In ABC ants move over a control network isomorphic to the one where the calls are established. In the adopted model the system evolves synchronously in discrete steps. Next hops are selected according to a random proportional or random uniform rule, as in AntNet, but taking into account only pheromone values; no heuristic correction is used. Arrived at a node, an ant waits ΔT steps, defined as a function of the spare node capacity ΔC,

$$\Delta T = Ke^{-a\Delta C}, \tag{5}$$

with K and a real constants, $K \gg a$, and increases in this way its *age*. This is equivalent to what happens in AntNet, where forward ants wait in the local data queues, with a consequent increase in their traveling time. Equivalently, the age is used in ABC to asses the quality of the ant path: an old ant is associated with a congested path. Pheromone entries are updated using the ant age T as follows. If s is the source and d the destination node of a traveling ant, after crossing the control link $i \rightarrow j$ the probabilistic pheromone table T^j at node j is immediately updated using the total ant age T. A reinforcement r inversely proportional to T is assigned to the normalized entry τ_{is} in T^j:

$$r = a/T + b, \tag{6}$$

where a and b are small constants dependent on network characteristics. The updating formula for the τ values is the same as in AntNet (Eq. 3). The main difference with AntNet in this respect consists in the fact that the pheromone table is updated during the forward journey in the backward direction of the source node s. This way of proceeding is justified by the fact that the network is cost-symmetric, such that the cost (level of congestion) of a path is the same in both directions. Therefore, at node j the ant age is a sound measure of the quality of the reverse ant path $j \rightarrow s$. In ABC ants do not need to retrace the path backward. Calls are routed according to a deterministic greedy policy that always selects the best next hop. If the destination can be reached, a circuit is established and the call can happen.

ABC's performance has been tested in simulation considering the real topology of the backbone of the British Telecom (BT) telephone network and a number of different call patterns. Reported results show that ABC outperforms an agent-based algorithm developed for BT by Appleby and Steward [1] and reacts better to changes in traffic.

5.4 Algorithms for Wired Connectionless Networks

In this section we review the main work concerning the application of AntNet, ABC, and, in general, ant colony ideas to wired best-effort routing in connectionless networks such as the Internet.

Subramanian, Druschel, and Chen (1997) [123]: Uniform Ant Algorithm

The authors consider generic cost-aysmmetric networks and provide an analysis of two algorithms; one is based on ABC, and the other is a very simple one that makes use of so-called *uniform ants*. In both algorithms, ants make routing table updates in the *reverse direction* of their motion: arriving at node j from node i, an ant originally launched from s updates the τ_{is} entry of j's routing table using some measure c_{ji} of link cost calculated in j. The difference between the two algorithms consists in the fact that uniform ants wonder in the network with no specific destination and make next hop selections blindly, without relying on pheromone. The core idea behind uniform ants is that *simple unbiased exploration* is a means to adapt to any change in the network, especially failure. Since they sample all the paths with equal probability, this results in their setting up a fully *multi-path* system. Moreover, the fact that they have no destination makes them potentially useful also in ad hoc networks in which node identifiers are not globally known in advance. The authors provide some theoretical proofs of asymptotic convergence of the two algorithms under stationary link costs. Simple simulation experiments considering small topologies show that the two approaches are more or less equivalent and comparable to simple link-state and distance-vector algorithms. The downside of the simple and general mechanism of uniform ants consists in the fact that its efficacy and efficiency is expected to dramatically decrease with the increase in network size. In some sense, the core idea behind ACO is precisely to find optimized ways to implement biased exploration and/or deal with failures, rather than rely on blind mechanisms.

Heusse et al. (1998) [65, 66]: Cooperative Asymmetric Forward (CAF)

CAF extends ABC's strategy for step-by-step updating in cost-asymmetric networks. In CAF, when a data packet arrives at node i, the arrival time t_i is written in the packet. After arriving at j from i at time t_j, the total time elapsed in going from i to j, $t_{ij} = t_j - t_i$, is written in j . An ant hopping from j to i reads the t_{ij} information in j and moves it to i, where it is used to update the local estimate for the time to travel from i to j. Since the ant is doing this for all the nodes along its path, the estimate of the travel cost from i to all the nodes the ant has visited so far can also be updated and used to update step-by-step the pheromone tables in the direction opposite to the ant motion, as in ABC. Clearly, if an ant arrives some time after the data packet, the information carried back by the ant might be out of date. The authors tested CAF under some static and dynamic conditions, using the average number of packets waiting in the queues and the average packet delay as performance measures. In [65] they compared CAF to an algorithm very similar to an earlier version of AntNet [26] and to Q-routing. Results were encouraging and under all the test situations CAF outperformed its competitors. In [66] the effectiveness of the approach for load balancing was favorably compared to that of more classical approaches.

Van der Put and Rothkrantz (1998) [132, 131]: ABC-backward

ABC-backward is designed as a combination of the basic ABC structure and formulas with the forward-backward updating strategy of AntNet. The authors have experimentally verified that ABC-backward has a better performance than ABC on both cost-symmetric and cost-asymmetric networks.

Oida and Kataoka (1999) [94]: DCY-AntNet, NFB-Ants

The authors improved an earlier version of AntNet [26] in which the heuristic term based on the instantaneous status of the data link queues was not included in the selection formula (Eq. 1). Without this dependency on the status of the queues, AntNet will suffer from what is termed *stagnation* in the ACO jargon: once the pheromone value τ_{nd} of any next hop link of a neighbor reaches 1 the routing tables get "locked". In ACO algorithms for combinatorial problems this problem is bypassed by applying at each time step t a sort of *pheromone evaporation* to all pheromone entries: $\tau_{nd}(t + 1) = \rho\tau_{nd}(t)$, $\rho \in [0, 1]$. The use of an evaporation mechanism allows us to keep good levels of exploration at any time. The authors of [94] modified pheromone table updating rules to avoid the locking behavior. Their algorithms, DCY-AntNet and NFB-Ants, upon comparison with the considered earlier version of AntNet performed much better under challenging situations.

Doi and Yamamura (2000) [40, 41]: BNetL

These authors also proposed a few additional heuristics to avoid the same locking problem addressed by Oida and Kataoka, but this time considering *AntNet-FA*, which is actually lock-free. Consistently, their algorithm showed a performance equivalent to that of AntNet-FA.

Baran and Sosa (2000) [3]: Improving AntNet-FA

These authors have introduced several modifications to AntNet-FA: (i) instead of starting from a uniform pheromone distribution among all the available next hops for all destinations, for the destinations coinciding with the actual neighbors, pheromone is explicitly initialized to give a much higher selection probability to the shortest, one-hop, route; (ii) assuming the existence of a mechanism that can locally detect and notify a link failure, the pheromone values for the next hop associated with the currently unavailable link are explicitly set to zero, which makes the algorithm explicitly *failure-resilient*; (iii) so-called *uniform ants* adopting a uniform random decision policy as in [123] are introduced to avoid the stagnation effect (however, as mentioned above, AntNet-FA does suffer from this, and therefore the introduced mechanism just helps to increase *exploration*); (iv) for the purpose of better exploiting the best paths, regular ants implement *greedy deterministic decisions* instead of random proportional ones, which reduces exploration (counterbalancing the effect of using uniform ants) and raises the probability that ants and data packets get trapped in long-lasting loops; (v) in order to limit routing overhead, the number of ants concurrently active in the network has been arbitrarily limited to four times the number of the links; unfortunately this can also impair the responsiveness of the algorithm and it is not precisely controllable in a distributed way.

Fenet and Hassas (2000) [54, 55]: Load-Balancing System

This work aimed at developing a novel multi-agent system for *multiple-criteria load-balancing* on a network of processors. The proposed system, which consists of both static and mobile agents, shows general characteristics similar to those of the previously mentioned ACR framework.

Michalareas and Sacks (2001) [85]: Deterministic Simplified AntNet

In this work the authors replaced the stochastic decision policy of AntNet with a *deterministic greedy policy* and did not use the heuristic based on queue lengths. This deterministic version of AntNet has been compared in simulation to OSPF on small tree, ring, and star topologies, and by considering FTP traffic using TCP Tahoe. According to the reported results, under stationary traffic conditions both the algorithms show equivalent performance.

Kassabalidis et al. (2002) [71]: Adaptive-SDR

This algorithm is derived from AntNet but makes use of a *hierarchical organization* by structuring the network into clusters using a centralized *K-means* algorithm. Once the partition process is completed, the algorithm maintains inter-clustering and intra-clustering routing tables at each node. *Multiple colonies* of ants are used to discover and maintain these different routing tables. In this manner the number of ants which need to be generated is significantly reduced because a node only maintains routes to the nodes inside the cluster and not to all the nodes in the network. The authors have compared Adaptive-SDR with a custom, non-standard, implementation of AntNet in which data are routed using a deterministic greedy policy, and with OSPF and RIP. Reported simulation results show that Adaptive-SDR achieves the best results with regard to throughput and average delay. The experiments were conducted on 16- and 48-node network topologies using the *NS-2* simulator [93]. The same authors provided in [70] a brief overview of swarm intelligence for routing, basically presenting ACO approaches.

Lang, Zincir-Heywood, and Heywood (2002) [76, 77]: AntNet vs. Distributed Genetic Algorithms

The authors have benchmarked AntNet and their GA-agent (2002) [78], based on a *distributed genetic algorithm* architecture, against several dynamic scenarios considering the 56-node topology of a former backbone of the Japanese company NTT Communications. AntNet was found to be able to deliver the best routing performance provided that complete and up-to-date global information on the number and identifiers of the reachable network nodes is given as an input to the algorithm. On the other hand, the GA-agent algorithm, which does not require a priori global knowledge, is shown to provide a performance which is between that of AntNet with and without global information.

Yang et al. (2002) [155]: AntNet on a Real Network

Differently from all previous works, which were based on simulation, these authors implemented and studied AntNet on a *real network*, a five-node LAN of Windows-based machines using the TCP/IP protocol. To shorten implementation time, the algorithm was actually implemented at the *application layer*, and not at the network layer. The authors made a study of the relative merits of different ways to define the reinforcement parameter r (see Eq. 2), which is central to the stable operation of the algorithm. They observed that the case of *constant reinforcements* leads to slow but dependable performance, whereas *adaptive reinforcements* might bring better performance but appear to be sensitive to the window length w used for statistics.

Doi and Yamamura (2004) [42]: Loop-Free AntNet

This work addresses two important aspects that have been neglected in most of the other mentioned works: (i) the fact that the Internet has a hierarchical structure and shows power-law properties regarding its topology, and (ii) a routing algorithm should provide some guarantees in terms of being loop-free. The authors proposed a loop-free variant of AntNet-FA in which forward ants explicitly avoid considering for next hop selection all the nodes previously visited. Both the original AntNet-FA and the loop-free variants have been tested on a set of hierarchical, scale-free, Internet-like topologies, and it has been found that the topological characteristics have a significant impact on the relative performance of the two algorithms.

Verstraete et al. (2006) [135]: AntNet on a Real Network

These authors have implemented AntNet on a *physical network* of five routers and two hosts. The authors ran extensive tests to tune AntNet's parameters and extend and modify the basic algorithm to make it work properly in a physical network. AntNet's performance has been compared to OSPF for throughput and failure adaptivity. In terms of throughput, AntNet largely outperformed OSPF in all the tested situations. On the other hand, since AntNet has no built-in mechanism to deal explicitly with topological failures, it recovers from failures slower than OSPF. The authors added a simple mechanism to overcome this problem, and were able to obtain significantly better performance than OSPF also with respect to topological failures.

Dhillon and Van Mieghem (2007) [23]: AntNet Performance Analysis

This work aimed at getting a deeper understanding of the properties of AntNet. The authors have made a performance analysis of AntNet comparing it with a centralized Dijkstra's shortest-path algorithm. The reported simulations show that the performance of AntNet is in general comparable to Dijkstra's algorithm. However, under varying traffic loads AntNet adapts better to the changing traffic and outperforms shortest-path routing.

Gadomska and Pacut (2007) [57]: AntNet with TCP and UDP

It is well known that the TCP, the Internet transport protocol, can show performance degradation in the case of arrival of out-of-order packets. This might happen because of packet losses, or when an adaptive multi-path routing algorithm is used at the network layer, or when the network is undergoing repeated topological modifications. In this work, the authors have studied the effect on performance of using an adaptive multi-path routing algorithm like AntNet at the network layer, together with either UDP or TCP at the transport layer, while the majority of the previously mentioned works are all based on the use of UDP. The authors have run a number of simulation experiments using different realistic network topologies, input traffic, and TCP implementations. Reported results show that while TCP sets higher demands than UDP on the adaptation processes, it is still possible to improve network performance with the use of an adaptive algorithm at the routing layer. In some cases the use of TCP can even improve adaptation time.

5.5 Algorithms for Wired Connection-Oriented Networks

In this section we review the main work concerning the application of ACO ideas to wired connection-oriented networks such as telephone networks and IP networks using virtual circuits (but not explicitly providing QoS).

Di Caro and Dorigo (1998) [31]: AntNet-FairShare (AntNet-FS)

Starting from their AntNet-FA, the authors have derived a novel model for *fair-share* routing and flow control in *virtual circuit* networks. In their model, for each flow a virtual circuit is allocated and bandwidth is reserved. However, the allocated bandwidth is not that requested by the session; it is the maximum bandwidth that can be provided at the moment the session is active and on the basis of a fair-share distribution of the bandwidth among the users. In AntNet-FS, on-demand mechanisms for session setup are added to the usual proactive ant generation. On the arrival of a new traffic session, a *forward setup ant* is *reactively* generated to find and reserve one or more paths for the session. During its journey toward the destination, it behaves like an AntNet-FA's forward ant, except for the fact that, if multiple equally good alternatives exist at a node, the ant is replicated and sent over all the equally good next hops. Moreover, the ant reads from the nodes the value of their residual available bandwidth. The first setup ant arriving at the destination goes back and allocates a virtual circuit with a reserved bandwidth that equals the minimum, bottleneck, bandwidth available along the path, and that does not exceed the bandwidth needed by the session. Further setup ants arriving at the destination are allowed to go back and add virtual circuits only if their trip time is comparable to that of the first ant and their path is sufficiently disjoint from those of the circuits allocated so far. Each session is forced to limit its data generation to not exceed its reserved bandwidth. On subsequent session arrival or departure, bandwidth allocation is dynamically recalculated and the sessions are notified in order for them to adjust their data rates.

White, Pagurek, and Oppacher (1998) [152, 153, 154]: ACO, Pheromone Evaporation, and Genetic Algorithms

These authors described several models and implementations for routing and path finding based on ACO [152, 153] or, more generally, on swarm intelligence [154]. The systems they proposed have an architecture which is very similar to the one of AntNet-FS [31] (described in Sect. 5.6) but make use of pheromone-updating formulas which are adapted from *Ant System* [47], one of the earlier ACO implementations for the traveling salesman problem. In particular, they imported from Ant System the notion of *pheromone evaporation* (see also Sect. 5.4) to sustain path exploration. The authors considered static and dynamic scenarios, as well as centralized and distributed ones. They conducted experiments on small topologies, and results show that the proposed algorithms are able to compute shortest paths in the considered situations. In [152] they used a *genetic algorithm* to dynamically adapt the parameters weighting the relative importance of pheromone and heuristic correction at routing decision time. The use of the genetic algorithm in their ASGA routing algorithm resulted in the improvement of performance.

Bonabeau et al. (1998) [8]: ABC and Dynamic Programming

This work extended ABC with *smart ants* derived from dynamic programming: an ant launched from s, updates at node i the pheromone values for all nodes visited during its trip, rather than just for the source node, as in ABC. That is, all the sub-paths of the $\mathcal{P}_{i \to s}$ path are updated. This is the same strategy adopted in AntNet and in many other algorithms. Compared to ABC ants, smart ants have a more complex behavior but, on the other hand, a better performance is achieved with less agents.

Sandalidis, Mavromoustakis, and Stavroulakis (2001) [110, 111]: Improving ABC with Anti-Pheromone

In their first work [111], these authors have studied the behavior of ABC on a few different network topologies and have confirmed the earlier results published by the authors of ABC. More recently, in [110] the same authors further improved the original ABC: if the age of an ant arrived at node i is greater than the maximum age calculated so far at i, then the pheromone entry related with the ant path is *decreased* instead of being increased. This is a form of so-called *anti-pheromone* similar to the repulsive pheromone used by ants in nature to block unfavorable paths (see 4.1): in the presence of experimental evidence that a sampled route is not good compared to other available routes, its probability of being selected is explicitly decreased. In the large majority of the other ACO implementations, after being sampled, the selection probability of a route is always increased. The performance of the algorithm has been compared to that of ABC for a topology of 25 nodes and has shown a slightly better performance.

Sim and Sun (2003) [120]: Multiple Ant Colony Optimization (MACO)

In their work, the authors first presented an overview of ACO for routing and load balancing and then proposed the MACO approach for *load balancing in connection-oriented networks*. MACO is based on the use of *multiple colonies*, where each colony lays its own type of pheromone. An ant is expected to select paths marked by high values of pheromone of the type laid by the colony the ant belongs to, and get repulsed by routes marked by high values of pheromone laid by ants of other colonies. This *anti-pheromone* mechanism is expected to be an efficient mechanism to find good *multiple disjoint paths*. The use of pheromone repulsion to favor the discovering of disjoint paths was earlier used by Navarro and Sinclair (1999) [91] to solve (static) problems of routing and wavelength allocation in all-optical networks.

Heegaard, Wittner, and Helvik (2003) [63]: Cross–Entropy Ants (CE–Ants)

CE-Ants shares the same forward-backward structure of AntNet but makes use of path-updating formulae derived from Rubinstein's *Cross-Entropy* (CE) optimization framework [107]. The CE method is based on the repeated sampling of paths and on the consequent adaptive adjustment of γ, a parameter that biases path sampling, to minimize the cross-entropy between the used generation probabilities and the optimal importance sampling probabilities. In the distributed version of the CE algorithm designed by the authors, path sampling is implemented by the ants and is biased by the pheromone values. CE formulae are used to define how pheromone values are updated. The authors have also introduced the notion of *elitist ants*: only the best ants are allowed to trace back and update pheromone tables (see [24], Sect. 4.3.2, for a general discussion on the use and efficacy of elitist strategies in general ACO implementations). CE-ants has been applied to *virtual-path discovery* and *failure management* in dynamic connection-oriented and label-switched IP networks offering some form of QoS. The authors have tested their approach considering the real backbone topology of Telenor, a major Norwegian network provider. In [62] Heegaard and Fuglem implemented and tested their system in a physical network using Linux routers.

5.6 Algorithms for Networks Providing Quality-of-Service

In this section we review the main work concerning the application of ACO ideas to wired networks providing QoS.

Di Caro and Vasilakos (2000) [24, 39]: AntNet and Stochastic Estimator Learning Automata (AntNet+SELA)

AntNet+SELA is intended for *QoS routing in ATM networks*. Ant path sampling is complemented by the presence of *node agents* designed after *stochastic estimator learning automata* (SELA) [89, 134]. Each node agent exploits the information gathered by the ants to adaptively learn an effective routing policy for QoS traffic based on the use of a *link-state* routing table in addition to the usual ant pheromone table. Stochastic learning automata have been used in early times [90, 92] to provide fully distributed adaptive routing. One of their main characteristics is that they learn by *induction*: no information is exchanged among the controllers. They only monitor local traffic and try to get an understanding of the effectiveness of the implemented routing choices. In AntNet+SELA, the static inductive learning component is enhanced by using the ants as active learners that gather also non-local information to keep up to date the link-state routing table to rapidly allocate resources for multi-path QoS routing when requested.

In addition to the proactive ant generation as in AntNet-FA, at the arrival of a new session, the node manager reactively generates a *setup ant* and a group of *path-probing ants*. The setup ant behaves similarly to the setup ants of AntNet-FS, with the difference that this time the ant searches for a path that strictly meets the QoS requirements. The path-probing ants are *source routed*: each node agent uses its link-state database to compute the first k paths with minimum hop count that satisfy the QoS requests of the session, and assigns each one of these paths to a different probing ant that will check at run-time its availability and QoS consistency. According to the results provided by the backward ants, the node agent decides whether or not to accept the session and how to possibly split it over multiple paths. Unfortunately, the authors ran only few preliminary tests to evaluate the efficacy of the proposed model.

Oida and Sekido (1999) [95, 96]: Agent-Based Routing System (ARS)

ARS is an enhancement of AntNet that supports both best-effort and QoS routing based on an *IntServ model* with resource reservation and admission control. A Weighted Fair Queueing algorithm distributes at the nodes the capacity between best-effort and QoS traffic. The QoS constraints considered are bandwidth and hop count. A real-time session can require one among n predefined levels of bandwidth and a number of hops less than a maximum value h. According to the basic AntNet scheme, from each node s ants are proactively generated and sent toward a sampled destination with the aim of finding a path with an available bandwidth that matches one of the n levels and with a hop count less than h. Links with more residual bandwidth are preferred when choosing the next hop. If a feasible path is found, it is reported back to the source which stores it in a local cache that is kept up-to-date. When a real-time session requires a QoS path, the session is admitted or not according to the path information held in the cache. If a path that can meet the QoS requirements is present, an ant is sent to probe it and reserve the necessary resources. If the path is not there anymore, the session is rejected. Simulation results on a 14-node network show a high efficiency using network resources.

Michalareas and Sacks (2001) [84, 86]: Multi-swarm

The authors have exploited the main features of both AntNet and ABC to design an algorithm for routing in *multi-constrained* QoS networks. The algorithm provides soft QoS guarantees on end-to-end delay and bandwidth constraints, or, in general, on additive (delay) and concave (bandwidth) constraints. Multi-Swarm deals with the two constraints adopting a *multi-colony*

approach based on the use of two different swarms of ants, one for each constraint. The ants dealing with delay are in practice the same as in AntNet.
On the other hand, since bandwidth is a non-additive metric and it cannot
be directly measured from the ant, the authors have introduced a resource
monitor that locally calculates the average spare bandwidth available at the
links. When a bandwidth ant arrives at a node, it is artificially delayed for a
time inversely proportional to the spare bandwidth, similarly to what happens
in ABC. In this way, the bandwidth estimate is reduced to a delay estimate.
Simulation experiments for three simple topologies under uniform TCP traffic
show that Multi-Swarm has performance comparable to OSPF.

Tadrus and Bai (2003) [126, 127, 128]: QColony

QColony is an algorithm for *QoS routing in multi-constrained networks* designed by extending and adapting AntNet behavior. QColony mostly addresses
the *IntServ QoS model* but its structure makes it suitable to be used with
other models such as DiffServ and MPLS. QColony categorizes network resources (e.g., bandwidth) in sets of adjacent *ranges*, where each range can
fit a different QoS request from a user flow. For instance, if the resource is
bandwidth, and, starting from the value of 0 Mbit/s, the network categorizes
bandwidth requests in ten ranges of 10 Mbit/s each, a user QoS request of 35
Mbit/s can be fit by all the seven upper ranges. At each node, learning and
using good paths for each range is realized by associating with each range a
unique vector of pheromone variables. In practice, this vector corresponds to
the pheromone table normally used by AntNet-like algorithms to deal with
the case of best-effort traffic, which can be seen as a special case of QoS traffic
with no traffic differentiation. Therefore, QColony, like Multi-Swarm, maintains *multi-pheromone tables*. This is reminiscent of what happens in nature,
where different resources and events in the environment are dealt with different types of pheromones (see Sect. 4.1). In addition to QoS tables, a best-effort
pheromone table is proactively maintained and used as in AntNet.

   Upon receiving a QoS request, the ingress node determines the range suitable to satisfy the required QoS and reactively launches an *allocator ant* to
find and reserve the resources. Allocator ants adopt a *greedy next hop selection*
based on the pheromone values associated with the range they are looking for.
If available, network resources are smartly allocated to accommodate the QoS
request while at the same time leaving space for future requests. In addition
to the allocator ants, QColony makes use of several other types of ants, all implementing greedy selections: (i) *explorer ants* are proactively generated and
have a behavior analogous to AntNet ants, but on their backward journey they
update pheromone entries associated with multiple ranges, (ii) *soldier ants*,
mimicking the behavior of soldier ants in nature that respond to potentially
harmful situations, are proactively generated to identify *short backup paths* to
be used in case of failures along the paths in use by running flows, (iii) *maintenance ants* are reactively generated when a *path failure* happens, in which
case they exploit the backup paths found by soldier ants to restore between

the ingress and egress nodes the broken path. Using a custom simulator, the authors have made a number of simulation experiments to test QColony's performance versus that of the previously mentioned ARS, a probing-based reactive algorithm based on selective flooding [18], and QOSPF, which is a reference algorithm in the QoS domain. For small topologies and under low network traffic load the performance of the four algorithms is comparable, while QColony's performance is significantly better for large networks and heavy traffic loads.

Carrillo et al. (2004) [16, 17]: AntNet-QoS

AntNet-QoS is based on a *multi-pheromone* extension of AntNet to support QoS in a *DiffServ* network with m different classes of service for end-to-end delay. For each class, every node holds a pheromone table, a data routing table, and a vector of statistics, replicating in this way m times the data structures held by best-effort AntNet nodes. Ants are generated per class of service: they follow and update the pheromone table associated with their specific class. Ants are routed with higher priority than data, but respecting class-based queuing, such that the quality of their path reflects the class-specific conditions. Preliminary results are promising.

5.7 Algorithms for Wireless Mobile Ad Hoc Networks

In this section we review ant-colony-inspired algorithms for MANETs. Most of the implementations focus on the optimization of throughput and end-to-end delays. On the other hand, we will see that the bee-inspired algorithm discussed later emphasizes battery optimization in addition to throughput and end-to-end delays.

Câmara and Loureiro (2000) [13, 14]: GPS/Ant-Like Algorithm (GPSAL)

These authors were among the first to propose an ACO algorithm for MANETs. GPSAL is a *location-based* algorithm. It assumes and exploits the presence of an on-board GPS device. The routing information is exchanged locally among neighbors, and globally by sending forward ants to distant nodes addressed geographically. Ants are propagated through a bandwidth-efficient flooding algorithm. The algorithm achieves a similar performance with less routing overhead compared to LAR [74], another location-based algorithm.

Matsuo and Mori (2001) [81]: Accelerated Ants Routing (AAR)

AAR is based on the work of Subramanian et al. (see Subsect. 5.4). In AAR, uniform ants are equipped with a stack where the last n visited nodes are stored. This allows them to update the pheromone tables for all the last n intermediate nodes. The authors have compared AAR with AntNet, Q-routing, and PQ-routing on a 56-node network and have shown its superior performance and faster convergence.

Guenes et al. (2002) [60]: Ant-Colony-Based Routing Algorithm (ARA)

ARA imports some basic aspects of AntNet into AODV. It is a *purely reactive* algorithm in which both forward and backward ants set up the paths to the nodes from which they arrive. Also *data packets* update the pheromone tables, reducing the number of ants needed to sample existing paths. According to simulation experiments, ARA's performance turns out to be slightly better than AODV's but worse than DSR's in highly dynamic environments.

Marwaha et al. (2002) [80]: Ant-AODV

In Ant-AODV, AODV is extended by a mechanism of proactive updating of the routing tables based on uniform ants. This increases the chance that a node or one of its neighbors will have a route to a destination when needed. The ants randomly traverse the network and keep track of the last n visited nodes. The results of simulation experiments indicate that Ant-AODV performs better than AODV and a simple ant-based algorithm.

Baras and Mehta (2003) [4]: Probabilistic Emergent Routing Algorithm (PERA)

These authors have introduced two routing algorithms for MANETs. The first algorithm is a *proactive* one very similar to AntNet. Nodes maintain pheromone entries for all destinations by periodically launching forward ants, which take random decisions for unbiased exploration, and data packets are deterministically routed over the paths with the highest quality. The large routing overhead and the inefficient route discovery of this algorithm led to PERA, which is a purely *reactive* algorithm not very different from AODV. The forward ants are now flooded through the network toward their destinations. This strategy leads to the dynamic discovery of multiple paths. However, data packets are routed over the single best path available. The presence of multiple paths is helpful in the quick recovery from link failures. The performance of the algorithm is comparable to that of AODV according to a limited set of simulation experiments.

Heissenbüttel and Braun (2003) [64]: Mobile Ant-Based Routing (MABR)

The algorithm proposed by these authors makes use of geographical partitioning of the node area and of pheromone exploiting geographical addressing. The algorithm is intended for large-scale MANETs and is purely proactive. Forward and backward ants are used to periodically check if the path to a randomly chosen destination is functional and reflects the current state of the network. Accordingly, paths followed by the ants are positively or negatively reinforced. In addition, pheromone evaporation favors further exploration and removal of out-of-date paths.

Roth and Wicker (2003) [104, 105]: `Termite`

The Termite algorithm was actually inspired by the behavior of termite colonies, which is indeed very similar to that of ant colonies. As a matter of fact, Termite retains most of the main features of the general ACO meta-heuristic such as pheromone tables, probabilistic decisions, and pheromone evaporation. In Termite, forward ants are unicast and follow a random walk. Backward ants do not necessarily follow the forward path backward, but are also routed stochastically. Each data packet follows the path to its destination according to stochastic decisions based on the pheromone values, and "drops" pheromone, indicating a path toward its source node. An exponential *pheromone evaporation* is introduced as a means of negative feedback to prevent old routes from remaining in the routing tables. Termite is a *hybrid* algorithm. Paths are discovered on demand by ants, but their goodness is implicitly sampled by data packets in a proactive fashion. The behavior and the properties of the algorithm have been studied using a *formal analysis* and by simulation, showing better performance than AODV.

Di Caro, Ducatelle, and Gambardella (2004) [33, 34, 35, 36, 51]: `AntHocNet`

AntHocNet combines the typical *path sampling* behavior of ACO algorithms with a *pheromone bootstrapping* mechanism analogous to that used in Bellman-Ford algorithms (see Sect. 2), to effectively and efficiently learn pheromone tables. This design results in superior performance at the expense of a relatively low routing overhead. AntHocNet is a *hybrid* algorithm. It is *reactive* in the sense that a node only starts gathering routing information for a specific destination when a local traffic session needs to communicate with the destination and no routing information is available. It is *proactive* because as soon as the communication starts, and for the entire duration of the communication, the nodes proactively keep the routing information related to the ongoing flow up to date with network changes for both topology and traffic.

To capture the complexity of MANET environments, pheromone values reflect the quality of next hop decisions in terms of a *composite metric* function of: number of hops, traffic congestion, and signal-to-noise ratio (see [49] for a study on the effectiveness of considering different sets of quality metrics to define pheromone variables). This means that the algorithm tries to find paths characterized by a minimal number of hops, low congestion, and good signal quality between adjacent nodes.

When a source node s starts a communication session with a destination node d, and no pheromone information is available about how to reach d, the node manager broadcasts a *reactive forward ant*. Ants are sent over high-priority queues. At each node, the ant is either unicast or broadcast, according to whether or not the current node has pheromone information for d. If pheromone information is available, the ant stochastic decision policy π_ϵ makes use of a random proportional rule as in AntNet to select its next hop. Selection probabilities at node i are defined as: $p_{nd} = \frac{(\tau_{nd})^\beta}{\sum_{j \in \mathcal{N}_d^i} (\tau_{jd}^i)^\beta}$, where $\mathcal{N}_d^i$

is the set of i's neighbors over which a path to d is currently known, and $\beta \geq 1$ is a parameter which controls the exploratory behavior of the ants. A node which receives multiple copies of the same ant only accepts the first and discards the others. When a forward ant arrives at the destination, it goes backward, updates the pheromone tables at the nodes, indicating a path between s and d, and triggers the sending of data packets from the traffic session. In this way, only one path is set up initially.

During the course of the communication session, additional paths are added and/or removed via a *proactive path maintenance and exploration* mechanism. This is implemented through a combination of ant path sampling and slow-rate *pheromone diffusion and bootstrapping* which mimics pheromone diffusion in nature. Each node n periodically and asynchronously broadcasts a sort of beacon message containing a list of destinations it has information about, including for each destination d its best pheromone value $\tau^n_{m^*d}$. A node i receiving the message from n, registers that n is its neighbor, and for each destination d listed in the message, it derives an estimate of the goodness of going from i to d over n, combining the cost of hopping from i to n with the reported pheromone value $\tau^n_{m^*d}$. The authors call the obtained estimate b^i_{nd} *bootstrapped pheromone*, since it is built by "bootstrapping" on the value of the path the quality estimate received from an adjacent node

If i already has a pheromone entry τ^i_{nd} in its table, b^i_{nd} is just treated as an update of the goodness estimate of a known, reliable path, and is used directly to replace τ^i_{nd} with an up-to-date estimate. This equals a *path maintenance* operation. If i does not have yet a value for τ^i_{nd}, b^i_{nd} could indicate a possible new path from i to d over n. However, this path has never been explicitly tried out by an ant from i, such that due to the slow multi-step process it could have disappeared, or it could contain undetected loops or dangling links. The path is therefore not safe to use for data forwarding before being checked. This is the task assigned to *proactive forward ants*, which behave similarly to reactive forward ants but make use of both regular and bootstrapped pheromone on their way to the destination. This way, promising pheromone is checked out, and if the associated path is there and has the expected good quality, it can be turned into a regular path available for data. This *guided exploration* mechanism increases the number of paths available for data routing, which grows to a full mesh, and allows the algorithm to exploit new opportunities in the ever-changing topology. *Stochastic decisions* are used to spread data packets over multiple paths with a strong preference for the best ones. *Link failures* are explicitly dealt with using a *local path repair* process that tries to exploit the additional paths made available by the proactive mechanism, or via the generation of ant agents carrying explicit notification information.

AntHocNet's performance has been extensively evaluated through simulations against state-of-the-art algorithms under a number of different MANET scenarios for both *open space* [2, 33, 34, 49, 35, 50, 51] and realistic *urban* conditions [36]. The authors studied the behavior of the algorithm under different conditions for network size (ranging from 50 to 1,000 nodes), connectivity, change rate, data traffic patterns, and node mobility. The performance of the algorithm has been assessed relative to two classical MANET routing algorithms, AODV and OLSR, using the QualNet commercial simulator [101]. In the reported experiments, AntHocNet robustly outperforms the two competitor algorithms in terms of general efficacy, and in terms of adaptivity, robustness, and scalability. Superior performances were obtained efficiently introducing only a relatively small overhead, usually smaller than that introduced by the two other algorithms.

Rajagopalan and Shen (2005) [102, 103]: Ad-Hoc Networking with Swarm Intelligence (ANSI)

This work is based on earlier works of Shen [116, 117, 118]. ANSI is a *reactive* algorithm. Forward ants are reactively generated to look for a route for a new session or to repair a route after a link failure. They are *deterministically flooded* toward the destination. Only the first ant arriving at the destination is converted to a source-routed backward ant that sets up the route. At each node i, the pheromone entry $\tau_{nd} \in \mathcal{T}^i$ represents a weighted measure of how many times the link $i \to n$ has been selected to go to d. Pheromone and routing tables are updated by both forward and backward ants, indicating and *reinforcing* the route for all nodes toward the starting node. Also the arrival of a data packet triggers an update, but only of the pheromone table. Contrary to what usually happens in ant algorithms, pheromone and routing tables are *not used for ant decisions*. Pheromone tables are used to derive *deterministic single-path* routing tables for data packets. At node i, the next hop r_d to be used for data bound for d is the next hop which has the highest value $p_{nd} = \tau_{nd}^{\alpha} \eta_{nd}^{\beta} \psi_{nd}$, $\forall n \in \mathcal{N}^i$, where η_{nd} is a heuristic measure of the inverse of the distance to d through n, ψ is an inverse heuristic measure of the congestion along the path, and α and β are appropriate weighting factors. Periodic sending of *Hello messages* is used to keep neighboring information up-to-date. The combination of Hello information and ant pheromone updates provides multiple paths for a destination, but only the best one is deterministically used to route data packets. *Pheromone evaporation* for all pheromone entries is triggered after each update to favor removal of unused and bad paths, which amounts to *negative reinforcement*. ANSI was shown to perform better than AODV in simulation experiments involving 50 mobile nodes.

6 Routing Protocols Inspired by Bee Colony Behaviors

Bee colony behaviors have driven the design of routing algorithms more recently than ant colony behaviors. Most of the work in this has been done by Farooq and colleagues. They developed two main algorithms, BeeHive, for wired IP networks, and BeeAdHoc, for MANETs. From these two reference algorithms they have further derived other algorithms, addressing a number of different network constraints and scenarios. In the following we describe in some detail the characteristics of the two main reference algorithms and provide a brief discussion of also the additional algorithms and studies based on them. A peculiar characteristic of all this work consists in the fact that the algorithms have been designed according to the guidelines of the so-called *natural engineering* framework [141]. That is, they have been designed keeping in mind the technical constraints of physical networks to prevent making design assumptions which hardly would hold once the algorithm is being implemented on a real network. The efficacy of this way of proceeding is witnessed by the fact that the implementation of these algorithms in physical networks of Linux routers does not significantly differ at the algorithmic level from their implementation in network simulators, and, even more importantly, their performance in physical networks is consistent with what is observed in simulation experiments [52, 61].

6.1 The BeeHive Algorithm for Wired Connectionless Networks

The BeeHive algorithm [52, 141, 142, 148] is based on a meta routing framework similar to that of ACO routing algorithms: paths are constantly tried out to discover new routes and adapt to changing network conditions, and data are spread over multiple paths to optimize network performance and resource utilization. This is achieved with a strategy that mimics bee foraging behaviors. More specifically, BeeHive is built around two types of agents, *short distance* and *long distance* bee agents, which are *proactively* generated at the nodes and are designed after the way bee foragers respond to waggle dances. Both types of agents undertake the same responsibility: *exploring* the network and *evaluating* the quality of paths that they traverse to update node routing tables. However, short distance bee agents are allowed to move only up to a restricted number of hops. On the other hand, long distance bees have to collect and disseminate routing information in the complete network. This two-level agent model is intended to quickly collect routing information while minimizing both processing and bandwidth overheads. BeeHive adopts a *hierarchical organization* of the network that matches the use of two different types of agents with different search ranges. The network is subdivided into *foraging zones* and *foraging regions*. A foraging zone is defined as the set of nodes around a given node from which short distance bee agents can reach the node. The same node may belong to foraging zones of many nodes. The network is also viewed as a collection of clusters of non-overlapping foraging regions, in which a node belongs to just one region. Each foraging region has a *representative node*, which is the node with the lowest IP address in the region. Its role is to launch long distance bee agents. Each node maintains routing information for all nodes within its foraging zone, and for representative nodes of the foraging regions. If the destination of a packet does not lie within the foraging zone of a node, then it is forwarded along a path leading to the representative node of the foraging region containing the destination node. Informally, the behavior of BeeHive and its main characteristics can be summarized as follows:

1. All nodes start the *foraging region formation process* during a start-up phase. They try to form a foraging region with the same address as their own and make themselves the representative node of the foraging region. They launch a *first generation of short distance bee agents* to propagate their identifier in their neighborhood.
2. If a node receives a short distance bee from a node whose representative node's address is smaller than that of the receiving node, then it discontinues its efforts to be a representative node and joins the foraging region of the node with the smaller address.
3. If a node later on learns that its representative node has joined another foraging region, then it repeats the region formation actions of Step 1.
4. Nodes keep on launching generations of short distance bee agents by following Steps 1–3 until the network is subdivided into *disjoint foraging regions* and *overlapping foraging zones*. Finally, each node informs all other nodes in the network to which it belongs. This step is repeated every time foraging regions are reshaped because of link or node failures.
5. At the end of Step 4, the algorithm enters into a *normal phase* in which each non-representative node periodically sends out a short distance bee agent by *broadcasting* it to each one of its neighbors.
6. When a replica of a bee agent is received at a node, it updates the local routing information and is broadcast again to all the neighbors except the one it was received from. This process continues until the lifetime of the agent has expired or the replica arrives at a node which has already received a copy of it.
7. Representative nodes only launch long distance bee agents that undergo the same process as the short distance ones but have longer lifetimes.
8. Each node dynamically maintains routing information for reaching nodes within its foraging zone and for reaching the representative nodes of foraging regions. According to this hierarchical organization, a node routes a data packet whose destination is beyond its foraging zone along a path toward the representative node of the foraging region containing the destination node. More specifically, each node i maintains three types of routing tables: the *Intra Foraging Zone* (IFZ), the *Inter Foraging Region* (IFR), and the *Foraging Region Membership* (FRM) table. The *Intra Foraging Zone* routing table $\mathcal{R}^i$ is organized as a matrix of size $|\mathcal{D}^i| \cdot |\mathcal{N}^i|$, where $\mathcal{D}^i$ is the set of destinations in the foraging zone of node i and $\mathcal{N}^i$ is the set of neighbors of i. Each entry r_{jd} is a pair of queuing delay and propagation delay (q_{jd}, p_{jd}) that a packet will experience traveling to destination d via neighbor j. The Inter Foraging Region routing table stores the queuing and propagation delay values for reaching the representative node of each foraging region through each of its neighbors. The Foraging Region Membership table provides the mapping of known destinations

to foraging regions. Thanks to the hierarchical organization, the overall memory occupancy of routing information at each node is reduced with respect to the case of a flat organization, as the one adopted in many ACO implementations, that would require $O(|\mathcal{N}| \cdot |\mathcal{D}|)$ entries.

9. A bee agent launched from s, traveling across the network incrementally, collects path information in the form of the *trip time estimate* t_{is} for reaching the source s from the current node i over the used link j. The main difference lies in the fact that path exploration by bee agents does not rely on a stochastic policy but is realized according to a *deterministic* scheme based on repeated broadcasting (i.e., *flooding*). Bee agents use *high-priority queues* for quick dissemination of routing information.

10. The core mechanism at work in BeeHive is the direct agent-to-agent communication model (see Fig. 2) inspired by the bee behavior. In this topology, three paths exist between node k and node s. Node s launches three replicas of the same agent on three paths and they arrive at node k through different paths. Each replica uses the estimation model described above to estimate the queuing delay and the propagation delay. The replica that arrived earliest is allowed to continue its exploration further while other replicas are killed. However, the other replicas do communicate their estimates to the replica that is allowed to continue the exploration. Using the communication paradigm explained in Fig. 2 the replica calculates p_{ks} and q_{ks}, which incorporate the estimates of all replicas proportional to the quality of the paths, g_{1s}, g_{2s}, g_{3s} (to be shortly defined in Step 11), which they traversed. Once this replica continues its exploration of the network, it tells the other nodes that there exists a path from k to s through which a packet could reach s with a propagation delay of p_{ks} and queuing delay of q_{ks}. The other nodes forward data packets to node k based on the quality, which is a function of p_{ks} and q_{ks}. Once the data packet is at node k it can take any one of the three paths based on their quality, which is calculated based on the delay estimates of *bee agents*.

11. The next hop for a data packet is selected according to a *stochastic rule* that depends on the quality of each next hop. The quality of the path is a function of the cumulative queuing and propagation delays to reach the desired destination. The cumulative values are a result of the bee-inspired direct agent-to-agent communication model as depicted in Fig. 2. The estimated goodness of a neighbor j of node i (i has N neighbors) for reaching a destination d is g_{jd} and defined as

$$g_{jd} = \frac{\frac{1}{p_{jd}+q_{jd}}}{\sum_{k=1}^{N}\left(\frac{1}{p_{kd}+q_{kd}}\right)}, \tag{7}$$

where p_{jd} and q_{jd} are respectively propagation and queuing delays estimated by the bee agents for reaching destination d via neighbor j of node i. The fundamental motivation behind this definition is to approximate the behavior of a real network. When the network is experiencing a heavy

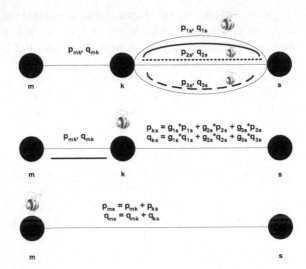

Fig. 2. Communication paradigm of *bee agents*

network traffic load, the queuing delay plays the primary role in the delay of a link. In this case it is trivial to say that $q_{jd} \gg p_{jd}$, and the goodness calculation becomes $g_{jd} \approx \frac{\frac{1}{q_{jd}}}{\sum_{k=1}^{N} \frac{1}{q_{kd}}}$. When the network is experiencing a low traffic load, it is the propagation delay that plays an important role in defining the latency of a link. Since $q_{jd} \ll p_{jd}$, we obtain $g_{jd} \approx \frac{\frac{1}{p_{jd}}}{\sum_{k=1}^{N} \frac{1}{p_{kd}}}$.

Figure 3 provides an exemplary working of the flooding mechanism. Short distance bee agents can travel up to two hops in this example. Each replica of the shown bee agent (launched by node 9) is specified with a different trail to identify its path unambiguously. The numbers on the paths show their cost. The flooding algorithm is a variant of the *breadth first* search algorithm. By following the above-mentioned Steps 1–4 the network is partitioned into two foraging regions with representative nodes 1 and 6 respectively. The foraging zone of node 9, which spans over both foraging regions, consists of nodes 2–8. The bee agents utilize the following estimation model for the trip time t_{is} that a packet would experience to reach s from current node i coming from j (protocol processing delays are ignored):

$$t_{is} \approx \frac{l_{ij}}{b_{ij}} + tx_{ij} + d_{ij} + t_{js} \tag{8}$$

where l_{ij} is the size of the queue (in bits) for neighbor j at node i, b_{ij} is the bandwidth of the link between node i and neighbor j, such as $l_{ij}/b_{ij} = q_{ij}$, tx_{ij} and pd_{ij} are respectively transmission and propagation delays of the link between node i and neighbor j (i.e., $tx_{ij} + pd_{ij} = p_{ij}$), and t_{js} is the trip time from j to s. Bandwidth and propagation delays of all links of a node are calculated at the beginning by transmitting back and forth so-called *hello packets*. Bee agents have a fixed size of 48 bytes and currently they are launched after every second or when a node has received a certain number of packets (240 in the performed experiments).

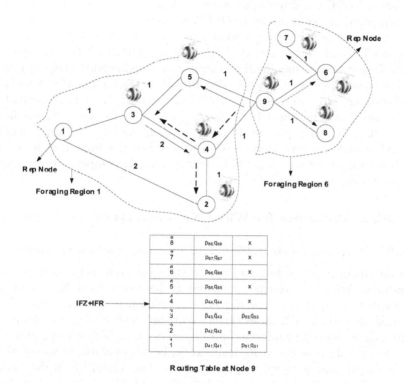

8	p_{88}, q_{88}	x
7	p_{87}, q_{87}	x
6	p_{86}, q_{86}	x
5	p_{85}, q_{85}	x
4	p_{84}, q_{84}	x
3	p_{83}, q_{83}	p_{53}, q_{53}
2	p_{82}, q_{82}	x
1	p_{81}, q_{81}	p_{51}, q_{51}

Routing Table at Node 9

Fig. 3. Illustration of the working of the BeeHive algorithm

BeeHive has been evaluated in a testing framework designed to provide a robust evaluation of generic SI-based routing algorithms over a large set of operational scenarios [52, 143]. The authors have shown that with the help of this framework, they were able to discover some previously unknown behavior of the considered routing algorithms. The framework considers a number of auxiliary parameters that provide valuable insight into the performance vs. cost benefits of the routing protocols. The authors compared BeeHive, AntNet, AntNet-FA, DGA [76], and OSPF. Reported simulation results (obtained by custom implementations of the algorithms using the OMNeT++ simulator) showed that BeeHive was able to deliver at least the same performance than AntNet, and solidly outperforms OSPF and DGA under heavy network traffic loads, while having a performance comparable to OSPF under low loads. An additional positive aspect of BeeHive is that it requires smaller routing tables and less computational resources than the considered competitors.

Harsch, Wedde, and Farooq (2005,2006) [52, 61]: BeeHive Implementation in Linux Routers

BeeHive has been also tested in a *physical network*. In fact, the authors implemented BeeHive inside the network stack of Linux routers and then tested and compared its performance to OSPF in a relatively small network of Linux routers using different types of synthetic and real-world traffic applications such as FTP and Voice over IP (VoIP) under UDP and TCP transport protocols. Results show that BeeHive can not only significantly help in reducing the download time of a file but also provide a better quality of service to VoIP sessions. Moreover, they have also shown that the performance of Bee-Hive in real-world networks is consistent with its performance in simulations. This work, confirming previously cited work on the physical implementation of ACO protocols, shows that an efficient and effective implementation of SI-based routing algorithms is possible, and it is highly competitive with current state-of-the-art algorithms.

6.2 Other Algorithms for Wired Networks Based on BeeHive

Wedde, Timm, and Farooq (2006) [149, 150]: BeeHiveGuard, BeeHiveAIS

The authors of these works have been the first to extensively analyze different types of security threats that malicious nodes can launch by manipulating the identity of agents or their routing information in networks controlled by SI-based algorithms. All these attacks can significantly degrade the network performance and compromise network operations. As a first step, they used a digital-signature-based security framework, BeeHiveGuard, to secure identity of a bee agent and of its routing information. The conclusion of the work was that BeeHiveGuard was able to counter different types of attacks but the processing complexity of the bee agents in BeeHiveGuard increased by more than 52,000% and the control overhead increased by more than 200% as compared to BeeHive. Then, the authors proposed a novel solution, BeeHiveAIS, based on Artificial Immune System (AIS) ideas, that provided the same security level as the digital-signature-based framework but with processing and control overheads which are approximately 200 and 20 times respectively smaller than BeeHiveGuard. This is an important achievement toward the definition of a SI-based security protocol with manageable processing and communication costs.

Brüntrup and Farooq (2006) [12]: BeeHivePlus, BeeHiveQoS

In BeeHive, data packets are spread stochastically over the available multiple paths according to their estimated quality. This is a behavioral trait common to many other algorithms reviewed in this chapter. While on the one hand this way of proceeding has some clear benefits, on the other hand, in some situations requiring strict and predictable performance, such as in QoS networks, this can result in unwanted side effects such as short-lived loops and jitter fluctuations. BeeHivePlus overcomes these potential problems by utilizing the concept of *waterfall routing*, in which only those neighbors that are nearer, in terms of hops, to the destination than the existing node are considered to be selected as a next hop . As a result, a packet always moves in the direction of the destination and loops are implicitly avoided. The authors have used the concept of *temporal stability* of routes, i.e., routing decisions for a destination remain fixed in the inter-arrival time period between two successive bee agents from the same destination. This feature ensures that packets for a short time period follow the same path. Consequently, it results in significant reduction in jitter fluctuations that make BeeHivePlus comparable to that of a single-path algorithm such as OSPF. However, in spite of having these desirable features of loop freedom and low jitter, BeeHivePlus still provides similar performance as BeeHive. The same authors derived from BeeHivePlus a novel algorithm, BeeHiveQoS, for QoS networks. The core mechanism in BeeHiveQoS is an intelligent hierarchical packet scheduler, which can be embedded in BeeHivePlus as well as in other schemes, and which provides soft guarantees to QoS sensitive applications. The results of the experiments conducted on network topologies of up to 150 nodes confirm that BeeHiveQoS is able to provide guarantees to QoS-sensitive applications.

Zahid, Shahzad, Ali, and Farooq (2007) [156]: Formal Framework for Performance Analysis

The authors have proposed a formal framework for analyzing the behavior of the BeeHive protocol. The framework utilizes relevant concepts of deductive mathematics and queuing theory. The framework also uses Markov transition matrices and probabilistic recursive functions that significantly augment the formal understanding about different design options adopted in BeeHive. The authors have formally modeled the goodness of a neighbor that represents the desirability for choosing it in order to reach a destination, end-to-end packet delay, throughput and the probability of packets following loops with the help of their model. The authors have empirically validated the results obtained form their formal model with the ones obtained from OMNeT++ simulations on a small network topology. The estimated performance values of their formal model closely follow patterns similar to the values obtained through the network simulator. This work will be a cardinal step in removing a serious shortcoming of SI-based algorithms: lack of formal understanding about their merits and behavior.

6.3 The BeeAdHoc Algorithm for Wireless Mobile Ad Hoc Networks

Wedde, Farooq, et al. (2004) [140, 144, 146, 147] designed BeeAdHoc with the aim of defining a MANET routing algorithm which at the same time is *energy-efficient* and provides performance comparable to that of existing state-of-the-art algorithms. Honey bee behavior served as the main inspiration to design the different types of agents at work in the system and their interaction. Adopting the bee metaphor, each node's routing controller is seen as an independent beehive where the bee agents live, act, and interact. BeeAdHoc is a relatively simple algorithm residing at the network layer that makes use of a *reactive* strategy for agent launching and of *source routing* to forward packets. In the following we discuss in separate subsections the multi-agent model, the beehive-like architecture of the routers, and the performance of the algorithm.

Multi-agent Model

BeeAdHoc is based on the use of four different bee-inspired types of agents: *packers, scouts, foragers*, and *beeswarms.*

Packers mimic the task of a food-storer bee. Packers reside inside a network node, and receive and store data packets from the upper transport layer (see Fig. 4). Their main task is to find a forager for the data packet at hand. Once the forager is found and the packet is handed over, the packer agents are removed from the system.

The task of *scouts* is to discover new routes from their launching node to their destination node. A scout is broadcast to all neighbors within range using an *expanding time to live (TTL) timer* heuristic analogous to that used in the AODV algorithm [75]. At the start of the route search, a scout agent is generated, its TTL is set to a small value (e.g., 3), and it is broadcast. If after a certain amount of time the scout is not back with a route, the strategy consists of the generation of a new scout and of the assignment of a TTL higher than that in the previous attempt. In this way the search radius of the generated scouts is incrementally enlarged, increasing the probability of reaching the searched destination. When a scout reaches the destination, it starts a backward journey on the same route that it has followed while moving forward toward the destination. Once the scout is back to its source node, it recruits foragers for its route by utilizing a mechanism derived from the waggle dance of scout bees in nature. A dance is abstracted into the number of clones that could be made of the same scout, which is encoded in their *dance number* (corresponding to recruiting forager bees in nature).

Foragers are the main workers in the BeeAdHoc algorithm. They are bound to the "bee hive" of a node. They receive data packets from packers and deliver them to their destination in a source-routed modality. To "attract" data packets foragers use the same metaphor of a waggle dance as scouts do. Foragers are of two types: delay and lifetime. From the nodes they visit, *delay foragers* gather end-to-end delay information, while *lifetime foragers* gather information about the remaining battery power. Delay foragers try to route packets along a minimum-delay path, while lifetime foragers try to route packets so that the lifetime of the network is maximized. A forager is transmitted from node to node using a unicast, point-to-point modality. Once a forager reaches the searched destination and delivers the data packets, it waits there until it can be piggybacked on a packet bounded for its original source node. In particular, since TCP acknowledges received packets, BeeAdHoc piggybacks the returning foragers in the TCP acknowledgments. This reduces the overhead generated by control packets, saving energy at the same time.

Beeswarms are the agents that are used to explicitly transport foragers back to their source node when the applications are using an unreliable transport protocol like UDP, such that no acknowledgments are sent for the received data packets. To optimize forager transport, one beeswarm agent can carry multiple foragers: one forager is put in the header of the beeswarm while the others are put in the agent payload.

Beehive-like Architecture of the Node Controllers

In BeeAdHoc, each MANET node contains at the network layer a software module called *hive*, which consists of three parts: the *packing floor*, the *entrance floor*, and the *dance floor*. The structure of the hive is shown in Fig. 4.

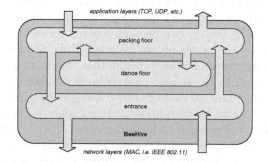

Fig. 4. Overview of the BeeAdHoc's *hive* architecture at a network node

The entrance floor is an interface to the lower MAC layer, while the packing floor is an interface to the upper transport layer. The dance floor contains the foragers and the routing information to route locally generated data packets. The functional characteristics of each floor composing the hive are as follows:

Packing floor. The packing floor is an interface to the upper transport layer (e.g., TCP or UDP). Once a data packet arrives from the transport layer, a matching forager for it is looked up on the dance floor. If one forager is found then the data packet is encapsulated in its payload. Otherwise, the data packet is temporarily buffered to wait for a returning forager. If no forager comes back within a certain predefined time, a scout is launched which is responsible for discovering new routes to the packet's destination.

Entrance floor. Actions at the dance floor depend on the type of packet that entered the floor from the MAC layer. If the packet is a forager and the current node is its destination, then the forager is forwarded to the packing floor; otherwise, it is directly routed to the MAC interface of the next hop node. If the packet is a scout agent, it is broadcast to the neighbor nodes if its TTL timer has not expired yet or if the current node is not its destination. The information about the ID of the scout and its source node is stored in a local list. If a replica of a previously received scout arrives at the entrance floor, the replica is removed from the system. If a forager with the same destination as the scout already exists in the dance floor, then the forager's route to the destination is given to the scout by appending it to the route held so far by the scout.

Dance floor. The dance floor is the heart of the hive because it maintains the *routing information* in the form of *foragers*. The dance floor is populated with routing information by means of a mechanism reminiscent of the waggle dance recruitment in natural bee hives: once a forager returns after its journey it recruits new foragers by "dancing" according to the quality of the path it traversed. A lifetime forager evaluates the quality of its route based on the average remaining battery capacity of the nodes along its route. Mimicking forager bees in nature that dance enthusiastically when they find a food source worth exploiting, recruiting in this way a number of foragers, a lifetime forager can be cloned many times in two distinct cases. In the first case, the nodes on the discovered route have a good amount of spare battery capacity, which means that this is a good route that can be well exploited. In the second case, a large number of data packets are waiting for the forager, so that the route needs to be exploited even though it might be having nodes with little battery capacity. On the other hand, if no data packets are waiting to be transported, then a forager with a very good route might even abstain from dancing because the other foragers are fully satisfying traffic requests. This concept is directly borrowed from the behavior of scout and forager bees in nature, and it helps to automatically regulate the number of foragers for each route.

The central activity of the dance floor module consists in sending a matching forager to the packing floor in response to a request from a packer. The foragers whose lifetimes have expired are not considered for matching. If multiple foragers could be identified for matching then a forager is selected in a random way. This helps in distributing the packets over *multiple paths*, which in turn serves two purposes: avoiding congestion under high loads and depleting batteries of different nodes at a comparable rate. A clone of the selected forager is sent to the packing floor and the original forager is stored on the dance floor after reducing its dance number, that is, the number of permitted clones. If the dance number is 0, then the original forager is sent to the packing floor, removed in this way from the dance floor. This strategy aims at favoring young over old foragers, since they represent fresher routes, which are expected to remain valid in the near future with higher chances than the older ones due to their greater mobility and lesser battery depletion. If the last forager for a destination leaves a hive, then the hive does not have anymore a route to the destination. Nevertheless, if a route to the destination still exists, then soon a forager will be returning to the hive, while if no forager comes back within a reasonable amount of time then the node has probably lost its connection to the destination node. This mechanism eliminates the need for explicitly monitoring the validity of the routes by using special Hello packets and informing other nodes through route error messages, as is done in several state-of-the-art algorithms such as AODV, as well as in several ACO implementations for MANETs. In this way fewer control packets are transmitted, resulting also in less energy expenditure.

Implementation and Performance Evaluation

BeeAdHoc has been implemented and evaluated both in simulation and in real networks. Results from extensive simulation tests show that BeeAdHoc delivers the same or better performance than that of the state-of-the-art algorithms like AODV, DSR and DSDV [98], but with a significantly smaller overall energy expenditure [140, 144, 146, 147]. To study its performance in more realistic and way more challenging physical networks, the authors have implemented BeeAdHoc inside the Linux network stack. They compared BeeAdHoc with AODV and OLSR proposing a three-step testing methodology with the intent of gradually moving toward a real MANET [140, 147]. First, in simulated reality, they tested the algorithms in a virtual network of five virtual machines connected through a software switch. They randomly changed the topology to simulate mobility. This scenario depicted an ideal MANET. Second, in quasi reality, the communication between laptops was established though 802.11 wireless network cards by placing laptops in the communication range of each other and the mobility was simulated by discarding packets through packet filtering at the link layer in case the laptop was not supposed to receive the packets. Finally, in real MANET, the authors conducted a MANET experiment on a 12-laptop network in which the nodes were moving at a walking speed on the north campus of University of Dortmund, Germany. Interestingly, performance values obtained in the simulated reality scenario did not show a correlation with those of the real MANET. However, the overall pattern of the values remained the same: BeeAdHoc was able to consistently a performance to that of OLSR and AODV, with the properties of using significantly fewer control packets and consuming less battery power.

Schletter, Fischer, et al. (2005) [147]: An Agent-Based Formal Investigation of BeeAdHoc

The formal analysis of routing protocols for MANETs is a challenging area of research. In [147] the authors have studied performance and behavior of

BeeAdHoc by adopting a *formal verification model*, which is a rather innovative approach in the domain of SI. They relied on the framework of Sumpter [124] to propose an agent interaction model based on the use of the Weight Synchronous Calculus of Communicating Systems (WSCCS). Using a probabilistic workbench to validate the formal model, the authors have shown that the formal model is able to predict the distribution of the foragers on multiple paths as a function of their quality. Moreover, they were also able to analyze the sending pattern of beeswarms and optimize it through the model. This work can play a vital role in developing a comprehensive formal model-checking framework for SI based routing algorithms in general.

6.4 Other Algorithms for MANETs Based on BeeAdHoc

Mazhar and Farooq (2007) [82, 83]: BeeSec, BeeAIS

The authors followed the same research methodology as was used in Bee-HiveGuard and BeeHiveAIS to analyze the security threats of BeeAdHoc and then proposed two solutions: BeeSec, which utilizes a digital signature based security framework, and BeeAIS, which utilizes the principles of AIS to provide security. However, in MANETs, providing an AIS-based security is more challenging because of the mobility of the nodes. As a result, it was difficult to identify whether the change in the path of an agent is due to the malicious activity of a node or to its internal mobility. This translates into the idea of a "self" which is changing. According to the reported results: (i) BeeAIS provides the same security level as compared to BeeSec but at significantly less processing and communication costs, which means a significant amount of energy and power saving compared to BeeSec; (ii) the performance of BeeAIS, even with the overhead for providing security, is significantly better than AODV and DSR and is approximately similar to that of the original BeeAdHoc algorithm. These results seem to indicate the efficacy of adopting an AIS component for SI-based security protocols in power-aware systems, considering the low processing complexity and the absence of additional communication costs.

7 Conclusions and Future Perspectives for SI Routing

In this chapter we addressed the problem of *routing* in current and next generation telecommunications networks, which are characterized by the fact of their being very complex, dynamic, large, and heterogeneous. Swarm intelligence design features a number of properties that are highly desirable with the challenges posed by these networks. In the chapter we reviewed the major routing protocols inspired by collective behaviors observed in *social insects* such as *ant* and *bee colonies*. This specific class of *SI algorithms* includes the majority of the most significant and promising applications of the SI paradigm to the solutions of adaptive network routing problems.

Social insects behaviors, mainly ant colonies, have fueled in the last ten years the design and implementation of a consistent number of protocols and algorithms for routing. We could not review here this whole body of work, but we provided anyway a quite comprehensive overview of the main algorithms, of their general design principles and general properties. More specifically, we discussed the *pheromone-driven shortest-path behavior* of foraging ant colonies, which has been reverse-engineered and put to work in the optimization framework of *ant colony optimization*, that, in turn, has guided the design of a relatively large number of routing algorithms. Analogously, we pointed out and abstracted those core mechanisms at work in foraging bee colonies, such as waggle dancing, which have recently driven the design of novel routing algorithms. We have considered different classes of networks characterized by different communication and transmission technologies and services, and for each class we have briefly discussed the characteristics of the main ant- and bee-colony inspired algorithms which can be found in the literature. We have pointed out their pros and cons and their distinctive features in relation to the classification features of routing algorithms given in Sect. 3. Tables 1 and 2 summarize these core features for a few of the most prominent among the reviewed algorithms, and compare them to established and classical state-of-the-art routing algorithms. From the tables it is apparent that, in general, SI algorithms have and make use of certain features that classical algorithms do not have and vice versa. On the other hand, since according to extensive comparative studies the performance of the reviewed algorithms, and in particular of those listed in the tables, seems to be significantly better than that of classical state-of-the-art algorithms, one might argue that the properties implied by the SI design are particularly suitable for facing the challenges of modern networks.

Generally speaking, a fundamental aspect of the SI paradigm is the fact that it emphasizes a particular *bottom-up design approach* which, for network routing results in the definition of protocols featuring, among others: *locality of interactions and self-organizing behaviors, availability of multiple paths for routing and failure backup, ability to adapt in a quick and robust way to topological and traffic changes and component failures, scalable performance, robustness to failures and losses internal to the protocol, easiness of design and tuning.* As is also confirmed by the data in the tables, most of the reviewed algorithms possess significant subsets of these important properties which are particularly appealing for current and future networking. In fact, with the impressive growth of the Internet, and the pervasive deployment of wireless and wired networking, these are the core properties which should characterize all modern protocols of the network protocol stack in order for us to be able to cope with the levels of complexity, heterogeneity, and dynamism of current and forthcoming networks. The relatively novel fields of *traffic engineering* and *autonomic communications* emphasize the need for both an efficient uti-

Table 1. General features of routing algorithms for wired networks. The considered features have been discussed in Sect. 3. The algorithms have been discussed in the previous sections. "y" and "n" stand respectively for "yes" and "no", while "p" means that the algorithm partly possesses the feature

	AntNet	Adaptive-SDR	BeeHive	OSPF	MDVA	QColony	QOSPF
Topology-adaptive	p	y	y	y	y	p	y
Traffic-adaptive	y	y	y	n	p	y	y
Router-Intelligent	y	y	y	y	y	y	n
Multi-path	y	y	y	n	y	n	n
Local representation	y	y	y	n	n	y	n
Hierarchical	n	y	y	y	n	y	y
Constructive	n	n	y	y	n	n	y
Loop-free	n	n	p	y	y	n	y
Proactive behavior	y	y	y	y	y	y	y
Reactive behavior	n	n	n	n	n	y	n
Stochastic exploration	y	y	n	n	n	p	n
Stochastic data routing	y	y	y	n	n	n	n
Formal properties	n	n	y	y	y	n	p
Physical implementation	y	n	y	y	n	n	y
Quality of service	n	n	n	n	n	y	y

Table 2. General features of routing algorithms for MANETs. The considered features have been discussed in Sect. 3. The algorithms are discussed in the previous sections. "y" and "n" stand respectively for "yes" and "no", while "p" means that the algorithm partly possesses the feature.

	AntHocNet	ANSI	Termite	BeeAdHoc	DSR	AODV	OLSR
Topology-adaptive	y	y	y	y	y	y	y
Traffic-adaptive	y	y	y	y	p	p	p
Router-Intelligent	y	y	y	n	n	y	y
Multi-path	y	p	y	y	n	n	n
Local representation	y	y	y	y	y	y	n
Hierarchical	n	n	n	n	n	n	n
Constructive	y	y	y	y	y	y	n
Loop-free	n	n	n	y	y	n	n
Proactive behavior	y	p	y	n	n	n	y
Reactive behavior	y	y	y	y	y	y	n
Stochastic exploration	y	n	y	n	n	n	n
Stochastic data routing	y	n	y	y	n	n	n
Formal properties	n	n	p	p	p	p	p
Physical implementation	p	n	n	y	y	y	y
Energy-aware	n	n	n	y	n	n	n

lization of network resources and the ability of the network to self-control and adapt over time both as a whole and at the level of the single components. "Classical" algorithms for network management and control have been designed top-down, not taking into account the explosion in complexity in all directions and dimensions faced by current networks. Again, this can be also understood looking at the tables. It is clear that established classical algorithms miss some core properties in terms of dynamic behavior, robustness, and locality.

We believe that routing algorithms inspired by social insect behaviors, and, more generally, by the SI paradigm, can play an important role in empowering future networks with an optimized, adaptive, robust, and scalable control system at the network layer. The downside of these novel approaches consists in the *current lack of extensive implementation and testing on physical networks*, and in the difficulty, somehow intrinsic to fully distributed and stochastic bottom-up approaches, of providing formal guarantees in terms of *dependability*. Solid work in these two aspects is still necessary to get a wider acceptance from the networking community and a deeper and solid understanding of the behavior and properties of these algorithms. If this will be done in the near future, we can expect a rapid deployment of SI algorithms in the control systems of forthcoming networks.

References

1. S. Appleby and S. Steward. Mobile software agents for control in telecommunications networks. *BT Technology Journal*, 18(1):68–70, 2000.
2. O. Babaoglu, G. Canright, A. Deutsch, G. A. Di Caro, F. Ducatelle, L. M. Gambardella, N. Ganguly, M. Jelasity, R. Montemanni, A. Montresor, and T. Urnes. Design patterns from biology for distributed computing. *ACM Transactions on Autonomous and Adaptive Systems*, 1(1):26–66, 2006.
3. B. Baran and R. Sosa. A new approach for AntNet routing. In *Proceedings of ICCCN*, pages 303–308, Las Vegas, NV, USA, 2000. IEEE Press.
4. J. S. Baras and H. Mehta. A probabilistic emergent routing algorithm (PERA) for mobile ad hoc networks. In *Proceedings of WiOpt*, 2003.
5. D. Bertsekas. *Dynamic Programming and Optimal Control*. Athena Scientific, USA, 1995.
6. D. Bertsekas and R. Gallager. *Data Networks*. Prentice Hall, Englewood Cliffs, NJ, USA, 1992.
7. E. Bonabeau, M. Dorigo, and G. Theraulaz. *Swarm Intelligence: From Natural to Artificial Systems*. Oxford University Press, Inc., New York, NY, USA, 1999.
8. E. Bonabeau, F. Henaux, S. Guérin, D. Snyers, P. Kuntz, and G. Theraulaz. Routing in telecommunications networks with ant-like agents. In *Proceedings of IATA*, pages 60–71, London, UK, 1998. Springer-Verlag.
9. J.A. Boyan and M.L. Littman. Packet routing in dinamically changing networks: A reinforcement learning approach. In J. D. Cowan, G. Tesauro, and J. Alspector, editors, *Proceedings of NIPS6*, pages 671–678. Morgan Kaufmann, San Francisco, CA, USA, 1994.

10. R. W. Brazier and M. D. Cookson. Intelligence design patterns. *BT Technology Journal*, 23(1):69–81, 2005.

11. J. Broch, D. A. Maltz, D. B. Johnson, Y. C. Hu, and J. Jetcheva. A performance comparison of multi-hop wireless ad hoc network routing protocols. In *Proceedings of MobiCom*, pages 85–97, New York, NY, USA, 1998. ACM Press.

12. R. Brüntrup. Quality of service in von der natur inspirierten routing-algorithmen (in German). Master thesis, LSIII, University of Dortmund, Germany, August 2006.

13. D. Câmara and A. F. Loureiro. A novel routing algorithm for ad hoc networks. In *Proceedings of HICSS*. IEEE Press, 2000.

14. D. Câmara and A. F. Loureiro. GPS/Ant-like routing in ad hoc networks. *Telecommunication Systems*, 18(1–3):85–100, 2001.

15. T. Camilo, C. Carreto, J. Sá Silva, and F. Boavida. An energy-efficient ant-based routing algorithm for wireless sensor networks. In *Proceedings ANTS*, volume 4150 of *Lecture Notes in Computer Science*, pages 49–59, Brussels, Belgium, 2006. Springer.

16. L. Carrillo, C. Guadal, J.-L. Marzo, G. A. Di Caro, F. Ducatelle, and L.M. Gambardella. Differentiated quality of service scheme based on the use of multiple classes of ant-like mobile agents. In *Proceedings of CoNEXT*, pages 234–235, Toulouse, France, October 24–27 2005. ACM Press.

17. L. Carrillo, J. L. Marzo, L. Fàbrega, P. Vilà, and C. Guadall. Ant colony behaviour as routing mechanism to provide quality of service. In *Proceedings of ANTS*, volume 3172 of *Lecture Notes in Computer Science*, pages 418–419, Berlin, 2004. Springer.

18. S. Chen and K. Nahrstedt. An overview of quality-of-service routing for the next generation high-speed networks: Problems and solutions. *IEEE Network Magazine, Special issue on Transmission and Distribution of Digital Video*, 12(6):64–79, 1998.

19. S. P. Choi and D.-Y. Yeung. Predictive Q-routing: A memory-based reinforcement learning approach to adaptive traffic control. In *Proceedings of NIPS8*, pages 945–951. MIT Press, 1996.

20. Cisco. Internetworking Technology Handbook, 2002.

21. T. Clausen, P. Jacquet, A. Laouiti, P. Muhlethaler, A. Qayyum, and L. Viennot. Optimized link state routing protocol. In *Proceedings of IEEE INMIC*, pages 62– 68. IEEE Press, 2001.

22. M. E. Csete and J. C. Doyle. Reverse engineering of biological complexity. *Science*, 295(5560):1664–1669, March 2002.

23. S. S. Dhillon and P. Van Mieghem. Performance analysis of the AntNet algorithm. *Computer Networks*, 51(8):2104–2125, 2007.

24. G. A. Di Caro. *Ant Colony Optimization and its application to adaptive routing in telecommunication networks*. PhD thesis, Faculté des Sciences Appliquées, Université Libre de Bruxelles, Brussels, Belgium, November 2004.

25. G. A. Di Caro and M. Dorigo. Adaptive learning of routing tables in communication networks. In *Proceedings of the Italian Workshop on Machine Learning (IWML)*, 1997.

26. G. A. Di Caro and M. Dorigo. AntNet: A mobile agents approach to adaptive routing. Technical Report 97–12, IRIDIA, Université Libre de Bruxelles, Brussels, Belgium, June 1997.

27. G. A. Di Caro and M. Dorigo. Ant colonies for adaptive routing in packet-switched communications networks. Technical Report 97-20.1, IRIDIA, Université Libre de Bruxelles, Brussels, Belgium, March 1997.
28. G. A. Di Caro and M. Dorigo. Mobile agents for adaptive routing. In *Proceedings of HICSS*, volume 7, pages 74–83. IEEE Computer Society Press, 1998.
29. G. A. Di Caro and M. Dorigo. AntNet: Distributed stigmergetic control for communications networks. *Journal of Artificial Intelligence Research (JAIR)*, 9:317–365, 1998.
30. G. A. Di Caro and M. Dorigo. Ant colonies for adaptive routing in packet-switched communications networks. In *Proceedings of PPSN-V*, volume 1498 of *LNCS*, pages 673–682. Springer, 1998.
31. G. A. Di Caro and M. Dorigo. Extending AntNet for best-effort Quality-of-Service routing. Proceedings of the First International Workshop on Ant Colony Optimization (ANTS'98), 1998.
32. G. A. Di Caro and M. Dorigo. Two ant colony algorithms for best-effort routing in datagram networks. In *Proceedings of the 11th International Conference on Parallel and Distributed Computing Systems (PDCS)*, pages 541–546, 1998.
33. G. A. Di Caro, F. Ducatelle, and L. M. Gambardella. AntHocNet: an ant-based hybrid routing algorithm for mobile ad hoc networks. In *Proceedings of PPSN-VIII*, volume 3242 of *LNCS*, pages 461–470. Springer, 2004.
34. G. A. Di Caro, F. Ducatelle, and L. M. Gambardella. AntHocNet: an adaptive nature-inspired algorithm for routing in mobile ad hoc networks. *European Transactions on Telecommunications*, 16(5):443–455, 2005.
35. G. A. Di Caro, F. Ducatelle, and L. M. Gambardella. Swarm intelligence for routing in mobile ad hoc networks. In *Proceedings of the IEEE Swarm Intelligence Symposium*, pages 76–83, Pasadena, USA, June 2005. IEEE Press.
36. G. A. Di Caro, F. Ducatelle, and L. M. Gambardella. Studies of routing performance in a city-like testbed for mobile ad hoc networks. Technical Report 07-06, IDSIA, Lugano, Switzerland, March 2006.
37. G. A. Di Caro, F. Ducatelle, and L. M. Gambardella. Swarm intelligence for routing in telecommunications networks. *Journal of Swarm Intelligence*, 2007. Submitted.
38. G. A. Di Caro, F. Ducatelle, and L. M. Gambardella. Theory and practice of Ant Colony Optimization for routing in dynamic telecommunications networks. In N. Sala and F. Orsucci, editors, *Reflecting interfaces: the complex coevolution of information technology ecosystems*. Idea Group, Hershey, PA, USA, 2007.
39. G. A. Di Caro and T. Vasilakos. Ant-SELA: Ant-agents and stochastic automata learn adaptive routing tables for QoS routing in ATM networks. In Proceedings of 2nd International Workshop on Ant Colony Optimization (ANTS'00), 2000.
40. S. Doi and M. Yamamura. BntNetL: Evaluation of its performance under congestion. *Journal of IEICE B* (in Japanese), pages 1702–1711, 2000.
41. S. Doi and M. Yamamura. BntNetL and its evaluation on a situation of congestion. *Electronics and Communications in Japan (Part I)*, 85(9):31–41, 2002.
42. S. Doi and M. Yamamura. An experimental analysis of loop-free algorithms for scale-free networks. In *Proceedings of ANTS*, volume 3172 of *Lecture Notes in Computer Science*, pages 278–285. Springer-Verlag, 2004.
43. M. Dorigo, E. Bonabeau, and G. Theraulaz. Ant algorithms and stigmergy. *Future Generation Computer Systems*, 16(8):851–871, 2000.

44. M. Dorigo and G. A. Di Caro. The ant colony optimization meta-heuristic. In D. Corne, M. Dorigo, and F. Glover, editors, *New Ideas in Optimization*, pages 11–32. McGraw-Hill, 1999.
45. M. Dorigo, G. A. Di Caro, and L. M. Gambardella. Ant algorithms for discrete optimization. *Artificial Life*, 5(2):137–172, 1999.
46. M. Dorigo and E. Sahin (Editors). Special issue on Swarm Robotics. *Autonomous Robots*, 17(2–3), 2004.
47. M. Dorigo, V. Maniezzo, and A. Colorni. The Ant System: Optimization by a colony of cooperating agents. *IEEE Transactions on Systems, Man, and Cybernetics—Part B*, 26(1):29–41, 1996.
48. M. Dorigo and T. Stützle. *Ant Colony Optimization*. MIT Press, Cambridge, MA, 2004.
49. F. Ducatelle, G. A. Di Caro, and L. M. Gambardella. An analysis of the different components of the AntHocNet routing algorithm. In *Proceedings of the 5th International Workshop on Ant Colony Optimization and Swarm Intelligence (ANTS'06)*, volume 4150 of *LNCS*, pages 37–48. Springer, 2006.
50. F. Ducatelle, G. A. Di Caro, and L. M. Gambardella. Ant agents for hybrid multipath routing in mobile ad hoc networks. In *Proceedings of WONS*, Switzerland, January 18–19, 2005. IEEE Press.
51. F. Ducatelle, G. A. Di Caro, and L. M. Gambardella. Using ant agents to combine reactive and proactive strategies for routing in mobile ad hoc networks. *International Journal of Computational Intelligence and Applications*, Special Issue on Nature-Inspired Approaches to Networks and Telecommunications, 5(2):169–184, 2005.
52. M. Farooq. *Bee-inspired Protocol Engineering: From Nature to Networks*. Natural Computing Series. Springer, (In Press).
53. L. M. Feeney. A taxonomy for routing protocols in mobile ad hoc networks. Technical Report ISRN:SICS-T-99/07-SE, Swedish Institute of Computer Science, Kista, Sweden, 1999.
54. S. Fenet and S. Hassas. An ant based system for dynamic multiple criteria balancing. Proceedings of the First International Workshop on Ant Colony Optimization (ANTS'98), 1998.
55. S. Fenet and S. Hassas. A.N.T.: a distributed network control framework based on mobile agents. In *Proceedings of the International ICSC Congress on Intelligent Systems and Applications*, 2000.
56. R. Freeman. *Telecommunication System Engineering.* John Wiley & Sons, 2004.
57. M. Gadomska and A. Pacut. Performance of ant routing algorithms when using TCP. In M. Giacobini et al., editors, *Applications of Evolutionary Computing, EvoWorkshops 2007*, volume 4448 of *Lecture Notes in Computer Science*, pages 1–10. Springer Verlag, 2007.
58. R. M. Garlick and R. S. Barr. Dynamic wavelength routing in WDM networks via Ant Colony Optimization. In M. Dorigo, G. A. Di Caro, and M. Sampels, editors, *Proceedings of ANTS*, volume 2463 of *Lecture Notes in Computer Science*, pages 250–255. Springer Verlag, 2002.
59. S. Goss, S. Aron, J. L. Deneubourg, and J. M. Pasteels. Self-organized shortcuts in the Argentine ant. *Naturwissenschaften*, 76:579–581, 1989.
60. M. Günes, U. Sorges, and I. Bouazizi. ARA—The ant-colony based routing algorithm for MANETS. In *Proceedings of the 2002 ICPP International Workshop on Ad Hoc Networks (IWAHN 2002)*, pages 79–85, 2002.

61. A. Harsch. Design and development of a network infrastructure for swarm routing protocols inside Linux. Master's thesis, LSIII, University of Dortmund, Germany, July 2005.

62. P. Heegaard and I. Fuglem. AntPing: prototype demonstrator of swarm based path management and monitoring (Poster). In *Proceedings of IWSOS*, 2006.

63. P. Heegaard, O. Wittner, and B. Helvik. Self-management of virtual paths in dynamic networks. In O. Babaoglu, M. Jelasity, A. Montresor, C. Fetzer, S. Leonardi, A. van Moorsel, and M. van Steen, editors, *Self-Star Properties in Complex Information Systems*, volume 3460 of *Lecture Notes in Computer Science*, pages 417–432. Springer-Verlag, 2005.

64. M. Heissenbüttel and T. Braun. Ants-based routing in large scale mobile ad-hoc networks. In *13. ITG/GI-Fachtagung Kommunikation in verteilten Systemen (KiVS 2003)*, pages 91–99, Leipzig, Germany, 2003.

65. M. Heusse, D. Snyers, S. Guérin, and P. Kuntz. Adaptive agent-driven routing and load balancing in communication networks. *Advances in Complex Systems*, 1(2):237–254, 1998.

66. M. Heusse, D. Snyers, and Y. Kermarrec. Adaptive agent driven routing in communication networks: comparison with a classical approach. *Advances in Complex Systems*, 2(3):209–219, 1999.

67. N. Hu and P. Steenkiste. Evaluation and characterization of available bandwidth probing techniques. *IEEE Journal on Selected Areas in Communications*, 21(6):879–894, 2003.

68. D. E. Jackson and F. L. Ratnieks. Communication in ants. *Current biology*, 16(15):570–574, 2006.

69. D. B. Johnson and D. A. Maltz. Dynamic source routing in ad hoc wireless networks. In T. Imielinski and H. F. Korth, editors, *Mobile Computing*, pages 153–181. Kluwer Academic Publishers, 1996.

70. I. Kassabalidis, M. A. El-Sharkawi, R. J. Marks, P. Arabshahi, and A. A. Gray. Swarm intelligence for routing in communication networks. In *Proceedings of GLOBECOM*, pages 3613–3617. IEEE Press, 2001.

71. I. Kassabalidis, M. A. El-Sharkawi, R. J. Marks II, P. Arabshahi, and A. A. Gray. Adaptive-SDR: Adaptive swarm-based distributed routing. In *Proceedings of IJCNN*, pages 351–354. IEEE Press, 2002.

72. J. Kephart and D. Chess. The vision of autonomic computing. *IEEE Computer Magazine*, 36(1):41–50, January 2003.

73. A. Khanna and J. Zinky. The revised ARPANET routing metric. *ACM SIG-COMM Computer Communication Review*, 19(4):45–56, 1989.

74. Y.-B. Ko and N. H. Vaidya. Location-aided routing (LAR) in mobile ad hoc networks. In *Proceedings of MOBICOM*, pages 66–75. ACM Press, 1998.

75. S.-J. Lee, E. M. Royer, and C. E. Perkins. Scalability study of the ad hoc on-demand distance vector routing protocol. *ACM/Wiley International Journal of Network Management*, 13(2):97–114, March 2003.

76. S. Liang, A. N. Zincir-Heywood, and M. I. Heywood. The effect of routing under local information using a social insect metaphor. In *Proceedings of the IEEE Congress on Evolutionary Computation (CEC)*, May 2002.

77. S. Liang, A. N. Zincir-Heywood, and M. I. Heywood. Adding more intelligence to the network routing problem: AntNet and GA-agents. *Applied Soft Computing*, 6(3):244–257, 2006.

78. S. Liang, A. N. Zincir-Heywood, and M. I. Heywood. Intelligent packets for dynamic network routing using distributed genetic algorithm. In *Proceedings of GECCO*, pages 88–96. ACM Press, 2002.
79. G. S. Malkin. *RIP: An Intra-Domain Routing Protocol*. Addison-Wesley, 1999.
80. S. Marwaha, C. K. Tham, and D. Srinavasan. Mobile Agents based routing protocol for mobile ad hoc networks. In *Proceedings of GLOBECOM*, pages 163–167, Taipei, Taiwan, November 2002. IEEE Press.
81. H. Matsuo and K. Mori. Accelerated ants routing in dynamic networks. In *Proceedings of SNPD*, pages 333–339, August 2001.
82. N. Mazhar and M. Farooq. BeeAIS: Artificial immune system security for nature inspired, MANET routing protocol, BeeAdHoc. In *Proceedings of the 6th International Conference on Artificial Immune Systems*, volume 4628 of *LNCS*, pages 370–381. Springer, 2007.
83. N. Mazhar and M. Farooq. Vulnerability analysis and security framework (BeeSec) for nature inspired MANET routing protocols. In *Proceedings of GECCO*, pages 102–109. ACM Press, 2007.
84. T. Michalareas and L. Sacks. Link-state and ant-like algorithm behaviour for single-constrained routing. In *Proceedings of HPSR*, pages 302–305. IEEE Press, May 2001.
85. T. Michalareas and L. Sacks. Stigmergic techniques for solving multi-constraint routing for packet networks. In *Proceedings of the First International Conference on Networking (ICN), Part II*, volume 2094 of *Lecture Notes in Computer Science*, pages 687–697. Springer-Verlag, 2001.
86. T. Michalareas and L. Sacks. Stigmergic techniques for solving multi-constraint routing for packet networks. In *Proceedings of ICN*, volume 2093 of *Lecture Notes in Computer Science*, pages 687—697. Springer Verlag, 2001.
87. J. Moy. *OSPF: Anatomy of an Internet Routing Protocol*. Addison-Wesley, 1998.
88. R. Muraleedharan and L. A. Osadciw. A predictive sensor network using ant system. In R. M. Rao, S. A. Dianat, and M. D. Zoltowski, editors, *Digital Wireless Communications VI, Proceedings of the SPIE*, pages 181–192, 2004.
89. K. S. Narendra and M. A. Thathachar. *Learning Automata: An Introduction*. Prentice-Hall, 1989.
90. K. S. Narendra and M. A. Thathachar. On the behavior of a learning automaton in a changing environment with application to telephone traffic routing. *IEEE Trans. on Systems, Man, and Cybernetics*, SMC-10(5):262–269, 1980.
91. G. Navarro-Varela and M. C. Sinclair. Ant colony optimisation for virtual-wavelength-path routing and wavelength allocation. In *Proceedings of the IEEE Congress on Evolutionary Computation (CEC)*, pages 1809–1816, 1999.
92. O. V. Nedzelnitsky and K. S. Narendra. Nonstationary models of learning automata routing in data communication networks. *IEEE Transactions on Systems, Man, and Cybernetics*, SMC-17:1004–1015, 1987.
93. The NS-2 network simulator. http://nsnam.isi.edu/nsnam/.
94. K. Oida and A. Kataoka. Lock-free AntNet and its evaluation adaptiveness. *Journal of IEICE B (in Japanese)*, J82-B(7):1309–1319, 1999.
95. K. Oida and M. Sekido. An agent-based routing system for QoS guarantees. In *IEEE International Conference on Systems, Man, and Cybernetics*, volume 3, pages 833–838, 1999.
96. K. Oida and M. Sekido. ARS: An efficient agent-based routing system for QoS guarantees. *Computer Communications*, 23:1437–1447, 2002.

97. S. Okdem and D. Karaboga. Routing in wireless sensor networks using Ant Colony Optimization. In *Proceedings of AHS*, pages 401–404, 2006.
98. C. E. Perkins and P. Bhagwat. Highly dynamic destination-sequenced distance-vector routing (DSDV) for mobile computers. In *Proceedings of SIGCOMM*, pages 234–244. ACM Press, 1994.
99. C. E. Perkins and E. M. Royer. Ad-hoc on-demand distance vector routing. In *Proceedings of WMCSA*, pages 90–100. IEEE Press, 1999.
100. L. Peshkin, N. Meuleau, and L. P. Kaelbling. Learning policies with external memory. In *Proceedings of ICML*, pages 307–314, 1999.
101. Qualnet Simulator, Version 3.9. Scalable Network Technologies, Inc., Culver City, CA, USA, 2005. http://www.scalable-networks.com.
102. S. Rajagopalan and C.-C. Shen. ANSI: A unicast routing protocol for mobile ad hoc networks using swarm intelligence. In *Proceedings of the International Conference on Artificial Intelligence (ICAI)*, pages 24–27, 2005.
103. S. Rajagopalan and C.-C. Shen. ANSI: a swarm intelligence-based unicast routing protocol for hybrid ad hoc networks. *Journal of System Architecture*, 52(8-9):485–504, 2006.
104. M. Roth and S. Wicker. Termite: Ad-hoc networking with stigmergy. In *Proceedings of IEEE GLOBECOM*, pages 2937–2941, 2003.
105. M. Roth and S. Wicker. Termite: Emergent ad-hoc networking. In *Proceedings of the 2nd Mediterranean Workshop on Ad-Hoc Networks (Med-Hoc-Net)*, 2003.
106. E. M. Royer and C.-K. Toh. A review of current routing protocols for ad hoc mobile wireless networks. *IEEE Personal Communications*, 6(2):46–55, 1999.
107. R. Y. Rubinstein. Combinatorial optimization, cross-entropy, ants and rare events. In S. Uryasev and P.M. Pardalos, editors, *Stochastic Optimization: Algorithms and Applications*, pages 304–358. Kluwer Academic Publisher, 2000.
108. M. Saleem and M. Farooq. Beesensor: A bee-inspired power aware routing protocol for wireless sensor networks. In M. Giacobini et al., editors, Lecture Notes in Computer Science, LNCS 4449, pages 81–90. Springer Verlag, 2007.
109. M. Saleem and M. Farooq. A framework for empirical evaluation of nature inspired routing protocols for wireless sensor networks. In *Proceedings of Congress on Evolutionary Computing (CEC)*, pages 751–758. IEEE, 2007.
110. H. G. Sandalidis, C. X. Mavromoustakis, and P. P. Stavroulakis. Ant based probabilistic routing with pheromone and antipheromone mechanisms. *Communication Systems*, 17:55–62, 2004.
111. H. G. Sandalidis, C. X. Mavromoustakis, and P. P. Stavroulakis. Performance measures of an ant based decentralised routing scheme for circuit switching communication networks. *Soft Computing*, 5(4):313–317, 2001.
112. R. Schoonderwoerd and O. Holland. Minimal agents for communications network routing: The social insect paradigm. *Software Agents for Future Communication Systems*, 1(1):1–2, 1999.
113. R. Schoonderwoerd, O. Holland, J. Bruten, and L. Rothkrantz. Ant-based load balancing in telecommunications networks. *Adaptive Behavior*, 5(2):169–207, 1996.
114. T. Seeley. *The Wisdom of the Hive*. Harvard University Press, London, 1995.
115. A. U. Shankar, C. Alaettinoğlu, I. Matta, and K. Dussa-Zieger. Performance comparison of routing protocols using MaRS: Distance-vector versus link-state. In *Proceedings of ACM SIGMETRICS/PERFORMANCE*, pages 181–192, 1992.

116. C.-C. Shen and C. Jaikaeo. Ad hoc multicast routing algorithm with swarm intelligence. *MONET*, 10(1-2):47–59, 2005.

117. C.-C. Shen, C. Jaikaeo, C. Srisathapornphat, Z. Huang, and S. Rajagopalan. Ad hoc networking with swarm intelligence. In M. Dorigo, M. Birattari, C. Blum, L. M. Gambardella, F. Mondada, and T. Stützle, editors, *Proceedings of ANTS*, volume 3172 of *Lecture Notes in Computer Science*, pages 262–269. Springer-Verlag, 2004.

118. C.-C. Shen, S. Rajagopalan, G. Borkar, and C. Jaikaeo. A flexible routing architecture for ad hoc space networks. *Computer Networks*, 46(3):389–410, 2004.

119. E. Sigel, B. Denby, and S. Le Heárat-Mascle. Application of ant colony optimization to adaptive routing in a LEO telecommunications satellite network. *Annals of Telecommunications*, 57(5–6):520–539, May-June 2002.

120. K. M. Sim and W. H. Sun. Ant colony optimization for routing and load-balancing: Survey and new directions. *IEEE Transactions on Systems, Man and Cybernetics-Part A*, 33(5):560–572, 2003.

121. K. M. Sim and W. H. Sun. Ant colony optimization for routing and load-balancing: Survey and new directions. *IEEE Transactions on Systems, Man, and Cybernetics–Part A*, 33(5):560–572, September 2003.

122. I. Stojmenović, editor. *Mobile Ad-Hoc Networks*. John Wiley & Sons, 2002.

123. D. Subramanian, P. Druschel, and J. Chen. Ants and reinforcement learning: A case study in routing in dynamic networks. In *Proceedings of IJCAI*, pages 832–838. Morgan Kaufmann, 1997.

124. D. J. T. Sumpter. From bee to society: An agent-based investigation of honey bee colonies. PhD thesis, University of Manchester, UK, 2000.

125. R. S. Sutton and A. G. Barto. *Reinforcement Learning: An Introduction*. Cambridge, MA: MIT Press, 1998.

126. S. Tadrus. *Generic multi-pheromone quality of service routing*. PhD thesis, Department of Computer Science, University of Nottingham, 2007.

127. S. Tadrus and L. Bai. A QoS network routing algorithm using multiple pheromone tables. In *Web Intelligence*, pages 132–138, 2003.

128. S. Tadrus and L. Bai. QColony: a multi-pheromone best-fit QoS routing algorithm as an alternative to shortest-path routing algorithms. *International Journal of Computational Intelligence and Applications*, 5(2):141–167, 2005.

129. G. Theraulaz and E. Bonabeau. A brief history of stigmergy. *Artificial Life, Special Issue on Stigmergy*, 5:97–116, 1999.

130. M. Thirunavukkarasu. Reinforcing reachable routes. Master's thesis, Virginia Polytechnic Institue and State University, 2004.

131. R. van der Put. Routing in packet switched networks using agents. Master thesis, KBS, Delft University of Technology, Netherlands, 1998.

132. R. van der Put. Routing in the faxfactory using mobile agents. Technical report, KPN Research, 1998.

133. S. Varadarajan, N. Ramakrishnan, and M. Thirunavukkarasu. Reinforcing reachable routes. *Computer Networks*, 43(3):389–416, 2003.

134. A. V. Vasilakos and G. A. Papadimitriou. A new approach to the design of reinforcement scheme for learning automata: Stochastic Estimator Learning Algorithms. *Neurocomputing*, 7(275), 1995.

135. V. Verstraete, M. Strobbe, E. Van Breusegem, J. Coppens, M. Pickavet, and P. Demeester. AntNet: ACO routing algorithm in practice. In *Proceedings of INFORMS Telecommunications Conference*, 2006.

136. K. von Frisch. *Tanzsprache und Orientierung der Bienen.* Springer-Verlag, Heidelberg, 1965.
137. K. von Frisch. *The Dance Language and Orientation of Bees.* Harvard University Press, Cambridge, 1967.
138. S. Vutukury. *Multipath routing mechanisms for traffic engineering and quality of service in the Internet.* PhD thesis, University of California, Santa Cruz, CA, USA, March 2001.
139. Z. Wang. *Internet QoS: Architectures and Mechanisms for Quality of Service.* Morgan Kaufmann, 2001.
140. H. F. Wedde and M. Farooq et al. BeeAdHoc—An Energy-Aware Scheduling and Routing Framework. Technical Report pg439, LSIII, School of Computer Science, University of Dortmund, 2004.
141. H. F. Wedde and M. Farooq. Beehive: New ideas for developing routing algorithms inspired by honey bee behavior. In Albert Y. Zomaya Stephan Olariu, editor, *Handbook of Bioinspired Algorithms and Applications,* chapter 21, pages 321–339. Chapman & Hall/CRC Computer and Information Science, 2005.
142. H. F. Wedde and M. Farooq. BeeHive: Routing algorithms inspired by honey bee behavior. *Künstliche Intelligenz,* Special Issue on Swarm Intelligence, 4:18–24, November 2005.
143. H. F. Wedde and M. Farooq. A performance evaluation framework for nature inspired routing algorithms. In *Applications of Evolutionary Computing,* volume 3449 of *LNCS,* pages 136–146. Springer, 2005.
144. H. F. Wedde and M. Farooq. The wisdom of the hive applied to mobile ad-hoc networks. In *Proceedings of the IEEE Swarm Intelligence Symposium,* pages 341–348, 2005.
145. H. F. Wedde and M. Farooq. A comprehensive review of nature inspired routing algorithms for fixed telecommunication networks. *Journal of System Architecture,* 52(8-9):461–484, 2006.
146. H. F. Wedde, M. Farooq, T. Pannenbaecker, B. Vogel, C. Mueller, J. Meth, and R. Jeruschkat. BeeAdHoc: an energy efficient routing algorithm for mobile ad-hoc networks inspired by bee behavior. In *Proceedings of GECCO,* pages 153–161, 2005.
147. H. F. Wedde, M. Farooq, C. Timm, J. Fischer, M. Kowalski, M. Langhans, N. Range, C. Schletter, R. Tarak, M. Tchatcheu, F. Volmering, S. Werner, and K. Wang. BeeAdHoc–An Efficient, Secure, Scalable Routing Framework for Mobile AdHoc Networks. Technical Report pg460, LSIII, School of Computer Science, University of Dortmund, 2005.
148. H. F. Wedde, M. Farooq, and Y. Zhang. BeeHive: An efficient fault-tolerant routing algorithm inspired by honey bee behavior. In *Ant Colony Optimization and Swarm Intelligence,* volume 3172 of *Lecture Notes in Computer Science,* pages 83–94. Springer Verlag, Sept 2004.
149. H. F. Wedde, C. Timm, and M. Farooq. BeeHiveAIS: A simple, efficient, scalable and secure routing framework inspired by artificial immune systems. In *Proceedings of the PPSN IX,* volume 4193 of *Lecture Notes in Computer Science,* pages 623–632. Springer Verlag, September 2006.
150. H. F. Wedde, C. Timm, and M. Farooq. BeeHiveGuard: A step towards secure nature inspired routing algorithms. In *Applications of Evolutionary Computing,* volume 3907 of *Lecture Notes in Computer Science,* pages 243–254. Springer Verlag, April 2006.

151. T. White. Swarm intelligence and problem solving in telecommunications. *Canadian Artificial Intelligence Magazine*, (41):14–16, 1997.

152. T. White, B. Pagurek, and F. Oppacher. ASGA: Improving the ant system by integration with genetic algorithms. In *Proceedings of the Third Annual Conference on Genetic Programming*, pages 610–617, 1998.

153. T. White, B. Pagurek, and F. Oppacher. Connection management using adaptive mobile agents. In H.R. Arabnia, editor, *Proceedings of the PDPTA*, pages 802–809. CSREA Press, 1998.

154. T. White, B. Pagurek, and F. Oppacher. Application oriented routing with biologically-inspired agents. In *Proceedings of GECCO*, pages 610–617, 1999.

155. Y. Yang, A. N. Zincir-Heywood, M. I. Heywood, and S. Srinivas. Agent-based routing algorithms on a LAN. In *IEEE Canadian Conference on Electrical and Computer Engineering*, 1442–1447 2002.

156. S. Zahid, M. Shehzad, S. Usman Ali, and M. Farooq. A comprehensive formal framework for analyzing the behavior of nature inspired routing protocols. In *Proceedings of Congress on Evolutionary Computing (CEC)*, pages 180–187. IEEE, 2007.

157. L. Zhang, S. Deering, and D. Estrin. RSVP: A new resource ReSerVation protocol. *IEEE Networks*, 7(5):8–18, September 1993.

158. R. Zhang and M. Bartell. *BGP Design and Implementation*. CISCO Press, 2003.

159. Z. Zhang, C. Sanchez, B. Salkewicz, and E. Crawley. Quality of service extensions to OSPF. Internet Draft draft-zhang-qos-ospf-00, Internet Engineering Task Force (IEFT), June 1996.

Part II

Applications

Evolution, Self-organization and Swarm Robotics

Vito Trianni[1], Stefano Nolfi[1], and Marco Dorigo[2]

[1] LARAL Research Group
   ISTC, Consiglio Nazionale delle Ricerche, Rome, Italy
   {vito.trianni,stefano.nolfi}@istc.cnr.it
[2] IRIDIA Research Group
   CoDE, Université Libre de Bruxelles, Brussels, Belgium
   mdorigo@ulb.ac.be

Summary. The activities of social insects are often based on a self-organising process, that is, "a process in which pattern at the global level of a system emerges solely from numerous interactions among the lower-level components of the system" (see [4], p. 8). In a self-organising system such as an ant colony, there is neither a leader that drives the activities of the group, nor are the individual ants informed about a global recipe or blueprint to be executed. On the contrary, each single ant acts autonomously following simple rules and locally interacting with the other ants. As a consequence of the numerous interactions among individuals, a coherent behaviour can be observed at the colony level.

A similar organisational structure is definitely beneficial for a swarm of autonomous robots. In fact, a coherent group behaviour can be obtained providing each robot with simple individual rules. Moreover, the features that characterise a self-organising system—such as decentralisation, flexibility and robustness—are highly desirable also for a swarm of autonomous robots. The main problem that has to be faced in the design of a self-organising robotic system is the definition of the individual rules that lead to the desired collective behaviour. The solution we propose to this design problem relies on artificial evolution as the main tool for the synthesis of self-organising behaviours. In this chapter, we provide an overview of successful applications of evolutionary techniques to the evolution of self-organising behaviours for a group of simulated autonomous robots. The obtained results show that the methodology is viable, and that it produces behaviours that are efficient, scalable and robust enough to be tested in reality on a physical robotic platform.

1 Introduction

Swarm robotics studies a particular class of multi-robot systems, composed of a large number of relatively simple robotic units, and it emphasises aspects like decentralisation of control, robustness, flexibility and scalability.[3] Swarm

---

[3] For an introduction to swarm robotics, see Chapter 4 in this book.

robotics is often inspired by the behaviour of social insects, such as ants, bees, wasps and termites. The striking ability of these animals consists in performing complex tasks such as nest building or brood sorting, despite the limited cognitive abilities of each individual and the limited information that each individual has about the environment. Many activities carried out by social insects are the result of self-organising processes, in which the system-level properties result solely from the interactions among the individual components of the system [4]. In a complex system like an ant colony, there is neither a leader that drives the activities of the group, nor are the individual ants informed of a global recipe or blueprint to be executed. On the contrary, each single ant acts autonomously following simple rules and locally interacting with the other ants. As a consequence of the numerous interactions among individuals, a coherent behaviour can be observed at the colony level.

A similar organisational structure is definitely beneficial for a swarm of autonomous robots. By designing for self-organisation, only minimal complexity is required for each individual robot and for its controller, and still the system as a whole can solve a complex problem in a flexible and robust way. In fact, the global behaviour results from the local interactions among the robots and between robots and the environment, without being explicitly coded within the rules that govern each individual. Rather, the global behaviour results from the interplay of the individual behaviours. Not all swarm robotic systems present self-organising behaviours, and self-organisation is not required for a robotic system to belong to swarm robotics. However, the importance of self-organisation should not be neglected: a high complexity at the system level can be obtained using simple rules at the individual level. It is therefore highly desirable to seek for self-organising behaviours in a swarm robotic system, as they can be obtained with minimal cost. However, because the relationship between simple local rules and complex global properties is indirect, the definition of the individual behaviour is particularly challenging.

> [The] problem is to determine how these so-called "simple" robots should be programmed to perform user-designed tasks. The pathways to solutions are usually not predefined but emergent, and solving a problem amounts to finding a trajectory for the system and its environment so that the states of both the system and the environment constitute the solution to the problem: although appealing, this formulation does not lend itself to easy programming [15].

The solution we propose to this *design problem* relies on artificial evolution as the main tool for the synthesis of self-organising behaviours. We discuss the evolutionary approach to swarm robotics in more detail in Sect. 2. In Sect. 3, we present three case studies in which self-organising behaviours have been evolved: synchronisation, coordinated motion and hole avoidance. With the obtained results, we show that the evolutionary methodology is viable and that it produces behaviours that are efficient, scalable and robust enough to be tested in reality on a physical robotic platform. Finally, Sect. 4 concludes the chapter.

2 Evolutionary Design of Self-organising Behaviours

As seen in the previous section, there is a fundamental problem—referred to as the *design problem*—that arises in the development of self-organising behaviours for a group of robots. As discussed in Sect. 2.1, this problem consists in defining the appropriate individual rules that will lead to a certain global pattern. In Sect. 2.2, we will discuss how collective behaviours can be obtained resorting to *evolutionary robotics*, an automatic technique for generating solutions for a particular robotic task, based on artificial evolution [7, 8]. Notwithstanding the many successful applications in the single robot domain [12, 20, 11], evolutionary robotics has been used only recently for the development of group behaviours. In Sect. 2.3, we review some of the most interesting achievements found in the literature about collective evolutionary robotics.

2.1 The Design Problem

The design of a control system that lets a swarm of robots self-organise requires the definition of those rules at the individual level that correspond to a desired pattern at the system level. This problem is not trivial. From an engineering perspective, it is necessary to discover the relevant interactions between the individual robots, which lead to the global organisation. In other words, the challenge is given by the necessity to decompose the desired global behaviour into simpler individual behaviours and into interactions among the system components. Furthermore, having identified the mechanisms that lead to the global organisation, we still have to consider the problem of encoding them into the controller of each robot, which is complicated by the non-linear, indirect relation between individual control rules and global behaviour: in fact, even a small variation in the individual behaviour might have large effects on the system-level properties. This two-step decomposition process—referred to as the *divide and conquer* approach to the design problem—is exemplified in Fig. 1. The self-organised system displays a global behaviour interacting with the environment (Fig. 1, left). In order to define the controller for the robots, two phases are necessary: first, the global behaviour is decomposed into individual behaviours and local interactions among robots and between robots and the environment (centre); then, the individual behaviour must be decomposed into fine-grained interactions between the robot and the environment, and these interactions must be encoded into a control program (right). Both these phases are complex because they attempt to decompose a process (the global behaviour or the individual one) that results from a dynamical interaction among its subcomponents (interactions among individuals or between the robots and the environment).

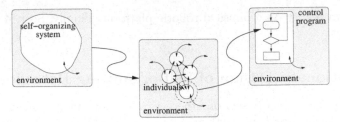

Fig. 1. The "divide and conquer" approach to the design problem. In order to have the swarm robotic system self-organise, we should first decompose the global behaviour of the system (left) into individual behaviours and local interactions among robots and between robots and environment (centre). Then, the individual behaviour must be in some way encoded into a control program (right)

The decomposition from the global to the individual behaviours could be simplified by taking inspiration from natural systems, such as insect societies, that could reveal the basic mechanisms which are to be exploited [3]. Following the observation of a natural phenomenon, a *modelling* phase is performed, which is of fundamental importance to "uncover what actually happens in the natural system" ([3], p. 8). The developed model can then be used as a source of inspiration for the designer, who can try to replicate certain discovered mechanisms in the artificial system, in order to obtain dynamics similar to the natural counterpart (see Fig. 2). However, it is not always possible to take inspiration from natural processes because they may differ from the artificial systems in many important aspects (e.g., the physical embodiment, the type of possible interactions between individuals and so forth), or because there are no natural systems that can be compared to the artificial one. Moreover, the problem of encoding the individual behaviours into a controller for the robots remains to be solved. Our working hypothesis is that both the decomposition problems discussed above can be efficiently bypassed relying on evolutionary robotics techniques [20], as discussed in the following section.

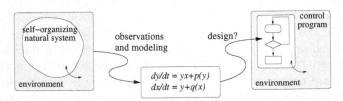

Fig. 2. The design problem solved by taking inspiration from nature: an existing self-organising system (left) can be observed and its global behaviour modelled (centre), obtaining useful insights on the mechanisms underlying the self-organisation process. The model can be used as a source of inspiration for the following design phase, which leads to the definition of the control program (right)

2.2 Evolution of Self-organising Behaviours

Evolutionary robotics represents an alternative approach to the solution of the design problem. By evaluating the robotic system as a whole (i.e., by testing the global self-organising behaviour starting from the definition of the individual rules), it eliminates the arbitrary decompositions at both the level of finding the mechanisms of the self-organising process and the level of implementing those mechanisms into the rules that regulate the interaction between robot and the environment. This approach is exemplified in Fig. 3: the controller encoded into each genotype is directly evaluated by looking at the resulting global behaviour. The evolutionary process autonomously selects the "good" behaviours and discards the "bad" ones, based on a user-defined evaluation function. Moreover, the controllers are directly tested in the environment; thus they can exploit the richness of solutions offered by the dynamic interactions among robots and between robots and the environment, which are normally difficult to be exploited by hand design.

The advantages offered by the evolutionary approach are not costless [16]. On the one hand, it is necessary to identify initial conditions that assure *evolvability*, i.e., the possibility to progressively synthesise better solutions starting from scratch. On the other hand, artificial evolution may require long computation time, so that an implementation on the physical robotic platform may be too demanding. For this reason, software simulations are often used. The simulations must retain as much as possible the important features of the robot-environment interaction. Therefore, an accurate modelling is needed to deploy simulators that well represent the physical system [14].

2.3 Collective Evolutionary Robotics in the Literature

As mentioned above, the use of artificial evolution for the development of group behaviours received attention only recently. The first examples of evolutionary techniques applied to collective behaviours considered populations of elementary organisms, evolved to survive and reproduce in a simulated scenario [31, 32]. Using a similar approach, flocking and schooling behaviours

Fig. 3. The evolutionary approach to the design problem: controllers (left) are evaluated for their capability to produce the desired group behaviour (right). The evolutionary process is responsible for the selection of the controllers and for evaluating their performance (*fitness*) within the environment in which they should work

were evolved for groups of artificial creatures [24, 30, 25]. Collective transport has also been studied using evolutionary approaches [9, 10].

The credit assignment problem in a collective scenario was studied by comparing homogeneous versus heterogeneous groups—composed of two simulated robots—evolved to display a coordinated motion behaviour [22]. Results indicate that heterogeneous groups are better performing for this rather simple task. However, the heterogeneous approach may not be suitable when coping with larger groups and/or with behaviours that do not allow for a clear role allocation [21]. In this case, homogeneous groups achieve a better performance, as they display altruistic behaviours that appear with low probability when the group is heterogeneous and selection operates at the individual level. Overall, the above-mentioned works confirm that artificial evolution can be successfully used to synthesise controllers for collective behaviours. However, whether these results can generalise to physical systems—i.e., real robots—remains to be ascertained. The three case studies presented in the following section are some examples—among few others, see [23, 19]—of evolutionary robotics techniques applied to group behaviours and successfully tested on physical robots.

3 Studies in Evolutionary Swarm Robotics

In this section, we present three case studies in which artificial evolution has been exploited to evolve collective self-organising behaviours. In Sect. 3.2, we consider the problem of synchronising the movements of a group of robots by exploiting a minimal communication channel. In Sect. 3.3, we present the problem of obtaining coordinated motion in a group of physically assembled robots. The obtained behaviour is extended in Sect. 3.4, in which the problem of avoiding holes is considered together with coordinated motion. Before reviewing these case studies, we present in Sect. 3.1 the robotic system used in our experiments.

3.1 A Swarm Robotics Artifact: The *Swarm-bot*

The experiments presented in this chapter have been mainly conducted within the SWARM-BOTS project,[4] which aimed at the design and implementation of an innovative swarm robotics artifact—the *swarm-bot*—which is composed of a number of independent robotic units—the *s-bots*—that are connected together to form a physical structure [18]. When assembled in a *swarm-bot*, the *s-bots* can be considered as a single robotic system that can move and reconfigure. Physical connections between *s-bots* are essential for solving many collective tasks, such as retrieving a heavy object or bridging a gap larger than a single *s-bot*. However, for tasks such as searching for a goal location

---

[4] For more details, see http://www.swarm-bots.org.

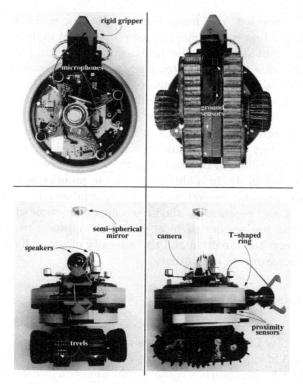

Fig. 4. View of the *s-bot* from different sides. The main components are indicated (see text for more details)

or tracing an optimal path to a goal, a swarm of unconnected *s-bots* can be more efficient.

An *s-bot* is a small mobile autonomous robot with self-assembling capabilities, shown in Fig. 4. It weighs 700 g and its main body has a diameter of about 12 cm. Its design is innovative with regard to both sensors and actuators. The traction system is composed of both tracks and wheels, called *treels*. The treels are connected to the chassis, which also supports the main body. The latter is a cylindrical turret mounted on the chassis by means of a motorised joint, that allows the relative rotation of the two parts. A gripper is mounted on the turret and it can be used for connecting rigidly to other *s-bots* or to some objects. The gripper does not only open and close, but it also has a degree of freedom for lifting the grasped objects. The corresponding motor is powerful enough to lift another *s-bot*. *S-bots* are also provided with a flexible arm with three degrees of freedom, on which a second gripper is mounted. However, this actuator has not been considered for the experiments presented in this chapter, nor was it mounted on the *s-bots* that have been used.

An *s-bot* is provided with many sensory systems, useful for the perception of the surrounding environment or for proprioception. Infrared proximity sensors are distributed around the rotating turret. Four proximity sensors placed under the chassis—referred to as *ground sensors*—can be used for perceiving holes or the terrain's roughness (see Fig. 4). Additionally, an *s-bot* is provided with eight light sensors uniformly distributed around the turret, two temperature/humidity sensors, a three-axis accelerometer and incremental encoders on each degree of freedom. Each robot is also equipped with sensors and devices to detect and communicate with other *s-bots*, such as an omnidirectional camera, coloured LEDs around the *s-bots'* turret, microphones and loudspeakers (see Fig. 4). In addition to a large number of sensors for perceiving the environment, several sensors provide information about physical contacts, efforts, and reactions at the interconnection joints with other *s-bots*. These include torque sensors on most joints as well as a *traction sensor*, a sensor that detects the direction and the intensity of the pulling and pushing forces that *s-bots* exert on each others.

3.2 Synchronisation

In this section, we provide the first case study in which self-organising behaviours are evolved for a swarm of robots. The task chosen is *synchronisation*: robots should exploit communication in order to entrain their individual movements. Synchronisation is a common phenomenon in nature: examples of synchronous behaviours can be found in the inanimate world as well as among living organisms. One of the most commonly cited self-organised synchronous behaviours is the one of fireflies from Southeast Asia: thousands of insects have the ability to flash in unison, perfectly synchronising their individual rhythm (see [4]). This phenomenon has been thoroughly studied and an explanation based on self-organisation has been proposed [17]. Fireflies are modelled as a population of pulse-coupled oscillators with equal or very similar frequency. These oscillators can influence each other by emitting a pulse that shifts or resets the oscillation phase. The numerous interactions among the individual oscillator fireflies are sufficient to explain the synchronisation of the whole population (for more details, see [17, 26]).

The above self-organising synchronisation mechanism was successfully replicated in a group of robots [33]. In this study, the authors designed a specialised neural module for the synchronisation of the group foraging and homing activities, in order to maximise the overall performance. Much like fireflies that emit light pulses, robots communicate through sound pulses that directly reset the internal oscillator designed to control the individual switch from homing to foraging and vice versa. Similarly, the case study presented in this section follows the basic idea that if an individual displays a periodic behaviour, it can synchronise with other (nearly) identical individuals by temporarily modifying its behaviour in order to reduce the phase difference with the rest of the group. However, while a firefly-like mechanism exploits the entrainment of the individual oscillators, in this work we do not postulate the need of internal dynamics. Rather, the period and the phase of the individual behaviour are defined by the sensory-motor coordination of the robot, that is, by the dynamical interactions with the environment that result from the robot embodiment. We show that such dynamical interactions can be exploited for synchronisation, allowing us to keep a minimal complexity of both the behavioural and the communication level (for more details, see [28]).

Experimental Setup

As mentioned above, in this work we aim at studying the evolution of behavioural and communication strategies for synchronisation. For this purpose, we define a simple, idealised scenario that contains all the ingredients needed for our study. The task requires that each s-bot in the group displays a simple periodic behaviour, that is, moving back and forth from a light bulb positioned in the centre of the arena. Moreover, s-bots have to synchronise their movements, so that their oscillations are in phase with each other.

The evolutionary experiments are performed in simulation, using a simple kinematic model of the s-bots. Each s-bot is provided with infrared sensors and ambient light sensors, which are simulated using a sampling technique. In order to communicate with each other, s-bots are provided with a very simple signalling system, which can produce a continuous tone with fixed frequency and intensity. When a tone is emitted, it is perceived by every robot in the arena, including the signalling s-bot. The tone is perceived in a binary way, that is, either there is someone signalling in the arena, or there is no one. The arena is a square of 6 × 6 meters. In the centre, a cylindrical object supports the light bulb, which is always switched on, so that it can be perceived from every position in the arena. At the beginning of every trial, three s-bots are initially positioned in a circular band ranging from 0.2 to 2.2 meters from the centre of the arena. The robots have to move back and forth from the light, making oscillations with an optimal amplitude of 2 meters.

Artificial evolution is used to synthesise the connection weights of a fully connected, feed-forward neural network—a perceptron network. Four sensory neurons are dedicated to the readings of four ambient light sensors, positioned in the front and in the back of the s-bot. Six sensory neurons receive input from a subset of the infrared proximity sensors evenly distributed around the s-bot's turret. The last sensory neuron receives a binary input corresponding to the perception of sound signals. The sensory neurons are directly connected to three motor neurons: two neurons control the wheels, and the third controls the speaker in such a way that a sound signal is emitted whenever its activation is greater than 0.5.

The evolutionary algorithm is based on a population of 100 binary-encoded genotypes, which are randomly generated. Each genotype in the population encodes the connection weights of one neural controller. Each real-valued connection weight is encoded by eight bits in the genotype. The population is evolved for a fixed number of generations, applying a combination of selection with elitism and mutation. Recombination is not used. At each generation, the 20 best individuals are selected for reproduction and retained in the subsequent generation. Each genotype reproduces four times, applying mutation with 5% probability of flipping a bit. The evolutionary process is run for 500 generations. During evolution, a genotype is mapped into a control structure that is cloned and downloaded in all the *s-bots* taking part in the experiment (i.e., we make use of a homogeneous group of *s-bots*). Each genotype is evaluated five times—i.e., five trials. Each trial differs from the others in the initialisation of the random number generator, which influences both the initial position and the orientation of the *s-bots* within the arena. Each trial lasts $T = 900$ simulation cycles, which corresponds to 90 seconds of real time.

The fitness of a genotype is the average performance computed over the five trials in which the corresponding neural controller is tested. During a single trial, the behaviour produced by the evolved controller is evaluated by a two-component fitness function. The first component rewards the periodic oscillations performed by the *s-bots*. The second component rewards synchrony among the robots, evaluated as the cross-correlation coefficient between the sequences of the distances from the light bulb. Additionally, an indirect selective pressure for the evolution of obstacle avoidance is given by blocking the motion of robots that collide. When this happens, the performance is negatively influenced. Furthermore, a trial is normally terminated after $T = 900$ simulation cycles. However, a trial is also terminated if any of the *s-bots* crosses the borders of the arena.

Results

We performed 20 evolutionary runs, each starting with a different population of randomly generated genotypes. After the evolutionary phase, we selected a single genotype per evolutionary run, chosen as the best individual of the final generation. We refer to the corresponding controllers as $c_i, i = 1, \ldots, 20$. Direct observation of the evolved behaviours showed that in some evolutionary runs—nine out of 20—communication was not evolved, and robots display a periodic behaviour without being able to synchronise. The remaining evolutionary runs produced simple behavioural and communication strategies in which signalling was exploited for synchronisation. All evolved solutions result in a similar behaviour, characterised by two stages, that is, phototaxis when the *s-bots* approach the light bulb, and antiphototaxis when the *s-bots* move away from it. Signalling is generally performed only during one of the two stages. We can classify the evolved controllers into three classes, according to the individual reaction to the perception of a sound signal.

The first two classes present a very similar behaviour, in which signalling strongly correlates with either phototaxis (controllers c_5, c_9, c_{13}, c_{15} and c_{16}) or antiphototaxis (controllers c_1, c_4, c_7, c_{19} and c_{20}). We describe here the behaviour using c_{13}, which can be appreciated by looking at the left part of

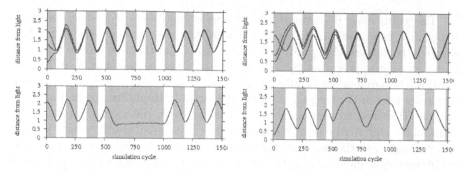

Fig. 5. The synchronisation behaviour of two controllers: c_{13} (left) and c_{14} (right). In the upper part, the *s-bots'* distances from the light bulb are plotted against the simulation cycles, in order to appreciate the synchronisation of the individual movements. The grey areas indicate when a signal is emitted by any of the *s-bots* in the arena. In the lower part, the distance and signalling behaviour of a single *s-bot* are plotted against the simulation cycles. From cycle 500 to 1000, a signal is artificially created, which simulates the behaviour of an *s-bot*. This allows us to visualise the reaction of an *s-bot* to the perception of a sound signal

Fig. 5. Looking at the upper part of the figure, it is possible to notice that whenever a robot signals, its distance from the light decreases and, vice versa, when no signal is perceived the distance increases. Synchronisation is normally achieved after one oscillation and it is maintained for the rest of the trial, the robots moving in perfect synchrony with each other. This is possible thanks to the evolved behavioural and communication strategy, for which a robot emits a signal while performing phototaxis and reacts to the perceived signal by reaching and keeping a specific distance close to the centre of the arena. As shown in the bottom part of Fig. 5, in presence of a continuous signal— artificially created from cycle 500 to cycle 1000—an *s-bot* suspends its normal oscillatory movement to maintain a constant distance from the centre. As soon as the sound signal is stopped, the oscillatory movement starts again. Synchronisation is possible because robots are homogeneous; therefore they all present an identical response to the sound signal that makes them move to the inner part of the arena. As soon as all robots reach the same distance from the centre, signalling ceases and synchronous oscillations can start. In conclusion, the evolved behavioural and communication strategies allow a fast synchronisation of the robots' activities, because they force all robots to perform synchronously phototaxis or antiphototaxis from the beginning of a trial, as a reaction to the presence or absence of a sound signal respectively. This also allows a fast synchronisation of the movements thanks to the reset of the oscillation phase. Finally, it provides a means to fine-tune and maintain through time a complete synchronisation, because the reset mechanism allows it to continuously correct even the slightest phase difference.

The third class is composed by a single controller—c_{14}—that produces a peculiar behaviour. In this case, it is rather the *absence* of a signal that strongly correlates with phototaxis. The individual reaction to the perceived signal can be appreciated by looking at the right part of Fig. 5. When the continuous signal is artificially created (see simulation cycles 500 to 1000 in the lower part of the figure), the *s-bot* performs both phototaxis and antiphototaxis. However, as soon as the signal is removed, the *s-bot* approaches the light bulb. Differently from the mechanism presented above, *s-bots* initially synchronise only the movement direction but not the distance at which the oscillatory movements are performed (see the top-right part of Fig. 5). Despite this limitation, this mechanism allows a very fast and precise synchronisation of the *s-bots*' phototaxis and antiphototaxis, which is probably the reason why it was evolved in the first place. In order to achieve a complete synchronisation, an additional mechanism was synthesised, which allows us to precisely entrain the movements of the robots on a fine-grained scale. This mechanism influences the distance covered by an *s-bot* during antiphototaxis: *s-bots* that are farther away from the light bulb slightly bend their trajectory and therefore cover a distance range shorter than the one covered by the other robots in the same time. In this way, the differences among *s-bots* are progressively reduced, until all *s-bots* are completely synchronised.

Scalability of the Evolved Behaviours

The above analysis clarified the role of communication in determining the synchronisation among the different robots. Here, we analyse the scalability of the evolved neural controllers when tested in larger groups of robots. For this

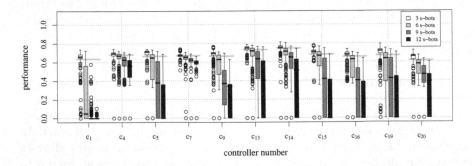

Fig. 6. Scalability of the successful controllers. Each controller was evaluated using 3, 6, 9 and 12 robots. In each condition, 500 different trials were executed. Each box represents the inter-quartile range of the corresponding data, while the black horizontal line inside the box marks the median value. The whiskers extend to the most extreme data points within 1.5 times the inter-quartile range from the box. The empty circles mark the outliers. The horizontal grey line shows the mean value over 500 trials measured in the evolutionary conditions, in order to better evaluate the scalability property

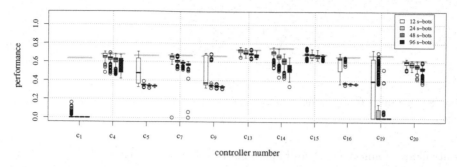

Fig. 7. Scalability of the synchronisation mechanism. Each controller was evaluated using 12, 24, 48 and 96 robots. In each condition, 500 different trials were executed

purpose, we evaluated the behaviour of the successful controllers using 3, 6, 9 and 12 *s-bots*. The obtained results are plotted in Fig. 6. It is easy to notice that most of the best evolved controllers have a good performance for groups composed of six *s-bots*. In such condition, in fact, *s-bots* are able to distribute in the arena without interfering with each other. Many controllers present a good behaviour also when groups are composed of nine *s-bots*. However, we also observe various failures due to interferences among robots and collisions. The situation gets worse when using 12 *s-bots*: the higher the density of robots, the higher the number of interferences that lead to failure. In this case, most controllers achieve a good performance only sporadically. Only c_4 and c_7 systematically achieve synchronisation despite the increased difficulty of the task.

In order to analyse the scalability property of the synchronisation mechanism only, we evaluate the evolved controllers by removing the physical interactions among the robots, as if each *s-bot* were placed in a different arena and perceived the other *s-bots* only through sound signals. Removing the robot-robot interactions allows us to test large groups of robots—we used 12, 24, 48 and 96 *s-bots*. The obtained results are summarised in Fig. 7. We observe that many controllers perfectly scale, having a performance very close to the mean performance measured with three *s-bots*. A slight decrease in performance is justified by the longer time required by larger groups to converge to perfectly synchronised movements (see for example c_7 and c_{20}).

Some controllers—namely c_4, c_5, c_9, c_{14} and c_{16}—present an interference problem that prevents the group from synchronising when a sufficiently large number of robots is used. In such a condition, the signals emitted by different *s-bots* at different times may overlap and may be perceived as a single, continuous tone (recall that the sound signals are perceived in a binary way, preventing an *s-bot* from recognising different signal sources). If the perceived signal does not vary in time, it does not bring enough information to be exploited for synchronisation. Such interference can be observed only sporadically for c_4 and and c_{14}, but it strongly affects the performance of the other

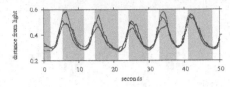

Fig. 8. Distances from the light bulb and collective signalling behaviour of the real *s-bots*

controllers—namely c_5, c_9 and c_{16}. This problem is the result of the fact that we used a "global" communication form in which the signal emitted by an *s-bot* is perceived by any other *s-bot* anywhere in the arena. Moreover, from the perception point of view, there is no difference between a single *s-bot* and a thousand signalling at the same time. The lack of locality and of additivity is the main cause of failure for the scalability of the evolved synchronisation mechanism. However, as we have seen, this problem affects only some of the analysed controllers. In the remaining ones, the evolved communication strategies present an optimal scalability that is only weakly influenced by the group size.

Tests with Physical Robots

We tested the robustness of the evolved controllers downloaded onto the physical robots. To do so, we chose c_{13} as it presented a high performance and good scalability properties. The neural network controller is used on the physical *s-bots* exactly in the same way as in simulation. The only differences with the simulation experiments are in the experimental arena, which is four times smaller in reality (1.5 × 1.5 meters), and accordingly the light bulb is approximately four times less intense. In these experiments, three *s-bots* have been used. A camera was mounted on the ceiling to record the movements of the robots and track their trajectories [5]. The behaviour of the physical robots presents a good correspondence with the results obtained in simulation. Synchrony is quickly achieved and maintained throughout the whole trial, notwithstanding the high noise of sensors and actuators and the differences among the three robots (see Fig. 8). The latter deeply influence the group behaviour: *s-bot* have different maximum speeds which let them cover different distances in the same time interval. Therefore, if phototaxis and antiphototaxis were very well synchronised, as a result of the communication strategy exploited by the robots, it was possible to notice some differences in the maximum distance reached.

3.3 Coordinated Motion

The second case study focuses on a particular behaviour, namely *coordinated motion*. In animal societies, this behaviour is commonly observed: we can think of flocks of birds coordinately flying, or of schools of fish swimming in perfect unison. Such behaviours are the result of a self-organising process, and various models have been proposed to account for them (see [4], chapter 11). In the *swarm-bot* case, coordinated motion takes on a particular flavour, due to the physical connections among the *s-bots*, which open the way to study novel interaction modalities that can be exploited for coordination. Coordinated motion is a basic ability for the *s-bots* physically connected in a *swarm-bot* because, being independent in their control, they must coordinate their actions in order to choose a common direction of movement. This coordination ability is essential for an efficient motion of the *swarm-bot* as a whole, and constitutes a basic building block for the design of more complex behavioural strategies, as we will see in Sect. 3.4. We review here a work that extends previous research conducted in simulation only [1]. We present the results obtained in simulation, and we show that the evolved controllers continue to exhibit a high performance when tested with physical *s-bots* (for more details, see [2]).

Experimental Setup

A *swarm-bot* can efficiently move only if the chassis of the assembled *s-bots* have the same orientation. As a consequence, the *s-bots* should be capable of negotiating a common direction of movement and then compensating possible misalignments that occur during motion. The coordinated motion experiments consider a group of *s-bots* that remain always connected in *swarm-bot* formation (see Fig. 9). At the beginning of a trial, the *s-bots* start with their chassis oriented in a random direction. Their goal is to choose a common direction of motion on the basis of only the information provided by their traction sensor, and then to move as far as possible from the starting position. The common direction of motion of the group should result from a self-organising process based on local interactions, which are shaped as traction forces. We exploit artificial evolution to synthesise a simple feed-forward neural network that encodes the motor commands in response to the traction force perceived by the robots.

Four sensory neurons encode the intensity of traction along four directions, corresponding to the directions of the semi-axes of the chassis' frame

Fig. 9. Left: four real *s-bots* forming a linear *swarm-bot*. Right: four simulated *s-bots*

of reference (i.e., front, back, left and right). The activation state of the two motor neurons controls the wheels and the turret-chassis motor, which is actively controlled in order to help the rotation of the chassis. The evolutionary algorithm used in this case differs from that described in Sect. 3.2 only in the mutation of the genotype, which is performed with 3% probability of flipping each bit. For each genotype, four identical copies of the resulting neural network controllers are used, one for each *s-bot*. The *s-bots* are connected in a linear formation, shown in Fig. 9. The fitness of the genotype is computed as the average performance of the *swarm-bot* over five different trials. Each trial lasts $T = 150$ cycles, which corresponds to 15 seconds of real time. At the beginning of each trial, a random orientation of the chassis is assigned to each *s-bot*. The ability of a *swarm-bot* to display coordinated motion is evaluated by computing the average distance covered by the group during the trials. Notice that this way of computing the fitness of the groups is sufficient to obtain coordinated motion behaviour. In fact, it rewards *swarm-bots* that maximise the distance covered and, therefore, their motion speed.

Results

Using the setup described above, 30 evolutionary runs have been performed in simulation. All the evolutionary runs successfully synthesised controllers that produced coordinated motion in a *swarm-bot*. The controllers evolved in simulation allow the *s-bots* to coordinate by negotiating a common direction of movement and to keep moving along in such a direction by compensating any possible misalignment. Direct observation of the evolved behavioural strategies shows that at the beginning of each trial the *s-bots* try to pull or push the rest of the group in the direction of motion in which they are initially placed. This disordered motion results in traction forces that are exploited for coordination: the *s-bots* orient their chassis in the direction of the perceived traction, which roughly corresponds to the average direction of motion of the group. This allows the *s-bots* to rapidly converge toward a common direction and to maintain it.

Behavioural Analysis

All the 30 controllers evolved in the different replications of the evolutionary process present similar dynamics. Hereafter, the controller synthesised by the 30th evolutionary run is considered, as it proved to have the best performance. In order to understand the functioning of the controller at the individual level, the activation of the motor units was measured in correspondence to a traction force whose angle and intensity were systematically varied. In this way, we can appreciate the behavioural strategy of each individual. When the intensity of traction is low, the *s-bot* moves forward at maximum speed (see the regions indicated by number 1 in Fig. 10). In fact, a low or null intensity of traction—i.e., no pulling or pushing forces—corresponds to the robots already

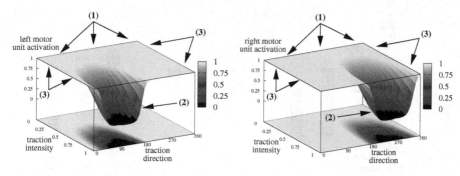

Fig. 10. Motor commands issued by the left and right motor units (left and right figure, respectively) of the best evolved neural controller in correspondence to traction forces having different directions and intensities. An activation of 0 corresponds to maximum backward speed and 1 to maximum forward speed. See text for the explanation of numbers in round brackets

moving in the same direction. Whenever a traction force is perceived from a direction different from the chassis' direction, the *s-bot* reacts by turning toward the direction of the traction force (see the regions indicated by number 2 in Fig. 10). For example, when the traction direction is about 90°—i.e., a pulling force from the left-hand side of the chassis' movement direction—the left wheel moves backward and the right wheel moves forward, resulting in a rotation of the chassis in the direction of the traction force. Finally, the *s-bot* keeps on moving forward if a traction force is perceived with a direction opposite to the direction of motion (see the regions indicated by number 3 in Fig. 10). Notice that this is an instable equilibrium point, because as soon as the angle of traction differs from 0°, for example due to noise, the *s-bot* rotates its chassis following the rules described above.

The effects of the individual behaviour at the group level can be described as follows. At the beginning of each test, all *s-bots* perceive traction forces with low intensity, and they start moving forward in the random direction in which they were initialised. However, being assembled together, they generate traction forces that propagate throughout the physical structure. Each *s-bot* perceives a single traction force, that is, the resultant of all the forces applied to its turret, which roughly indicates the average direction of motion of the group. Following the simple rules described above, an *s-bot* rotates its chassis in order to align to the perceived traction force. In doing so, some *s-bots* will be faster than the others, therefore reinforcing the traction signal in their direction of motion. As a consequence, the other *s-bots* perceive an even stronger traction force, which speeds up the alignment process. Overall, this positive feedback mechanism makes all *s-bots* quickly converge toward the same direction of motion.

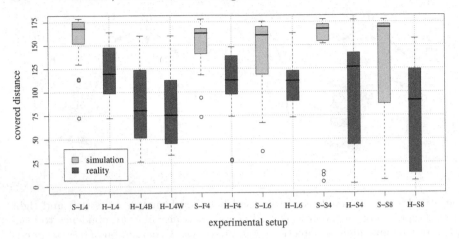

Fig. 11. Performance of the best evolved controller in simulation and reality (distance covered in 20 trials, each lasting 25 s). Labels indicate the experimental setup: 'S' and 'H' indicate tests performed respectively with simulated and physical *s-bots*; 'L4' indicates tests involving four *s-bots* forming a linear structure; 'L4B' and 'L4W' indicate tests performed on rough terrain, respectively brown and white terrain (see text for details). 'F4' indicates tests involving four *s-bots* forming a linear structure not rigidly connected. 'L6' indicates tests involving six *s-bots* forming a linear structure. 'S4' indicates tests involving four *s-bots* forming a square shape; 'S8' indicates tests involving eight *s-bots* forming a "star" shape

Scalability and Generalisation with Simulated and Physical Robots

The self-organising behaviour described above is very effective and scalable, leading to coordinated motion of *swarm-bots* of different sizes and shapes, despite its being evolved using a specific configuration for the *swarm-bot* (i.e., four *s-bots* in linear formation). Tests with real robots showed a good performance as well, confirming the robustness of the evolved controller. In Fig. 11, we compare the performance of the evolved controller in different tests with both simulated and real robots. In all tests performed, *s-bots* start connected to each other, having randomly assigned orientations of their chassis. Each experimental condition is tested for 20 trials, each lasting 25 seconds (250 cycles). In the following, we briefly present the tests performed and we discuss the obtained results.

The reference test involves four simulated *s-bots* forming a linear structure. The *swarm-bot* covers on average about 160 cm in 25 seconds. The performance decreases of 23%, on average, when tested with the real *s-bots* (see Fig. 11, conditions *S-L4* and *H-L4*). The lower performance of the real *swarm-bot* with respect to the simulated *swarm-bot* is due to the longer time required by real *s-bots* to coordinate. This is caused by many factors, among which is the fact that tracks and toothed wheels of the real *s-bots* sometimes get stuck during the initial coordination phase, due to a slight bending of the structure that caused an excessive thrust on the treels. This leads to a sub-optimal motion of the *s-bots*, for example while turning on the spot. However, coordination is always achieved and the *s-bots* always move away from the initial position. This result proves that the controller evolved in simulation can effectively produce coordinated motion when tested in real *s-bots*, notwithstanding the fact that the whole process takes some more time compared with simulation.

The evolved controller is also able to produce coordinated movements on two types of rough terrain (see Fig. 11, conditions *H-L4B* and *H-L4W*). The brown rough terrain is a very regular surface made of brown plastic isolation foils. The white rough terrain is an irregular surface made of plaster bricks that look like stones. In these experimental conditions, the *swarm-bot* is always able to coordinate and to move from the initial position, having a performance comparable to what was achieved on flat terrain. However, in some trials coordination is achieved only partially, mainly due to a more difficult grip of the treels on the rough terrain.

Another test involves a *swarm-bot* in which connections among *s-bots* are "semi-rigid" rather than completely rigid (see Fig. 11, conditions *S-F4* and *H-F4*). In the case of semi-rigid links the gripper is not completely closed and the assembled *s-bots* are partially free to move with respect to each other. In fact, a partially open gripper can slide around the turret perimeter, while other movements are constrained. One interesting aspect of semi-rigid links is that they potentially allow *swarm-bots* to dynamically rearrange their shape in order to better adapt to the environment [1, 29]. Despite the different connection mechanism, which deeply influences the traction forces transmitted through the physical links, the obtained results show that the evolved controller preserves its capability of producing coordinated movements both in simulation and in reality. The performance using semi-rigid links is only 4% and 11% lower than using rigid links, respectively in tests with simulated and real *swarm-bots*.

The best evolved controller was tested with linear *swarm-bots* composed of six *s-bots*. The results showed that larger *swarm-bots* preserve their ability to produce coordinated movements both in simulation and in reality (see Fig. 11, conditions *S-L6* and *H-L6*). The performance in the new experimental condition is 10% and 8% lower than what was measured with *swarm-bots* formed by four *s-bots*, respectively in tests in simulation and in reality. This test suggests that the evolved controller produces a behaviour that scales well with the number of individuals forming the group both in simulated and real robots (for more results on scalability with simulated robots, see [1, 6]).

Finally, we tested *swarm-bots* varying both shape and size. We tested *swarm-bots* composed of four *s-bots* forming a square structure and *swarm-bots* composed of eight *s-bots* forming a "star" shape (see Fig. 12). The results show that the controller displays an ability to produce coordinated movements independently of the *swarm-bot*'s shape, although the tests that use real *s-bots* show a higher drop in performance (see Fig. 11, conditions *S-S4* and *H-S4*

Fig. 12. *Swarm-bots* with different shapes. Left: a *swarm-bot* composed of four *s-bots* forming a square shape. Right: a *swarm-bot* composed of eight *s-bots* forming a "star" shape

for the square formation, and conditions *S-S8* and *H-S8* for the "star" formation). This is due to a high chance for the *swarm-bot* to achieve a rotational equilibrium in which the structure rotates around its centre of mass, therefore resulting in a very low performance. This rotational equilibrium is a stable condition for central-symmetric shapes, but it is never observed in the experimental conditions used to evolve the controller. Additionally, increasing the size of the *swarm-bots* leads to a slower coordination. This not only lowers the performance, but also increases the probability that the group falls into rotational equilibrium. As a consequence, the performance of square and "star" formation in reality is 27% and 40% lower than that in the corresponding simulated structures.

Overall, the tests with simulated and physical robots prove that the evolved controllers produce a self-organising system able to achieve and maintain coordination among the individual robots. The evolved behaviour maintains its properties despite the particular configuration of the *swarm-bot*. It also constitutes an important building block for *swarm-bots* that have to perform more complex tasks such as coordinately moving toward a light target [1], and coordinately exploring an environment by avoiding walls and holes [1, 29]. In the following section, we analyse in detail one of these extensions of the coordinated motion task, that is, *hole avoidance*.

3.4 Hole Avoidance

The third case study presents a set of experiments that build upon the results on coordinated motion described above. Also in this case, we study a coordination problem among the *s-bots* forming a *swarm-bot*. Additionally, *s-bots* are provided with a sound-signalling system, that can be used for communication. The task we study requires the *s-bots* to explore an arena presenting holes in which the robots may fall. Individual *s-bots* cannot avoid holes due to their limited perceptual apparatus. In contrast, a *swarm-bot* can exploit the physical connections and the communication among its components in order to safely navigate in the arena. Communication is an important aspect in a social domain: insects, for example, make use of different forms of communication, which serves as a regulatory mechanism of the activities of the colony [13]. Similarly, in swarm robotics communication is often required for the coordination of the group.

The experiments presented here bring forth a twofold contribution. We examine different communication protocols among the robots (i.e., no signalling, handcrafted and evolved signalling), and we show that a completely evolved approach achieves the best performance. This result is in accordance with the assumption that evolution potentially produces a system more efficient than those obtained with other conventional design methodologies (see Sect. 2.2). Another important contribution of these experiments consists in the testing of the evolved controllers on physical robots. We show that the evolved controllers produce a self-organising system that is robust enough to be tested on real *s-bots*, notwithstanding the huge gap between the simulation model used for the evolution and the physical *s-bot* (for more details, see [27]).

Experimental Setup

The hole avoidance task has been defined for studying collective navigation strategies for a *swarm-bot* that moves in environments presenting holes in which it risks remaining trapped. For a *swarm-bot* to perform hole avoidance, two main problems must be solved: (i) coordinated motion must be performed in order to obtain coherent movements of the *s-bots*; (ii) the presence of holes must be communicated to the entire group, in order to trigger a change in the common direction of motion. We study and compare three different approaches to communication among the *s-bots*. In a first setup, referred to as *Direct Interactions* setup (*DI*), *s-bots* communicate only through the pulling and pushing forces that one *s-bot* exerts on the others. The second and third setups make use of direct communication through binary sound signals. In the second setup, referred to as *Direct Communication* setup (*DC*), the *s-bots* emit a tone as a handcrafted reflex action to the perception of a hole. In the third setup, referred to as *Evolved Communication* setup (*EC*), the signalling behaviour is not a priori defined, but it is left to evolution to shape the best communication strategy.

We decided to let evolution shape the neural controller testing the *swarm-bot* both in environments with and without holes. In this way, we focus on the ability of both efficiently performing coordinated motion and avoiding falling into holes. In all cases, the *s-bots* start connected in a *swarm-bot* formation, and the orientation of their chassis is randomly defined, so that they need to coordinate in order to choose a common direction of motion. Also in this case, the *s-bots* are controlled by a simple perceptron network, whose parameters are set by the same evolutionary algorithm described in Sect. 3.2. In all three setups (*DI*, *DC* and *EC*), *s-bots* are equipped with traction and ground sensors. In *DC* and *EC*, microphones and speakers are also used. In the *DC* setup, the activation of the loudspeaker has been handcrafted, simulating a sort of reflex action: an *s-bot* activates the loudspeaker whenever one of its ground sensors detects the presence of a hole. Thus, the neural network does not control the emission of a sound signal. However, it receives the information coming from the microphones, and evolution is responsible for shaping the correct reaction to the perceived signals. In contrast, in the *EC* setup the speaker is controlled by an additional neural output. Therefore, the complete communication strategy is under the control of evolution.

Each genotype is evaluated in 12 trials, each lasting $T = 400$ control cycles, corresponding to 40 seconds in real time. Similarly to the previous experiments, we make use of homogeneous robots: each genotype generates a single neural controller that is cloned and downloaded in all the *s-bots*. In each trial, the behaviour of the *s-bots* is evaluated rewarding fast and straight motion. Moreover, *s-bots* are asked to minimise the traction force perceived— in order to perform coordinated motion—and the activation of the ground sensors—in order to avoid holes. Finally, *s-bots* are strongly penalised for every fall out of the arena in order to obtain a robust avoidance behaviour.

Results

For each setup—*DI*, *DC* and *EC*—the evolutionary experiments were replicated ten times. All evolutionary runs were successful, each achieving a good performance. Looking at the behaviour produced by the evolved controllers, we observe that the initial coordination phase that leads to the coordinated motion is performed with rules very similar to those described in Sect. 3.3. The differences between the three setups appear once the hole avoidance behaviour is considered.

DI setup: *s-bots* can rely only on direct interactions, shaped as traction forces. Here, the *s-bots* that detect a hole invert the direction of motion, therefore producing a traction force that is perceived by the rest of the group as a signal to move away from the hole. The interactions through pushing and pulling forces are sufficient to trigger collective hole avoidance. However, in some cases the *swarm-bot* is not able to avoid falling because the signal encoded in the traction force produced may not be strong enough to trigger the reaction of the whole group.

DC setup: *s-bots* can rely on both direct interactions shaped as traction forces and direct communication through sound signals. The *s-bots* that detect a hole invert their direction of motion and emit a continuous tone. In contrast, the *s-bots* that perceive a sound signal stop moving. Signalling ceases when no *s-bots* perceive the hole, and coordinated motion can start again. In this setup, direct communication reinforces the interactions through traction forces, achieving a faster collective reaction to the perception of the hole.

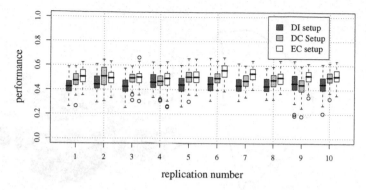

Fig. 13. Post-evaluation analysis of the best controller produced by all evolutionary runs of the three different setups

EC setup: Similarly to the *DC* setup, *s-bots* can exploit both traction and sound signals. However, here, evolution is responsible for shaping both the signalling mechanisms and the response to the perceived signals. This results in complex signalling and reaction strategies that exploit the possibility to control the speaker. In general, signalling is associated with the perception of a hole, but it is also inhibited in certain conditions. For example, signals are not emitted if a strong traction force is perceived or if a sound signal was previously emitted: in both cases, in fact, an avoidance action was already initiated, and further signalling could only interfere with the coordination effort.

The results obtained using direct communication seem to confirm our expectations: direct communication allows a faster reaction to the detection of a hole and therefore a more efficient avoidance behaviour is obtained. Additionally, the evolved communication strategy appears more adaptive than the handcrafted solution. This intuition is also confirmed by a quantitative analysis we performed in order to compare the three setups.

For each evolutionary run, we selected the best individual of the final generation and we re-evaluated it 100 times. A box-plot summarising the performance of these individuals is shown in Fig. 13. It is easy to notice that *EC* generally performs better than *DC* and *DI*, while *DC* seems to be generally better than *DI*. On the basis of these data, we performed a statistical analysis, which allowed us to state that the behaviours evolved within the *EC* setup performs significantly better than those evolved within both the *DI* and the *DC* setups. The latter in turn results in being significantly better than the *DI* setup. We can conclude that the use of direct communication is clearly beneficial for hole avoidance. In fact, it speeds up the reaction to the detection of a hole, and it makes the avoidance action more reliable. Moreover, we demonstrated, evolving the communication protocol leads to a more adapted system.

Fig. 14. Hole avoidance performed by a physical *swarm-bot*. Left: view of the arena taken with the overhead camera. The dark line corresponds to the trajectory of the *swarm-bot* in a trial lasting 900 control cycles. Right: a physical *swarm-bot* while performing hole avoidance. It is possible to notice how physical connections among the *s-bots* can serve as support when a robot is suspended out of the arena, still allowing the whole system to work. Notwithstanding the above difficult situation, the *swarm-bot* was able to successfully avoid falling

Tests with Physical Robots

One controller per setup was selected for tests with physical robots. Each selected controller was evaluated in 30 trials. The behaviour of the *swarm-bot* was recorded using an overhead camera, in order to track its trajectory with a tracking software [5] (see the left part of Fig. 14). Qualitatively, the behaviour produced by the evolved controllers tested on the physical *s-bots* is very good and closely corresponds to that observed in simulation. *S-bots* coordinate more slowly in reality than in simulation, taking a few seconds to agree on a common direction of motion. Hole avoidance is also performed with the same modalities as observed in simulation.

From a quantitative point of view, it is possible to recognise some differences between simulation and reality, as shown in Fig. 15. We compare the performance recorded in 100 trials in simulation with the one obtained from the 30 trials performed in reality. Generally, we observe a decrease in the maximum performance, mainly due to a slower coordination among the *s-bots*. This means that real *s-bots* start moving coordinately later than the simulated ones, both at the beginning of a trial and after the perception of a hole. This influences the performance, as the *swarm-bot* cannot cover large distances until coordination among the *s-bots* is achieved. With the *DI* controller, the combination of tracks and wheels of the traction system brings an advantage in hole avoidance as the *s-bot* that perceives the hole can produce a traction force even if it is nearly completely suspended out of the arena. Moreover, the high friction provided by the tracks produces higher traction forces that can

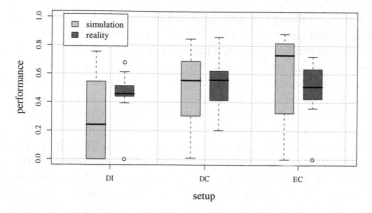

Fig. 15. Comparison of the performance produced in the different settings by the selected controllers tested in both simulation and reality

have a greater influence on the behaviour of the rest of the group. Similarly, the treels system is advantageous for the *DC* controller, in which the *s-bot* perceiving the holes pushes the other *s-bots* away from the arena border while emitting a sound signal. Concerning the *EC* controller, in contrast, the treels system does not lead to a clear advantage from a qualitative point of view.

On the whole, the neural controllers synthesised by artificial evolution proved to be robust enough to be tested on physical robots, notwithstanding the huge gap between the simulation model used for the evolution and the actual *s-bot*. The performance of the controllers tested in the real world was somewhat affected by various factors, but the difference with simulation was never higher than 20% on average. We can therefore conclude that the transferring of the evolved self-organising behaviour from simulated to physical *s-bots* was successful.

4 Conclusions

In this chapter, we have argued that self-organising behaviours represent a viable solution for controlling a swarm robotic system, and that evolutionary robotics techniques are a valuable design tool. There are multiple reasons why self-organisation should be aimed at. Among these are the properties of decentralisation, flexibility, and robustness that pertain to self-organising systems and that are highly desirable for a swarm of autonomous robots. However, if everything seems to fit in nicely, some problems arise when trying to design a self-organising behaviour. In fact, the features that determine the behaviour of a self-organising system are not explicitly coded anywhere, while the design of a control system requires exactly the definition of the control rules for each robot of the system. The *design problem*—treated in detail in

Sect. 2—consists in filling the gap between the desired global behaviour of the robotic system and the control rules that govern each single robot.

The three case studies presented here present a possible solution to the design problem based on evolutionary robotics. All experiments share the same methodology, which consists in evolving neural controllers for homogeneous groups of simulated robots. The free parameters that are varied during the evolutionary process encode the connection weights of the neural controllers that regulate the fine-grained interactions between the robots and the environment. Variations of the free parameters are retained or discarded on the basis of their effect at the level of the global behaviour exhibited by the swarm of robots. The evolved controllers are afterwards tested in simulation and, whenever possible, also with physical robots. The analysis of the behaviours produced by the evolutionary process is useful to assess the quality of the obtained results. However, the same analysis can be seen from a different, equally important, perspective, that is, the discovery and the understanding of the basic principles underlying self-organising behaviours and collective intelligence. The analysis of the evolved behaviours presented in this chapter shows how complex behavioural, cognitive and social skills might arise from simple control mechanisms. These results are important to assess evolutionary robotics not only as a design tool, but also as a methodology for modelling and understanding intelligent adaptive behaviours.

Acknowledgements

This work was supported by the SWARM-BOTS project and by the ECAgents project, two projects funded by the Future and Emerging Technologies programme (IST-FET) of the European Commission, under grant IST-2000-31010 and 001940 respectively. The information provided is the sole responsibility of the authors and does not reflect the Community's opinion. The Community is not responsible for any use that might be made of data appearing in this publication. The authors thank Nikolaus Correll and Alcherio Martinoli for providing the tracking software used in the experiments presented in this paper. Marco Dorigo acknowledges support from the Belgian FNRS, of which he is a research director, through the grant "Virtual Swarm-bots", contract no. 9.4515.03, and from the ANTS project, an Action de Recherche Concertée funded by the Scientific Research Directorate of the French Community of Belgium.

References

1. G. Baldassarre, D. Parisi, and S. Nolfi. Distributed coordination of simulated robots based on self-organisation. *Artificial Life*, 12(3):289–311, 2006.

2. G. Baldassarre, V. Trianni, M. Bonani, F. Mondada, M. Dorigo, and S. Nolfi. Self-organised coordinated motion in groups of physically connected robots. *IEEE Transactions on Systems, Man and Cybernetics—Part B: Cybernetics*, 37(1):224–239, 2007.

3. E. Bonabeau, M. Dorigo, and G. Theraulaz. *Swarm Intelligence: From Natural to Artificial Systems*. Oxford University Press, New York, NY, 1999.

4. S. Camazine, J.-L. Deneubourg, N. Franks, J. Sneyd, G. Theraulaz, and E. Bonabeau. *Self-organization in Biological Systems*. Princeton University Press, Princeton, NJ, 2001.

5. N. Correll, G. Sempo, Y. Lopez de Meneses, J. Halloy, J.-L. Deneubourg, and A. Martinoli. SwisTrack: A tracking tool for multi-unit robotic and biological systems. In *Proceedings of the 2006 IEEE/RSJ International Conference in Intelligent Robots and Systems (IROS '06)*, pages 2185–2191. IEEE Press, Piscataway, NJ, 2006.

6. M. Dorigo, V. Trianni, E. Şahin, R. Groß, T. H. Labella, G. Baldassarre, S. Nolfi, J.-L. Deneubourg, F. Mondada, D. Floreano, and L. M. Gambardella. Evolving self-organizing behaviors for a *swarm-bot*. *Autonomous Robots*, 17(2–3):223–245, 2004.

7. L. J. Fogel, A. J. Owens, and M. J. Walsh. *Artificial Intelligence through Simulated Evolution*. John Wiley & Sons, New York, NY, 1966.

8. D. E. Goldberg. *Genetic Algorithms in Search, Optimization and Machine Learning*. Addison-Wesley, Reading, MA, 1989.

9. R. Groß and M. Dorigo. Group transport of an object to a target that only some group members may sense. In X. Yao, E. Burke, J. A. Lozano, J. Smith, J. J. Merelo-Guervós, J. A. Bullinaria, J. Rowe, P. Tiño, A. Kabán, and H.-P. Schwefel, editors, *Parallel Problem Solving from Nature – 8th International Conference (PPSN VIII)*, volume 3242 of *Lecture Notes in Computer Science*, pages 852–861. Springer Verlag, Berlin, Germany, 2004.

10. R. Groß, F. Mondada, and M. Dorigo. Transport of an object by six pre-attached robots interacting via physical links. In *Proceedings of the 2006 IEEE Interantional Conference on Robotics and Automation (ICRA '06)*, pages 1317–1323. IEEE Computer Society Press, Los Alamitos, CA, 2006.

11. I. Harvey, E. A. Di Paolo, R. Wood, M. Quinn, and E. Tuci. Evolutionary robotics: A new scientific tool for studying cognition. *Artificial Life*, 11(1–2):79–98, 2005.

12. I. Harvey, P. Husbands, and D. Cliff. Issues in evolutionary robotics. In J.-A. Meyer, H. Roitblat, and S. W. Wilson, editors, *From Animals to Animats 2. Proceedings of the Second International Conference on Simulation of Adaptive Behavior (SAB 92)*, pages 364–373. MIT Press, Cambridge, MA, 1993.

13. B. Hölldobler and E. O. Wilson. *The Ants*. Belknap Press, Harvard University Press, Cambridge, MA, 1990.

14. N. Jakobi. Evolutionary robotics and the radical envelope of noise hypothesis. *Adaptive Behavior*, 6:325–368, 1997.

15. C. R. Kube and E. Bonabeau. Cooperative transport by ants and robots. *Robotics and Autonomous Systems*, 30(1–2):85–101, 2000.

16. M. J. Matarić and D. Cliff. Challenges in evolving controllers for physical robots. *Robotics and Autonomous Systems*, 19(1):67–83, 1996.

17. R. E. Mirollo and S. H. Strogatz. Synchronization of pulse-coupled biological oscillators. *SIAM Journal on Applied Mathematics*, 50(6):1645–1662, 1990.

18. F. Mondada, G. C. Pettinaro, A. Guignard, I. V. Kwee, D. Floreano, J.-L. Deneubourg, S. Nolfi, L. M. Gambardella, and M. Dorigo. SWARM-BOT: A new distributed robotic concept. *Autonomous Robots*, 17(2–3):193–221, 2004.

19. A. L. Nelson, E. Grant, and T. C. Henderson. Evolution of neural controllers for competitive game playing with teams of mobile robots. *Robotics and Autonomous Systems*, 46:135–150, 2004.

20. S. Nolfi and D. Floreano. *Evolutionary Robotics: The Biology, Intelligence, and Technology of Self-Organizing Machines*. MIT Press/Bradford Books, Cambridge, MA, 2000.

21. A. Perez-Uribe, D. Floreano, and L. Keller. Effects of group composition and level of selection in the evolution of cooperation in artificial ants. In W. Banzhaf, T. Christaller, P. Dittrich, J. T. Kim, and J. Ziegler, editors, *Advances in Artificial Life. Proceedings of the 7th European Conference on Artificial Life (ECAL 2003)*, volume 2801 of *Lecture Notes in Artificial Intelligence*, pages 128–137. Springer Verlag, Berlin, Germany, 2003.

22. M. Quinn. A comparison of approaches to the evolution of homogeneous multi-robot teams. In *Proceedings of the 2001 Congress on Evolutionary Computation (CEC 01)*, pages 128–135. IEEE Press, Piscataway, NJ, 2001.

23. M. Quinn, L. Smith, G. Mayley, and P. Husbands. Evolving controllers for a homogeneous system of physical robots: Structured cooperation with minimal sensors. *Philosophical Transactions of the Royal Society of London, Series A: Mathematical, Physical and Engineering Sciences*, 361(2321-2344), 2003.

24. C. W. Reynolds. An evolved, vision-based behavioral model of coordinated group motion. In J.-A. Meyer, H. Roitblat, and S. W. Wilson, editors, *From Animals to Animats 2. Proceedings of the Second International Conference on Simulation of Adaptive Behavior (SAB 92)*, pages 384–392. MIT Press, Cambridge, MA, 1993.

25. L. Spector, J. Klein, C. Perry, and M. Feinstein. Emergence of collective behavior in evolving populations of flying agents. *Genetic Programming and Evolvable Machines*, 6(1):111–125, 2005.

26. S. H. Strogatz and I. Stewart. Coupled oscillators and biological synchronization. *Scientific American*, 269(6):102–109, 1993.

27. V. Trianni and M. Dorigo. Self-organisation and communication in groups of simulated and physical robots. *Biological Cybernetics*, 95:213–231, 2006.

28. V. Trianni and S. Nolfi. Minimal communication strategies for self-organising synchronisation behaviours. In *Proceedings of the 2007 IEEE Symposium on Artificial Life*, pages 199–206. IEEE Press, Piscataway, NJ, 2007.

29. V. Trianni, S. Nolfi, and M. Dorigo. Cooperative hole avoidance in a *swarm-bot*. *Robotics and Autonomous Systems*, 54(2):97–103, 2006.

30. C. R. Ward, F. Gobet, and G. Kendall. Evolving collective behavior in an artificial ecology. *Artificial Life*, 7(1):191–209, 2001.

31. G. M. Werner and M. G. Dyer. Evolution of communication in artificial organisms. In C. Langton, C. Taylor, D. Farmer, and S. Rasmussen, editors, *Artificial Life II*, volume X of *SFI Studies in the Science of Complexity*, pages 659–687. Addison-Wesley, Redwood City, CA, 1991.

32. G. M. Werner and M. G. Dyer. Evolution of herding behavior in artificial animals. In J.-A. Meyer, H. Roitblat, and S. W. Wilson, editors, *From Animals to Animats 2. Proceedings of the Second International Conference on Simulation of Adaptive Behavior (SAB 92)*, pages 393–399. MIT Press, Cambridge, MA, 1993.

33. S. Wischmann, M. Huelse, J. F. Knabe, and F. Pasemann. Synchronization of internal neural rhythms in multi-robotic systems. *Adaptive Behavior*, 14(2):117–127, 2006.

Particle Swarms
for Dynamic Optimization Problems

Tim Blackwell[1], Jürgen Branke[2], and Xiaodong Li[3]

[1] Department of Computing
  Goldsmiths College, London, UK
  t.blackwell@gold.ac.uk
[2] Institute AIFB
  University of Karlsruhe, Karlsruhe, Germany
  branke@aifb.uni-karlsruhe.de
[3] School of Computer Science and Information Technology
  RMIT University, Melbourne, Australia
  xiaodong.li@rmit.edu.au

1 Introduction

Many practical optimization problems are dynamic in the sense that the best solution changes in time. An optimization algorithm, therefore, has to both find and subsequently track the changing optimum. Examples include the arrival of new jobs in scheduling, changing expected profits in portfolio optimization, and fluctuating demand.

Clearly, if the changes in the problem instance are radical, the best one can do is to repeatedly solve the optimization problem from scratch. However, in most practical applications the changes are gradual. If this is the case, it should be possible to speed up optimization after a problem change by utilizing some of the information on the fitness landscape gathered during the optimization process so far. In recent years, appropriately modified evolutionary algorithms (EAs) have been shown to achieve this successfully; see, e.g., [11, 23]; the focus of this chapter is to present similar advances within the context of particle swarm optimization.

Particle swarm optimization (PSO) is a versatile population-based optimization technique, in many respects similar to evolutionary algorithms. Basically, particles "fly" above the fitness landscape, while a particle's movement is influenced by its attraction to its neighborhood best (the best solution found by members of the particle's social network), and its personal best (the best solution the particle has found so far). PSO has been shown to perform well for many static problems [32], and is introduced in more detail in Section 2.

The application of PSO to dynamic problems has been explored by various authors [9, 7, 14, 17, 19, 20, 21, 32, 29, 38]. The overall consequence of this work is that PSO, just like EAs, must be modified for optimal results in dynamic environments. There are two main difficulties that need to be addressed:

1. **Outdated memory:** If the problem changes, a previously good solution stored as neighborhood or personal best may no longer be good, and will mislead the swarm towards false optima if the memory is not updated.
2. **Diversity loss:** In normal operation, the swarm contracts around the best solution found during the optimization. As has been demonstrated, the time taken for a partially converged swarm to re-diversify, find the shifted peak, and then re-converge is quite deleterious to performance [4].

A number of adaptations have been applied to PSO in order to solve these difficulties; memories can be refreshed or forgotten and swarms may be re-diversified through randomization, repulsion, and dynamic information exchange and with the use of multi-populations. An account of these adaptations, and a summary of how PSO can detect change (this is especially important when change is unpredictable), is presented in Section 3. A more detailed review of our own work on PSO algorithms, the species PSO and the multi-swarm PSO, is described and extended in Section 4. These approaches are empirically tested and compared in Section 5. The chapter concludes with a summary and some ideas for future work.

2 Particle Swarm Optimization

Optimization with particle swarms (see Chap. 2 for a detailed introduction) has two major ingredients, the particle dynamics and the particle information network. The particle dynamics are derived from swarm simulations in computer graphics [34], and the information sharing component is inspired by social networks [18, 25]. These ingredients make PSO a robust and efficient optimizer of real-valued objective functions (although PSO has also been successfully applied to combinatorial and discrete problems). PSO is an accepted computational intelligence technique, sharing some qualities with evolutionary computation [1]. For an introduction to PSO see also Chap. 2 of this book.

In PSO, population members (particles) move over the search space according to

$$\mathbf{v}(t+1) = \mathbf{v}(t) + \mathbf{a}(t+1) \tag{1}$$
$$\mathbf{x}(t+1) = \mathbf{x}(t) + \mathbf{v}(t+1) \tag{2}$$

where $\mathbf{a}$, $\mathbf{v}$, $\mathbf{x}$, and t are acceleration, velocity, position and time (iteration counter) respectively.

A particle's acceleration $\mathbf{a}$ is primarily governed by attraction to two solutions: its personal best and its neighborhood best. Particles possess a memory of the best (with respect to an objective function) location that they have visited in the past, their *personal best* or *pbest*, and of its fitness. In addition, particles have access to the best location of any other particle in their neighborhood, usually denoted as *neighborhood best* or *gbest*. Naturally, these two locations will coincide if the particle has the best local best in its neighborhood. Several neighborhood topologies have been tried, with the fully connected network remaining a popular choice for unimodal problems. With this neighborhood structure, every particle will share information with every other particle in the swarm so that there is a single *gbest* global best attractor representing the best location found by the entire swarm.

The particles experience a linear or spring-like attraction, weighted by a random number, towards each attractor. Convergence towards a good solution will not follow from these dynamics alone; the particle movement must progressively contract. This contraction is implemented by Clerc and Kennedy with a constriction factor χ, $\chi < 1$, [16]. For our purposes here, the Clerc-Kennedy PSO will be taken as the canonical swarm; χ replaces other energy draining factors such as decreasing 'inertial weight' and velocity clamping.

Overall, the acceleration of particle i in Eq.1 is given by

$$\mathbf{a}_i = \chi[c\epsilon_1 \cdot (\mathbf{p}_g - \mathbf{x}_i) + c\epsilon_2 \cdot (\mathbf{p}_i - \mathbf{x}_i)] - (1 - \chi)\mathbf{v}_i \qquad (3)$$

where ϵ_1 and ϵ_2 are vectors of random numbers drawn from the uniform distribution $\mathcal{U}[0, 1]$, $c > 2$ is the spring constant and $\mathbf{p}_i$, $\mathbf{p}_g$ are personal and neighborhood attractors. This formulation of the particle dynamics emphasizes constriction as a frictional force, opposite in direction and proportional to velocity. Clerc and Kennedy derive a relation for $\chi(c)$: standard values are $c = 2.05$ and $\chi = 0.729843788$. The complete PSO algorithm for maximizing an objective function f is summarized as Algorithm 2.

3 Addressing the Challenges in Dynamic Environments

As has been mentioned in Section 1, in dynamic environments, PSO suffers from outdated memory and lost diversity. This section summarizes the approaches proposed in the literature to address these challenges. Because many of these approaches explicitly react to a change in the landscape, we start by discussing ways to detect a change.

3.1 Detecting a Change

In many applications, a change is known to the system, e.g., in scheduling, when a new job arrives and has to be integrated into the schedule. If the time of a change is not known, it has to be detected. In the literature, this is usually done by simply re-evaluating one or more solutions, and concluding that a change has occurred if the fitness value of at least one of these solutions has changed [14, 20]. How many solutions and which ones to re-evaluate

---

Algorithm 2 Canonical PSO

FOR EACH particle i
    Randomly initialize $\mathbf{v}_i, \mathbf{x}_i = \mathbf{p}_i$
    Evaluate $f(\mathbf{p}_i)$
    $g = \arg\max f(\mathbf{p}_i)$
REPEAT
    FOR EACH particle i
        Update particle position $\mathbf{x}_i$ according to Eqs. 1, 2 and 3
        Evaluate $f(\mathbf{x}_i)$
        //Update personal best
        IF $f(\mathbf{x}_i) > f(\mathbf{p}_i)$ THEN
            $\mathbf{p}_i = \mathbf{x}_i$
        //Update global best
        IF $f(\mathbf{x}_i) > f(\mathbf{p}_g)$ THEN
            $\mathbf{p}_g = \arg\max f(\mathbf{p}_i)$
UNTIL termination criterion reached

---

may depend on the particular application. Usually, the best solution found so far is re-evaluated, which prevents the algorithm from converging around a previously good solution which is no longer good.

In [22], a change in the environment is detected from observing the algorithm behavior, which has the advantage of also working in noisy environments when the above re-evaluation scheme might lead to false alarms.

3.2 Memory Update

If the environment changes, the particle memory (namely the best location visited in the past, and its corresponding fitness) may no longer be valid, with potentially disastrous effects on the search.

This problem is typically solved in one of two ways: re-evaluating the memory or forgetting the memory [14]. In the latter, each particle's memory is simply set to the particle's current position, and the global best is updated making sure that $\mathbf{p}_g = \arg\max f(\mathbf{p}_i)$.

3.3 The Problem of Lost Diversity

If the environment changes after the swarm has converged to a peak, it takes time for the population to re-diversify and re-converge, making it slow in tracking a moving optimum.

The diversity loss was examined in [3] based on the swarm diameter $|S|$, defined as the largest distance, along any axis, between any two particles. When a change occurs and the new optimum location is within the collapsing swarm, there is a good chance that a particle will find itself close to the new optimum within a few iterations and the swarm will successfully track the moving target. The swarm as a whole has sufficient diversity. However, if the new optimum is outside the swarm's expansion, the low velocities of the particles (which are of order $|S|$) will inhibit re-diversification and tracking, and the swarm can even oscillate about a false attractor and along a line perpendicular to the true optimum, in a phenomenon known as linear collapse [6]. [4] uses these considerations to examine under what conditions a swarm can track a single moving optimum.

Because of the problem of convergence, either a diversity increasing mechanism should be invoked at change (or at predetermined intervals), or sufficient diversity has to be ensured at all times [8]. There are four principle mechanisms proposed in the literature for either re-diversification or diversity maintenance: randomization [20], repulsion [6], dynamic networks [21, 37] and multi-populations [30, 7]. They will be discussed in turn in the following.

3.4 Re-diversification

Hu and Eberhart [20] study a number of re-diversification mechanisms. They all involve randomization of the entire or part of the swarm after a change. Since randomization implies information loss, there is a danger of erasing too much information and effectively re-starting the swarm. On the other hand, too little randomization might not introduce enough diversity to cope with the change. In the multi-swarm approaches (see below), it has been suggested to always keep one swarm searching, and randomize the least-fit swarm whenever all swarms have converged [8]. The way quantum particles are used in this chapter, as described in Section 4.1, can also be seen as a form of local re-diversification.

3.5 Repulsion

A constant degree of swarm diversity can be maintained at all times through some type of repulsive mechanism. Repulsion can be either between particles, or from an already-detected optimum. For example, Krink et al. [36] study finite-size particles as a means of preventing premature convergence. The hard sphere collisions produce a constant diversification pressure. Alternatively, Parsopoulos and Vrahatis [32] place a repeller at an already-detected optimum, in an attempt to divert the swarm and find new optima. Neither technique, however, has been applied to the dynamic scenario.

An example of repulsion that has been tested in a dynamic context is the atom analogy [6, 5, 9, 7]. In this model, a swarm is comprised of 'charged' and 'neutral' particles. The charged particles repel each other, leading to a cloud of charged particles orbiting a contracting, neutral, PSO nucleus (as shown in Figure 1). Charge enhances diversity in the vicinity of the converging PSO subswarm, so that optimum shifts within this cloud should be traceable. Good tracking (outperforming canonical PSO) has been demonstrated for unimodal dynamic environments of varying severities [4].

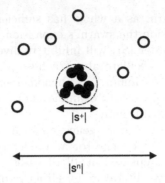

Fig. 1. An example of a swarm containing both neutral and charged particles. A solid circle denotes a neutral particle, whereas a hollow circle denotes a charged particle. $|S^+|$ and $|S^n|$ denote the size of the charged and neutral swarms, respectively.

In [7, 8], the charged particle idea has been simplified to the quantum particle, which does not follow the usual particle movement laws, but instead is re-generated in each iteration at a random position in a vicinity around the swarm's global best. Quantum particles have been shown to be easier to control, to be computationally faster and to perform better than charged particles in [7, 8]. More details on quantum particles are discussed in Section 4.1.

3.6 Dynamic Network Topology

Adjustments to the information-sharing topology can be made with the intention of reducing, maybe temporarily, the desire to move towards the global best position, thereby enhancing population diversity. Li and Dam [37] use a grid-like neighborhood structure, while Janson and Middendorf [21] apply a tree-like structure. Both papers report improvements over unmodified PSO for unimodal dynamic environments. In the latter approach, the particles can change places in the hierarchy, In [22], it has been additionally suggested to break up the tree into sub-trees after a change, so that they can independently search for a new optimum for a while, until they are joined again.

A division of the swarm into subswarms is intuitively helpful in dynamic multi-modal environments, where several promising regions of the search space can be tracked simultaneously. This is the core idea of the speciation PSO (SPSO) proposed in [27, 30], and the multi-swarm PSO (MPSO) proposed in [7, 8]. SPSO uses a fixed swarm size and dynamically divides the swarm into subswarms based on a technique known as clearing [33]. MPSO, on the other hand, has a set of swarms of predetermined size, and adjusts the number of such swarms during the run, thereby also changing the overall population size. These two approaches are examined in more detail below.

4 Multi-swarms and Speciation

The authors of this chapter have independently proposed two different approaches which divide the swarm into a number of subswarms: the multi-swarm PSO (MPSO, [7, 8]) and the speciation-based PSO (SPSO, [27, 31, 28]). The motivation for both approaches is that in a dynamic environment, it is useful to maintain information about several promising regions of the search space. This proposition was already the motivation behind the self-organizing scouts approach [11] and has recently been confirmed also in [12]. By dividing up the swarm, the subswarms may simultaneously track different promising regions of the search space. This is particularly helpful if the environment changes in a way that makes a previous local optimum the new global optimum. If one of the subswarms was tracking the local optimum or a nearby region, the new global optimum is immediately found.

In this section, we describe the previously proposed MPSO and SPSO in detail and also present some new extensions. The approaches are then compared empirically in Section 5. Because both approaches now use quantum particles, these are discussed first.

4.1 Quantum Particles

The quantum particles have been proposed in [7] as a means to maintaining a certain level of diversity within a swarm. They have been inspired by atomic models. In a classical atom, a number of electrons orbit, at various distances, a small ball of nucleons. The picture is altered in the quantum atom; the electrons do not orbit in deterministic paths but are distributed in a probability 'cloud' around the nucleus. The PSO atom consists of a nucleus of normal PSO particles moving under the normal update rules. Typically this nucleus will be shrinking in size as it converges on an optimum. The nucleus is surrounded by quantum particles. Quantum particles do not follow PSO dynamics but are placed at positions around the center of the nucleus, defined by $\mathbf{p}_g$, according to a probability distribution. As a result, they do not converge, but maintain a constant level of diversity.

Different probability distributions are conceivable. It is reasonable to assume that the quantum probability distribution should be spherically symmetric, i.e., shells of constant density centered on $\mathbf{p}_g$. In the following, three different distributions are considered: The Gaussian distribution, the uniform volume distribution (UVD), and a non-uniform volume distribution (NUVD); see Figure 2 for examples in a two-dimensional space. The UVD in d dimensions can be generated as follows [15]:

1. Generate a point $\mathbf{x_i}$ from $\mathcal{N}[0, 1]$ for $1 \leq i \leq d$.
2. Calculate the distance of x_i to the origin $dist = \sqrt{\sum_{i=1}^{i=d} x_i^2}$.
3. Select a value u from $\mathcal{U}[0, 1]$.
4. The new point will be $r_{cloud} \cdot \mathbf{x_i} \frac{\sqrt[d]{u}}{dist}$.

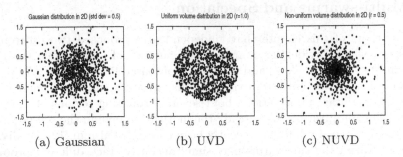

(a) Gaussian   (b) UVD   (c) NUVD

Fig. 2. 1,000 sampling points for a Gaussian distribution, a uniform volume distribution (UVD), and a non-uniform volume distribution (NUVD), in two dimensions.

r_{cloud} is a parameter defining the radius of the distribution.

The above procedure can be modified to generate other distributions by simply changing the distribution for u. For example, instead of selecting u from $\mathcal{U}[0,1]$ in Step 3, Gaussian $\mathcal{N}[0,\sigma]$ can be used with σ set to roughly $\frac{1}{3}$, since in a Gaussian distribution with 3σ away from the mean would cover 99% of all possible samples. This would create a distribution with higher density closer to mean. When changing the calculation of the new solution in Step 4 to $r_{cloud} \cdot \mathbf{x_i}\frac{u}{dist}$, one obtains the non-uniform volume distribution (NUVD) where the density decreases linearly with distance from the center.

Quantum particles are somehow related to the bare-bones PSO proposed by Kennedy, where each dimension of the new position of a particle is randomly selected from a Gaussian distribution with the mean being the average of $\mathbf{p_i}$ and $\mathbf{p_g}$ and the standard deviation σ being the distance between $\mathbf{p_i}$ and $\mathbf{p_g}$ [24]:

$$\mathbf{x_i} \leftarrow \mathcal{N}(\frac{\mathbf{p_i} + \mathbf{p_g}}{2}, \|\mathbf{p_i} - \mathbf{p_g}\|) \tag{4}$$

Richer and Blackwell also reported work on PSO variants employing Gaussian distribution [35], as well as the more general Lévy distribution[4]. Algorithms employing a Lévy or Cauchy distribution, which both have a long fat tail, are more capable of escaping local optima than the Gaussian counterpart. Escape from local optima is profitable in circumstances in which a single global optimum must be found. In the context of tracking moving peaks in a multi-modal dynamic environment, we are more interested in finding multiple peaks in parallel so that when peaks move the optimizer still has a chance to relocate them. Distributions that can provide good sampling in an adjacent area of the peaks would be more suitable; hence the Gaussian distribution is preferred.

---

[4] The shape of the Lévy distribution can be controlled by a parameter α. For $\alpha = 2$ it is equivalent to Gaussian distribution, whereas for $\alpha = 1$ it is equivalent to the Cauchy distribution [35].

4.2 Multi-swarm PSO

The multi-swarm PSO (MPSO) was originally proposed in [7] and then extended in [8] and [2]. It maintains diversity on two levels: the swarm is divided into a number of subswarms which are forced to different areas of the search space (diversity between swarms), and each swarm has some quantum particles to ensure diversity within the swarm.

Basic Features

The MPSO proposed in [8] uses the following mechanisms:

- **Change Detection and Outdated Memory:** For change detection, in each iteration, a subswarm's global best is re-evaluated. If the fitness value has changed, a change is detected and all of the subswarm's personal best are re-evaluated before commencing.
- **Multiple Swarms:** The particles in MPSO are divided into a number of M independent subswarms. Each subswarm in MPSO has a fixed number of particles. Information sharing within a swarm is global, i.e., any good position found by any particle (neutral or quantum) is available to any other. It is known that a global information topology between particles produces better optimization of a uni-modal environment, whereas a local topology is preferred in the multi-modal situation. Even if the landscape is multi-modal, the role of the neutral PSO is to climb up a single peak. The diversity needed to find peaks stems from a dynamic interaction between separate swarms (exclusion, see below) rather than information transfer between social neighborhoods of a single swarm.
- **Exclusion:** Intuitively, several swarms sitting on the same local optimum are not very helpful; they should explore different promising regions of the search space. To this end, a so-called *exclusion* operator is employed. Swarm exclusion forbids two swarms moving to within r_{excl} of each other, where the distance between swarms is defined as the distance between their $\mathbf{p}_g$'s. When exclusion is invoked, the worse swarm, as judged by the current best values $f(\mathbf{p}_g)$, is randomized in the entire search space.
- **Anti-convergence:** Exclusion makes sure that the different swarms converge to different local optima. In order to be able to detect also new, emerging peaks, MPSO contains an additional mechanism termed *anti-convergence* in [8]: Whenever all swarms have converged, the least-fit swarm (as judged again by the current best values $f(\mathbf{p}_g)$) is randomized in the entire search space. Thereby, a swarm is considered as converged when its expansion, i.e., the radius of the smallest circle encompassing the neutral particles, is less than r_{conv}. Anti-convergence is particularly important if the number of swarms is significantly smaller than the number of local optima, in which case each swarm might converge to and get stuck on a different local optimum.

- **Quantum Particles:** Each swarm in MPSO consists of a number N^0 of neutral particles and a number N^Q of quantum particles. The quantum particles are generated in each iteration according to a UVD distribution with parameter r_{cloud} around the swarm's p_g.

Parameter Settings

MPSO as described so far introduces a number of new parameters: the number M of swarms, the number of quantum particles N^Q, the exclusion distance r_{excl}, and the quantum cloud radius r_{cloud}. Several guidelines for setting these parameters are provided in [8] and shall be briefly summarized here. Intuitively, there is a relationship between the distance a local optimum shifts, s, and the quantum cloud parameter r_{cloud}. It is expected that these factors are of the same order of magnitude so that a quantum particle might be found close to the new optimum. Previous experiments have shown that $r_{cloud} = 0.5s$ is a good default setting. r_{excl} can be estimated by assuming that all p peaks occupy the same portion of the search space X^d. The radius r_{boa} of the basin of attraction of a peak is then $p \cdot r_{boa}^d = X^d$, or $r_{boa} = X/p^{1/d}$. The exclusion radius r_{excl} is thus set to r_{boa}. The multi-swarm cardinality M can be estimated from the number of optima, p. If possible we would expect that $M > p$ is undesirable since free swarms absorb valuable function evaluations and there is no need to have many more swarms than peaks. Similarly, many fewer swarms than peaks means the multi-swarm is in danger of missing good locations.

Self-adaptation of the Number of Swarms

Because usually the number of peaks is not known beforehand, the number of swarms, M, might be difficult to set. For this reason, in [2], a self-adaptation mechanism for the number of swarms has been proposed. The mechanism for this is quite simple. The multi-swarm will need a new, patrolling swarm, if all current swarms are converging. On the other hand, too many free swarms will absorb function evaluations and one should be removed. If there is more than one free swarm, the choice for removal is arbitrary and the worst of the free swarms is removed, as judged by $f(\mathbf{p}_g)$. This birth/death mechanism removes the need for the anti-convergence operator, and for specifying the multi-swarm cardinality M.

The number of swarms $M(t)$ is then dynamic and at any iteration t given by

$$M(0) = 1$$
$$M(t) = \begin{cases} M(t-1) + 1, & M_{free} = 0 \\ M(t-1) - 1, & M_{free} > n_{excess} \end{cases} \tag{5}$$

where n_{excess} is a parameter specifying the desired number of free swarms, and a free swarm is a swarm whose expansion is larger than r_{conv}. An intuitive choice is to allow just a single free swarm, $n_{excess} = 1$, as one swarm roaming around and searching for new peaks should be sufficient. Setting $n_{excess} = \infty$ means that no swarm can ever be removed and the multi-swarm can only grow.

When the number of swarms is adapted, r_{excl} can also be adapted by assuming that M corresponds to the number of peaks:

$$r_{excl} = \frac{X}{2M^{1/d}} \tag{6}$$

where X is the extent of the search space in each dimension.

Note that this adaptation scheme, by changing the number of swarms, also changes the overall number of particles. In practice, the number of swarms should be bounded. Too many particles will slow down the PSO, as each particle is processed less frequently. But in the empirical tests reported below, no such bound was used.

Particle Conversion

Finally, let us propose another modification of the original MPSO. In the original MPSO, each population has a fixed number of neutral and quantum particles. Intuitively, quantum particles are most useful at or just after an environmental change, where they provide the tracking that a tightly converged swarm cannot perform. Their role during environmentally stable periods is less clear. Thus, we propose here to convert all neutral particles to quantum particles for one iteration immediately after a change has been detected. After this iteration, they are reverted back to neutral. We expect that this mechanism might allow us to reduce the number of permanent quantum particles in a population.

The overall algorithm is summarized in Algorithm 3.

4.3 Speciation-Based PSO

Other than the just-discussed MPSO which uses fixed swarms with a fixed number of particles each, the speciation-based PSO (SPSO) is able to dynamically distribute particles to species. It was inspired by a clearing procedure proposed in [33]. SPSO was developed based on the notion of species [27]. The definition of species depends on a parameter r_s, which denotes the radius measured in Euclidean distance from the center of a species to its boundary. The center of a species, the so-called *species seed*, is always the best-fit particle in the species. All particles that fall within distance r_s from the species seed are classified as the same species.

Algorithm 4 summarizes the steps for determining species seeds [26]. By performing this algorithm at each iteration step, each different species seed can be identified for a different species and the seed's p_i can be used as the p_g for particles belonging to that species accordingly [27, 31].

Algorithm 3 Multi-Swarm

//Initialization
  Begin with a single free swarm, $M = 1$
FOR EACH particle ni
  Randomly initialize $\mathbf{v}_{ni}, \mathbf{x}_{ni} = \mathbf{p}_{ni}$
  Evaluate $f(\mathbf{p}_{ni})$
FOR EACH swarm n
  $\mathbf{p}_{ng} := argmax\{f(\mathbf{p}_{ni})\}$

REPEAT
  // Adapt number of swarms
  IF all swarms have converged THEN
    Generate a new swarm.
  ELSE IF $(M_{free} > n_{excess})$ THEN
    Remove worst swarm.
  FOR EACH swarm n
    // Test for Change
    Evaluate $f(\mathbf{p}_{ng})$.
    IF new value is different from last iteration THEN
      Convert all particles to quantum particles for one iteration.
      Re-evaluate each particle attractor.
      Update swarm attractor.
    FOR EACH particle i of swarm n
      // Update Particle
      Move particle depending on particle type.
      // Update Attractor
      Evaluate $f(\mathbf{x}_{ni})$.
      IF $f(\mathbf{x}_{ni}) > f(\mathbf{p}_{ni})$ THEN
        $\mathbf{p}_{ni} := \mathbf{x}_{ni}$.
      IF $f(\mathbf{x}_{ni}) > f(\mathbf{p}_{ng})$ THEN
        $\mathbf{p}_{ng} := \mathbf{x}_{ni}$
    // Exclusion.
    FOR EACH swarm $m \neq n$
      IF swarm attractor p_{ng} is within r_{excl} of p_{mg} THEN
        IF $f(\mathbf{p}_{ng}) \leq f(\mathbf{p}_{mg})$ THEN
          Re-initialize swarm n
        ELSE
          Re-initialize swarm m
      FOR EACH particle in re-initialized swarm
        Re-evaluate function value.
        Update swarm attractor.
UNTIL number of function evaluations performed $> max$

Algorithm 4 Algorithm for determining species seeds

$//L_{sorted}$ - a list of particles sorted in their decreasing $f(\mathbf{p_i})$ values
$//S$ - a list of dominating particles identified as species seeds
$S = \emptyset$
REPEAT
    Get the best unprocessed $p \in L_{sorted}$
    $found \leftarrow$ FALSE
    FOR all $s \in S$
        IF $d(s,p) \leq r_s$ THEN
            $found \leftarrow$ TRUE
            break
        IF $not found$ THEN
            let $S \leftarrow S \cup \{p\}$
UNTIL reaching the end of L_{sorted}

Basically Algorithm 4 sorts all particles in decreasing order of the fitness values of their personal bests $\mathbf{p_i}$[5]. The species seed set S is initially set to $\emptyset$. All particles' $\mathbf{p_i}$ are checked in turn (from best to least fit) against the species seeds found so far. If a particle's $\mathbf{p_i}$ does not fall within the radius r_s of all the seeds of S, then this particle will become a new seed and be added to S. Figure 3 provides an example to illustrate the working of this algorithm. In this case, applying the algorithm will identify s_1, s_2 and s_3 as the species seeds. Note also that if seeds have their radii overlapped (e.g., s_2 and s_3 here), the first identified seed (such as s_2) will dominate over those less fit seeds in the list L_{sorted}. For example, s_2 dominates s_3; therefore p should belong to the species led by s_2. This has the nice side-effect of helping SPSO to locate the fitter peaks before the less fit ones.

The identified species seeds represent particles that are highly fit and at least distance r_s away from each other. Since a species seed is the best-fit particle's $\mathbf{p_i}$ within a species, all particles belonging to the same species can be made to follow their species seed's $\mathbf{p_i}$ as their leader (i.e., neighborhood best). Each species acts as a separate PSO with all its particles moving according to the conventional particle velocity rules. The sorting of particles, determination of species seeds, and allocation of species seeds as leaders to particles are performed at each iteration, which have the effect of moving particles within the same species to positions that make them even fitter. Because species are formed around different peaks in parallel, species seeds will provide the right guidance for particles to converge towards different peaks that exist on the fitness landscape. Comparing with the multi-swarm concept introduced in the earlier sections, a species in SPSO is equivalent to an individual swarm of the

---

[5] Note that in our previous implementation, the fitness values of current particle positions $\mathbf{x_i}$ were sorted [27, 28]. We changed to use particle's personal best $\mathbf{p_i}$ because it is a more stable point compared with $\mathbf{x_i}$

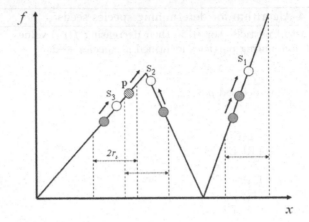

Fig. 3. An example of determining species seeds from a population of particles. s_1, s_2, and s_3 are chosen as the species seeds; for the other particles, arrows indicate which species seed they are attracted to. Note that p falls within the radius r_s of two seeds but follows s_2 as the seed which was identified earlier.

multi-swarm model, but the particles are re-distributed to species dynamically in each iteration, resulting in a variable number of species with different sizes.

To detect whether a change has occurred, SPSO simply re-evaluates the recorded personal best positions of the top t best species seeds at each iteration. A change is considered to have occurred if any of these t re-evaluations is different from its corresponding personal best's recorded fitness. All particles' personal bests are then reset to their associated current positions since these recorded personal bests are outdated.

Species Cap, Quantum Swarm, and Anti-convergence

This section describes several useful techniques incorporated into SPSO to enhance its performance in dynamic environments [28].

Since the algorithm for identifying species seeds favors those seeds with higher fitness values resulting in more particles being allocated to fitter species than to less fit ones, on a multi-modal fitness landscape this may result in too many particles assigned to just a few very best peaks while leaving other lower peaks unoccupied. In order to distribute more evenly the number of particles across different species, a parameter p_{max} can be used to set a maximum number of particles that a species is allowed [30]. This means that only the best-fit p_{max} particles will be allocated as members of a species. Least-fit members that cause the species population to exceed p_{max} are reinitialized as randomly generated new particles into the search space, as a side-effect helping SPSO better explore the search space.

The quantum atom model described in Section 4.1 is also adopted in SPSO, but as described in [28], only triggered by convergence: When the neutral particles belonging to a species converge below a pre-specified threshold (largest distance between any pair of particles), half of the neutral particles are converted to quantum particles around the species seed to form a 'quantum cloud'.

Anti-convergence was first proposed and used as an effective mechanism for global information sharing in the multi-swarm model [8] (see also Section 4.2). This idea can be adopted in SPSO to reduce the number of less fit species [28] and to improve information sharing among species. Anti-convergence is carried out simply by replacing the least-fit species and reinitializing them into the search space at each iteration step. While this seems drastic, note that the speciation procedure usually results in some isolated particles forming species of size 1. Anti-convergence as implemented here randomizes these until they either discover a promising region or are close enough to join a larger species.

4.4 Improving Local Convergence

SPSO is shown to be an effective optimizer for solving static multi-modal problems [31], and with enhancements described in the previous section, it can be used for handling multi-modal problems in a dynamic environment [28]. In order to track moving peaks, SPSO must be able to locate peaks and follow them as closely as possible if they have moved. It is observed that SPSO can consistently locate the majority of the peaks most of the time. However, relocating moved peaks with a satisfactory convergence speed (so as to reduce the offline error) still remains as a challenge. Techniques promoting faster local convergence would be desirable to tackle this problem.

The convergence-triggered strategy as described above and in [28] keeps the particles in a species spread out so that the species will have a better chance to recapture the peak if it has moved again. The downside is that these quantum particles contribute little to the local convergence of the species most of the time. Although changes occur only occasionally, at each iteration quantum particles are generated to form a 'cloud' spreading out around the species seed, rather than being used to converge towards the species attractor, like the neutral particles.

In this study, instead of using the usual quantum 'cloud' approach, for one iteration after a change has been detected, all particles are moved as quantum particles, i.e., to a randomly selected position according to a given distribution. Also, we found that centering the distribution not around the species' seed, but around the center of species seed and particle position, has slight advantages. After this, all particles will move again as neutral particles according to their allocated seed and personal best positions, following the standard PSO velocity update rules. This new variant of SPSO is summarized in Algorithm 5.

Algorithm 5 SPSO with local sampling

//Initialization
FOR EACH particle i
    Randomly initialize $\mathbf{v}_i, \mathbf{x}_i, \mathbf{p}_i = \mathbf{x}_i$
REPEAT
    FOR EACH particle i
        Evaluate $f(\mathbf{x_i})$
        //Update personal best
        IF $f(\mathbf{x_i}) > f(\mathbf{p_i})$ THEN
            $\mathbf{p_i} \leftarrow \mathbf{x_i}$
    IF change is detected THEN
        $\mathbf{p_i} \leftarrow \mathbf{x_i}$
    //Following steps are carried out at each iteration
    Sort particles in descending order of their $\mathbf{p_i}$ fitness values
    Identify species seeds from the above sorted list based on r_s
    $\mathbf{p_i}$ of each species seed is assigned as the leader (neighborhood best) to
    all particles belonging to the same species
    IF $numParticles > p_{max}$ THEN
        Anti-convergence to replace the excess particles with new particles
    Replace particles in the least-fit species by initializing them
    Adjust all particle positions according to Eqs. (1) and (2)
    //Invoke local sampling only if change is detected
    IF change is detected THEN
        Generate a new particle for each particle by local sampling
UNTIL number of function evaluations performed $> max$

---

5 Empirical Results

5.1 Moving Peaks Benchmark and Experimental Setup

For empirical tests, we used the publicly available moving peaks benchmark (MPB) [10]. It consists of p peaks changing in height and width, and moving by a fixed shift length s in random directions every K evaluations. The peaks are constrained to move in a search space of extent X in each of the d dimensions, $[0, X]^d$.

Unless stated otherwise, the parameters have been set as follows: the search space has five dimensions $X^5 = [0, 100]^5$, there are $p = 10$ peaks, the peak heights vary randomly in the interval $[30, 70]$, and the peak width parameters vary randomly within $[1, 12]$. The peaks change position every $K = 5000$ evaluations by a distance of $s = 1$ in a random direction, and their movements are uncorrelated (the MPB coefficient $\lambda = 0$). These parameter settings are summarized in Table 1. They correspond to Scenario 2 in [10] and have been used to facilitate comparison with other published results. The termination condition for each experiment is 100 peak changes, corresponding to 500,000 function evaluations.

Table 1. Standard settings for the moving peaks benchmark

Parameter	Setting
Number of peaks p	10
Number of dimensions d	5
Peak heights	$\in [30, 70]$
Peak widths	$\in [1, 12]$
Evals between changes	5000
Change severity s	1.0
Correlation coefficient λ	0

Scenario 2 actually specifies a family of benchmark functions, since the initial location, initial height and width of the peaks, and their subsequent development is determined by a pseudorandom number generator. All our results are based on averages over 50 runs, where each run uses a different random number seed for the optimization algorithm as well as the MPB. The primary performance measure is the offline error [13] which is the average over, at every point in time, the error of the best solution found since the last change of the environment. This measure is always greater than or equal to 0 and would be 0 for perfect tracking.

Unless specified otherwise, PSO acceleration parameters χ and c are set to standard values 0.72984 and 2.05, respectively. For MPSO, parameters were set according to the guidelines from Section 4.2, and n_{excess} was set to 1. For SPSO, the overall population size was set to 100, p_{max} was set to 10, and $t = 5$ best species seeds were re-evaluated to detect a change.

5.2 Optimal Swarm Size

We begin by determining the optimum neutral swarm size for a single stationary peak in five dimensions. Since canonical PSO does not use the local shape of the function, all spherically symmetric peaks are equivalent. Table 5.2 reports on tests of the neutral subswarm, which is a canonical PSO. The results demonstrate that a small, five-particle swarm is the best hill climber in 5 dimensions. We will therefore set the number of neutral particles in MPSO swarms to five.

5.3 Quantum Particles in MPSO

In previous experiments [8], it was shown that a swarm's particles should be divided equally into neutral and quantum particles performed, and swarms with five neutral and five quantum particles yielded best results. However, in Section 4.2, we have proposed converting the neutral particles to quantum particles for one iteration after a change has been detected, hoping this would allow us to reduce the number of quantum particles in a swarm.

Table 2. Performance of canonical PSO for a single cone, $f(\mathbf{x}) = |\mathbf{x}^* - \mathbf{x}|$. The table shows the best f attained after 2,500 evaluations, averaged over 100 runs with differing initial configurations and cone position

Number of particles	f (std error)
1	65.62 (2.37)
2	6.86(1.00)
3	0.0072 (0.0066)
4	5.43E-8 (4.17)
5	3.64E-10 (1.52E-10)
6	1.18E-9 (3.59E-10)
7	9.40E-9 (1.90E-9)
8	6.64E-8 (1.52E-8)
9	2.87E-7 (5.34E-8)
10	9.13E-7 (1.12E-7)

Table 5.3 shows results for MPSO on the MPB ($p = 10$, $s = 1.0$) for various swarm configurations. A configuration with a neutral particles and b quantum particles is denoted as $(a + b)$. As can be seen, even the $(5 + 0)$ configuration, which has no permanent quantum particles (only the converted particles at the iteration immediately following function change), performs remarkably well. The optimum configuration appears to have just one permanent quantum particle for both values of n_{excess}. Without particle conversion after a change, a $(5 + 1)$ MPSO obtains an offline error of only 2.05 (0.08). This confirms our hope that we might be able to reduce the overall number of particles due to the conversion method. It also demonstrates the importance of the exclusion operator which continuously repositions swarms until they find a peak to settle on. After this happens, swarm diversity is not required and the neutral particles will rapidly converge to the center of the peak. The presence of a small amount of diversity, i.e., just one quantum particle, presumably helps tracking in the few iterations just after the function change.

Comparing the results for $n_{excess} = 1$ and $n_{excess} = 3$, we see that differences are small, but slightly better results are obtained with three rather than one patrolling swarms. So, n_{excess} can be used for fine-tuning if necessary, but $n_{excess} = 1$ seems a good and intuitively justifiable default setting. This was also confirmed in some additional tests with only one peak or 200 peaks, where $n_{excess} = 1$ performed slightly better than $n_{excess} = 3$.

5.4 Quantum Distribution in SPSO

The following experiments look at the influence of the quantum distribution in scenarios with $p = 10$ peaks and with shift severity $s \in \{1, 5\}$.

Table 5.4 provides the results on offline errors using Gaussian, UVD and NUVD distribution respectively. These results are also visualized in Figure 4. The best offline error for Gaussian distribution is 1.73, at $\sigma = 0.3$. The best

Table 3. Variation of average offline error with swarm configuration and excess parameter for MPSO. The data is for 50 runs of MPB, scenario 2, p = 10, and s = 1.0 and with 500,000 evaluations per run

Configuration	oe (std error), $n_{excess} = 1$	oe (std error), $n_{excess} = 3$
5 + 0	1.80 (0.08)	1.85 (0.07)
5 + 1	1.73 (0.08)	1.69 (0.07)
5 + 2	1.85 (0.09)	1.69 (0.06)
5 + 3	1.82 (0.09)	1.83 (0.07)
5 + 4	1.77 (0.07)	1.88 (0.07)
5 + 5	1.85 (0.09)	1.92 (0.08)

Table 4. Offline errors for sampling using UVD and NUVD distribution with radius r_{cloud}, and Gaussian with standard deviation σ (set to r_{cloud}). The shift severity value $s = 1.0$ and $p - 10$

r_{cloud}	Gaussian	UVD	NUVD
0.05	1.86 (0.06)	2.24 (0.07)	2.31 (0.07)
0.1	1.82 (0.06)	1.97 (0.07)	2.02 (0.07)
0.2	1.84 (0.07)	1.72 (0.06)	1.78 (0.04)
0.3	1.73 (0.05)	1.74 (0.07)	1.74 (0.08)
0.4	1.85 (0.06)	1.64 (0.06)	1.73 (0.06)
0.5	2.00 (0.08)	1.77 (0.06)	1.62 (0.05)
0.6	1.94 (0.06)	1.84 (0.08)	1.69 (0.06)
0.7	2.23 (0.06)	1.76 (0.06)	1.66 (0.05)
0.8	2.29 (0.07)	1.79 (0.06)	1.87 (0.07)
0.9	2.42 (0.06)	1.78 (0.05)	1.98 (0.06)
1.0	2.55 (0.06)	1.79 (0.06)	1.89 (0.05)
1.2	2.75 (0.07)	2.01 (0.07)	1.89 (0.05)
1.5	3.07 (0.09)	2.09 (0.06)	1.95 (0.06)
2.0	3.48 (0.08)	2.40 (0.09)	2.12 (0.06)

result for UVD is 1.64 at $r_{cloud} = 0.4$, and for NUVD is 1.62 at $r_{cloud} = 0.5$. The differences between the results of UVD and NUVD are insignificant, but both the best results for UVD (1.64) and NUVD (1.62) are better than the best for Gaussian (1.73). These results show that sampling more frequently closer to the mean is not always beneficial for the purpose of the quantum particles. The reason may be that their task is not local improvement (as, e.g., for mutations), but exploration. In any case, the results are are better than the 1.98 reported previously with the convergence-triggered particle diversification scheme [28].

The scenario with $s = 5.0$ should be more difficult to track than the $s = 1.0$ counterpart. Consequently, the performance values summarized in Table 5.4 report higher offline errors. Note that r_{cloud} values in the range [1.0, 5.0] were

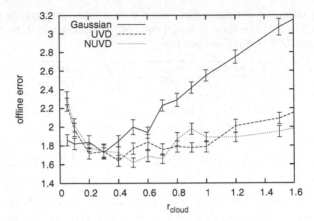

Fig. 4. Offline errors depending on radius r_{cloud} and distribution used for sampling.

Table 5. Offline errors for sampling using Gaussian, UVD, and NUVD distribution, on scenario 2. The shift severity $s = 5.0$ and $p = 10$.

r_{cloud}	Gaussian	UVD	NUVD
1.0	4.53(0.10)	4.43(0.10)	4.53(0.12)
1.5	4.62(0.09)	4.20(0.11)	4.36(0.08)
2.0	5.06(0.09)	4.15(0.08)	4.47(0.09)
2.5	5.54(0.11)	4.28(0.09)	4.56(0.10)
3.0	5.68(0.13)	4.43(0.09)	4.62(0.08)
3.5	6.29(0.11)	4.58(0.13)	4.79(0.11)
4.0	6.40(0.11)	4.61(0.12)	4.97(0.10)
4.5	7.15(0.13)	4.85(0.11)	4.86(0.11)
5.0	7.85(0.17)	4.93(0.12)	5.20(0.11)

used to reflect our assumption that $r_{cloud} \approx s/2$ should be a good default value, which is also confirmed with the empirical data.

Again, UVD performs better than NUVD and Gaussian. The best offline error was obtained by UVD with an offline error 4.15, at $r_{cloud} = 2.0$.

5.5 Adapting the Number of Swarms

Both MPSO and SPSO have mechanisms to adapt the number of swarms over time. While MPSO starts with a single swarm and adds additional swarms as needed, SPSO usually starts with many swarms, slowly converging to the required number of swarms to cover all peaks.

Fig. 5 shows a typical SPSO simulation run for an MPB problem with $p = 10$ and $s = 1$. As expected, SPSO starts with a large number of species seeds, but over iterations this number decreases to a value close to the number of existing peaks (if they are found). Anti-convergence is particularly effective

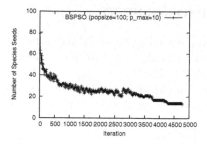

Fig. 5. The number of species seeds is decreasing over the run of a (basic) SPSO model on an MPB problem with ten peaks.

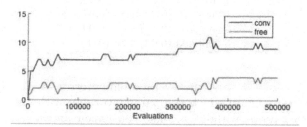

Fig. 6. The number of swarms in MPSO for a ten peak MPB problem.

during the early stage of the run in replacing the least-fit species with new particles that can be better used to explore other parts of the search space. Although it is possible that anti-convergence may remove a species that has already occupied a peak, the chance of this being the best-fit peak is small since the removed species is always the least-fit species. Furthermore, the best-fit peak is most likely watched and tracked by the fittest species.

The number of swarms in a typical MPSO run for a ten-peak MPB problem and with $n_{excess} = 3$ is shown in Figure 6. The figure displays both the number of converged swarms at any stage of the optimization, and the number of free swarms, where a swarm is deemed to be converging if its spatial extent is less than a convergence radius. This distance is dynamically determined by Eq. 6.

In the environment with 200 peaks and $s = 5$, SPSO with a population size of 100 is no longer able to cover all the peaks, and fluctuates between 20 and 30 species. MPSO, although potentially able to add an arbitrary number of sub-populations, converges only slightly higher to around 32 subpopulations. One reason for this convergence is certainly that of the 200 peaks, several smaller peaks are "covered" by higher peaks and thus not visible to the swarm, or they are too close to be regarded as separate peaks. Another reason may be that increasing the number of subpopulations also increases the total number of particles and thus slows down convergence, which leads to a slower convergence of swarms and thus fewer new swarms being spawned.

5.6 Comparing MPSO and SPSO

In this section, we compare MPSO and SPSO on four scenarios: shift severity $s \in \{1, 5\}$ and number of peaks $p \in \{10, 200\}$. Default parameters are used i.e., $r_{cloud} = 0.5s$ (SPSO, MPSO), $n_{excess} = 1$ (MPSO), $p_{max} = p$ (SPSO) and the UVD distribution (MPSO, SPSO). The exclusion radius and number of swarms (MPSO) are set dynamically according to Eqs. 5 and 6.

Table 6. Comparison of MPSO and SPSO on four test scenarios. Offline error and standard error

s	p	MPSO	SPSO
1	10	1.73 (0.08)	1.77 (0.06)
1	200	2.18 (0.02)	2.88 (0.04)
5	10	3.52 (0.11)	4.28 (0.09)
5	200	3.93 (0.03)	4.36 (0.05)

For $s = 1$ and $p = 200$, the average offline error of MPSO was found to be 2.12(0.02), a figure which compares favorably with the best previous adaptive MPSO result without particle conversion and $n_{excess} = 4$ of 2.37 (0.03) [2]. This is also slightly better than the 10 $(5 + 5)$ MPSO result of 2.26 (0.03) of the original version in [8], although it does not require us to specify the number of swarms. For SPSO, the offline error obtained is 2.88 (0.04), i.e., slightly worse. One possible explanation is that SPSO can not adjust the overall number of particles, and that the individual species are becoming too small to successfully track the moving peaks.

Both approaches suffer significantly as the shift severity is increased to $s = 5$, although it seems that MPSO is slightly better in this scenario.

6 Conclusions

This chapter has reviewed the application of particle swarms to dynamic optimization. The canonical PSO algorithm must be modified for good performance in dynamic environments. In particular the problems of outdated memory information and diversity loss must be addressed. Two promising PSO variants are the multi-swarm PSO (MPSO) and the speciation-based PSO (SPSO). They both maintain diversity by dividing the swarm into several subswarms. While MPSO starts with a single swarm and adds additional swarms as needed (all consisting of a predefined number of particles), SPSO has a fixed overall number of particles and dynamically distributes particles to swarms, usually starting with many swarms, slowly converging to the required number of swarms to cover all peaks. Also, they both use a mechanism to maintain diversity within a swarm. We have described MPSO and SPSO in detail and suggested new variations of both. For both approaches, we suggest a reduction of the number of permanent quantum particles and a conversion of all neutral particles to quantum particles for one iteration only after a change has been detected. This approach seems to be more efficient as it speeds up convergence towards local peaks by maximizing its use of neutral particles. On the other hand, it still provides the diversity necessary to recapture a peak after it has moved. Also, we looked at the influence of the probability distribution for the quantum particles on performance.

As the empirical results showed, both variants outperform their previously published originals. Among the examined distributions, the uniform volume distribution outperformed the distributions with higher sampling probability around the species' best. On the other hand, the best possible radius is significantly less than the actual distance a peak shifts. The reason for the benefit of the uniform distribution is probably that the task for the quantum particles is exploration and tracking, rather than fine-tuning, which is better done by the neutral particles. The reason for the smaller radius is probably that as particles generated still possess velocities, placing them too close to the new peak might not be necessarily better, as the particles tend to 'overshoot' the optimum frequently.

Overall, the performance of MPSO and SPSO is comparable on slowly changing problems with fewer peaks, but MPSO seemed to be able to better cope with many peaks, while SPSO seems to be better when the changes are more severe.

Future work will aim at making both approaches more flexible, allowing MPSO to adapt the number of particles within a species, and SPSO to adapt the overall number of particles.

References

1. A. Engelbrecht. *Computational Intelligence.* John Wiley and sons, 2002.
2. T. Blackwell. Particle swarm optimization in dynamic environments. In: S. Yang et al., editors, *Evolutionary Computation in Dynamic and Uncertain Environments.* Springer, Berlin, Germany, 2007.
3. T. M. Blackwell. Particle swarms and population diversity I: Analysis. In: J. Branke, editor, *GECCO Workshop on Evolutionary Algorithms for Dynamic Optimization Problems*, pages 9–13, 2003.
4. T. M. Blackwell. Particle swarms and population diversity II: Experiments. In: J. Branke, editor, *GECCO Workshop on Evolutionary Algorithms for Dynamic Optimization Problems*, pages 14–18, 2003.
5. T. M. Blackwell and P. Bentley. Don't push me! Collision avoiding swarms. In: *Proc. of the 2002 Congress on Evolutionary Computation*, pages 1691–1696. IEEE Press, 2002.
6. T. M. Blackwell and P. J. Bentley. Dynamic search with charged swarms. In: W. B. Langdon et al., editor, *Proc. of the 2002 Genetic and Evolutionary Computation Conference*, pages 19–26. Morgan Kaufmann, 2002.
7. T. M. Blackwell and J. Branke. Multi-swarm optimization in dynamic environments. In: G. R. Raidl, editor, *Applications of Evolutionary Computing*, volume

3005 of *Lecture Notes in Computer Science*, pages 489–500. Springer, Berlin, Germany, 2004.

8. T. M. Blackwell and J. Branke. Multi-swarms, exclusion and anti-convergence in dynamic environments. *IEEE Transactions on Evolutionary Computation*, 10(4):459–472, 2006.

9. T. M. Blackwell. Swarms in dynamic environments. In: E. Cantu-Paz, editor, *Genetic and Evolutionary Computation Conference*, volume 2723 of *Lecture Notes in Computer Science*, pages 1–12. Springer, Berlin, Germany, 2003.

10. J. Branke. The Moving Peaks Benchmark Website. *http://www.aifb.uni-karlsruhe.de/jbr/movpeaks*.

11. J. Branke. *Evolutionary Optimization in Dynamic Environments*. Kluwer, 2001.

12. J. Branke, E. Salihoglu, and S. Uyar. Towards an analysis of dynamic environments. In: Beyer H.-G. et al., editor, *Proc. of the Genetic and Evolutionary Computation Conference, GECCO-2005*, pages 1433–1440. ACM Press, 2005.

13. J. Branke and H. Schmeck. Designing evolutionary algorithms for dynamic optimization problems. *Theory and Application of Evolutionary Computation: Recent Trends*, pages 239–262. Springer, Berlin, 2002. S. Tsutsui and A. Ghosh, editors.

14. A. Carlisle and G. Dozier. Adapting particle swarm optimisation to dynamic environments. In: *Int. Conference on Artificial Intelligence*, pages 429–434. CSREA Press, 2000.

15. M. Clerc. *Particle Swarm Optimization*. ISTE Publishing Company, 2006.

16. M. Clerc and J. Kennedy. The particle swarm: explosion, stability and convergence in a multi-dimensional space. *IEEE Transactions on Evolutionary Computation*, 6:158–73, 2000.

17. J. P. Coelho, P. B. De Moura Oliveira, and J. Boaventura Cunha. Non-linear concentration control system design using a new adaptive particle swarm optimiser. In: *Proc. of the 5th Portuguese Conference on Automatic Control*, pages 132–137, 2002.

18. R. Eberhart and Y. Shi. *Swarm Intelligence*. Morgan Kaufmann, 2001.

19. R. C. Eberhart and Y. Shi. Tracking and optimizing dynamic systems with particle swarms. In: *Proc. of 2001 Congress on Evolutionary Computation*, pages 94–100. IEEE Press, 2001.

20. X. Hu and R.C. Eberhart. Adaptive particle swarm optimisation: detection and response to dynamic systems. In: *Proc. of the 2002 Congress on Evolutionary Computation*, pages 1666–1670. IEEE Press, 2002.

21. S. Janson and M. Middendorf. A hierachical particle swarm optimizer for dynamc optimization problems. In: G. R. Raidl, editor, *Applications of Evolutionary Computing*, volume 3005 of *Lecture Notes in Computer Science*, pages 513–524. Springer, Berlin, Germany, 2004.

22. S. Janson and M. Middendorf. A hierarchical particle swarm optimizer for noisy and dynamic environments. *Genetic Programming and Evolvable Machines*, 7(4):329–354, 2006.

23. Y. Jin and J. Branke. Evolutionary optimization in uncertain environments – a survey. *IEEE Transactions on Evolutionary Computation*, 9(3):303–317, 2005.

24. J. Kennedy. Bare bones particle swarm. In: *Proc. of the 2003 Conference on Evolutionary Computation*, pages 80–87. IEEE Press, 2003.

25. J. Kennedy and R.C. Eberhart. Particle swarm optimization. In: *International Conference on Neural Networks*, pages 1942–1948. IEEE Press, 1995.

26. J.-P. Li, M.E. Balazs, G.T. Parks, and P.J. Clarkson. A species conserving genetic algorithm for multimodal function optimization. *Evolutionary Computation*, 10(3):207–234, 2002.
27. X. Li. Adaptively choosing neighborhood bests in a particle swarm optimizer for multimodal function optimization. In: K. Deb et al., editor, *Proc. of the 6th Genetic and Evolutionary Copmutation Conference*, volume 3102 of *Lecture Notes in Computer Science*, pages 105–116. Springer, Berlin, Germany, 2004.
28. X. Li, J. Branke, and T. Blackwell. Particle swarm with speciation and adaptation in a dynamic environment. In: M. Keijzer et al., editor, *Proc. of the 8th Genetic and Evolutionary Computation Conference*, volume 1, pages 51–58. ACM Press, 2006.
29. G. Pan, Q. Dou, and X. Liu. Performance of two improved particle swarm optimization in dynamic optimization environments. In: *Proc. of the 6th International Conference on Intelligent Systems Design and Applications*, pages 1024–1028. IEEE Press, 2006.
30. D. Parrott and X. Li. A particle swarm model for tracking multiple peaks in a dynamic environment using speciation. In: *Proc. of the 2004 Congress on Evolutionary Computation*, pages 98–103, 2004.
31. D. Parrott and X. Li. Locating and tracking multiple dynamic optima by a particle swarm model using speciation. *IEEE Transactions on Evolutionary Computation*, 10(4):440 – 458, 2006.
32. K. E. Parsopoulos and M.N. Vrahatis. Recent approaches to global optimization problems through particle swarm optimization. *Natural Computing*, pages 235–306, 2002.
33. A. Petrowski. A clearing procedure as a niching method for genetic algorithms. In: *Proc. of the 2003 Conference on Evolutionary Computation*, pages 798–803. IEEE Press, 2003.
34. C. Reynolds. Flocks, herds and schools: a distributed behavioral model. *Computer Graphics*, 21:25–34, 1987.
35. T. Richer and T. Blackwell. The Lévy particle swarm. In: *Proc. of the 2006 Congress on Evolutionary Computation*, pages 808– 815. IEEE Press, 2006.
36. J. Vesterstrom T. Krink and J. Riget. Particle swarm optimisation with spatial particle extension. In: *Proc. of the 2002 Congress on Evolutionary Computation*, pages 1474–1479. IEEE Press, 2002.
37. X. Li and K. H. Dam. Comparing particle swarms for tracking extrema in dynamic environments. In: *Proc. of the 2003 Congress on Evolutionary Computation*, pages 1772–1779. IEEE Press, 2003.
38. X. Zhang et al. Two-stage adaptive PMD compensation in a 10 Gbit/s optical communication system using particle swarm optimization algorithm. *Optics Communications*, 231:233–242, 2004.

An Agent-Based Approach to Self-organized Production

Thomas Seidel[1], Jeanette Hartwig[2], Richard L. Sanders[3], and
Dirk Helbing[4,5,6]

[1] Institute for Transport & Economics
    Dresden University of Technology, Dresden, Germany
    seidel@vwi.tu-dresden.de
[2] SCA Packaging Ltd.
    Wigan, United Kingdom
    jeanette.hartwig@sca.com
[3] Institute of Economic Research
    Lund University, Lund, Sweden
    dick.sanders@ics.lu.se
[4] Institute for Transport & Economics
    Dresden University of Technology, Dresden, Germany
[5] Collegium Budapest–Institute for Advanced Study, Budapest, Hungary
[6] Department of Humanities and Social Sciences
    ETH Zurich, Switzerland
    dhelbing@ethz.ch

Summary. The chapter describes the modeling of a material handling system with the production of individual units in a scheduled order. The units represent the agents in the model and are transported in the system which is abstracted as a directed graph. Since the hindrances of units on their path to the destination can lead to inefficiencies in the production, the blockages of units are to be reduced. Therefore, the units operate in the system by means of local interactions in the conveying elements and indirect interactions based on a measure of possible hindrances. If most of the units behave cooperatively ("socially"), the blockings in the system are reduced.

A simulation based on the model shows the collective behavior of the units in the system. The transport processes in the simulation can be compared with the processes in a real plant, which draws conclusions about the consequences of production based on superordinate planning.

1 Introduction

Since the world is becoming more and more complex, linear models developed in the past are increasingly failing to produce effective management tools. While the forecast for the behavior of systems such as production networks is of crucial interest to those who have to make strategic, planning, and operational decisions within a plant, current approaches are often insufficient to cope with the occuring dynamics.

 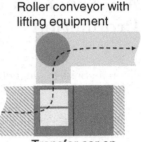

Roller conveyor with lifting equipment

Transfer car on its track system

Fig. 1. Left: Two stacks on a deck position of an automated guided transfer car at an intersection to a roller conveyor [33]. Right: Schematic representation of the element in the left figure. The change of the movement direction on the conveyor is performed by a short roller conveyor with lifting equipment (so-called chain crossover) and is marked by a circle.

In the following, the modeling of complex production networks – in particular of the packaging industry – will be described. Figure 1 shows an element of the transport and buffer system in one of the modeled production plants and its presentation within the simulation software.

When modeling a general transport and buffer system within a multistage production network, one has to consider that the units (i.e. the intermediate products or work in process [43]) leave a machine in the order of their production, but are often scheduled for the next production step in a different order. Thus, a sorting of the units within the system is necessary (see Fig. 2). Since the planning of the production program for all machines is done centrally and in advance, the model has to take into account the given production programs. The model has to describe both, the characteristics of the transport and buffer system and the movement of the production units within the system.

The implications resulting from the required sorting of units in the system are described in Fig. 2. There, the workstation processes four units. The units X_1 and X_2 are authorized to enter the input buffer of the workstation first. Since X_1 and X_2 belong to the same job within the production program, i.e. they are the same type of product, they are accepted at the workstation in any order. In our example, however, both units X_1 and X_2 are blocked by the units Y_1 and Y_2. Thus, Y_1 and Y_2 have to be removed from the lanes first. Y_1 will relocate to the next buffer area on its way to the destination. Y_2 can just be transferred to the same buffer area again and, therefore, has to execute a relocation cycle.

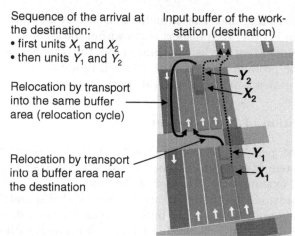

Sequence of the arrival at the destination:
• first units X_1 and X_2
• then units Y_1 and Y_2

Input buffer of the work-station (destination)

Relocation by transport into the same buffer area (relocation cycle)

Relocation by transport into a buffer area near the destination

Y_2
X_2
Y_1
X_1

Fig. 2. To meet the scheduled arrival sequence, relocations may be required to process or finish units in the right order: In this illustration, the units Y_1 and Y_2 have to relocate, as they obstruct the units X_1 and X_2.

1.1 Problem Definition

The aim of this chapter is the modeling of a material handling system as an integral part of a multistage production network. Since the hindrances of units during their transport between the production stages lead to inefficiencies like shortfalls, undesired machine stops, and later completion times, blockages of units are to be reduced. Therefore, the reduction of units' blocking is a central aspect in our model.

Due to the possibility of their mutual blocking, units have to locally *interact* and *react* accordingly. In order to avoid potential blockings, units also have to *act in an anticipatory way* with the help of indirect interactions. This involves avoiding critical buffer areas, in which the material flow is likely to be hindered by a high buffer utilization. Thus, the aim is the design of a transport and buffer system where local und indirect interactions help to reduce mutual hindrances of the units.

In the following, the conveying elements (e.g. roller conveyors or plastic chain conveyors) of material handling systems are called *lanes*. Intersections in factories that are represented by chain transfers or turntables will generally be called *turntables*. The transport and buffer system composed of lanes, turntables, and transport systems (e.g. transfer cars, see Fig. 1) will furthermore be described as a mathematical graph with nodes and directed edges.

The *units* to be transferred are boardstacks, production waste, and auxiliary material such as ink boxes and cutting tools etc. They represent the agents in our model and operate on the nodes and the edges of the graph.

Our model is the basis for the simulation of transport processes in a plant, while a production of the units and their material handling is simulated with given production programs. The ultimate goal is to avoid hindrances to units in the real plant in order to transport them within the system to their destination on time. Simulation can help optimize both, the transport processes given a superordinate production plan of the overall production and the planning itself, if it integrates a model of the transport and buffer system.

1.2 Organization of This Chapter

Within the first part of the chapter, the theoretical model of the transport and buffer system is introduced. The second part addresses applications of the model and examines the behavior of the units.

Section 2 provides an overview of agent-based modeling and its application to the description of transport and buffer systems. In Sect. 3, an overview of the methods used to model the production system is given, without presenting the mathematical and algorithmic details. These are provided in [65].

The mathematical abstraction of the functional interrelationship in a real factory is described in Sect. 4, in particular the illustration of the transport and buffer system as a mathematical graph with nodes and directed edges. Afterwards, Sect. 5 summarizes the treatment of transport processes described in the previous sections.

The implementation of our model in a simulation environment is described in the second part, which begins with Sect. 6. In Sect. 7, we examine the factors contributing to a cooperative behavior of the units. Our contribution concludes with a discussion of the pros and cons of the modeling approach and evaluates the practical relevance for production systems.

2 Relation to the Previous Literature

2.1 Material Handling Elements as Part of Production Networks

Production systems are generally modeled at different levels of aggregation. Frequently, the description of individual material handling elements and their interaction in larger systems is done within the framework of queuing theory [31, 27, 2, 28, 4]. However, the necessary sequence of the entrance of single units into the workstation is difficult to handle. In addition, the analytical and numerical effort increases with the degree of complexity of the system and the level of detail of the elements being described.

Therefore, entire production networks or supply chains are often modeled by means of macroscopic approaches, e.g. fluid-dynamic ones [36, 37, 3]. These do not distinguish individual units, which are rather modeled by event-driven simulations [7, 46, 19]. Other modeling approaches are Petri nets and the max-plus algebra [6, 61, 47, 18]. One interesting feature of the latter is the possibility of analytical calculations, but a disadvantage is the effort required to adapt the description to new or modified setups.

2.2 Agent-Based Modeling of the Transport and Buffer System

Agent-based modeling allows one to describe the complex interactive behavior of many individual units [17, 73, 72, 25]. Wooldridge [74] describes an "intelligent agent" as a computer that has the ability to perform flexible and autonomous actions in a certain environment, in order to achieve its planned goals. Agents show

- *reactive* behavior in relation to the environment,
- *proactive* behavior (by showing initiative and acting anticipatively), and
- *social* (e.g. cooperative) behavior.

These kinds of behavior are realized by the description of an energy and a utility function respectively, that are optimized in a distributed way. Furthermore, such an agent has the ability to forecast its future state or the state of the other implemented agents [16]. But suitable organizational structures and communication strategies are necessary.

A further important aspect of multi-agent systems (MAS) is the environment in which the agents interact. The given production system consisting of the machines, connected by a transport and buffer system, constitutes the environment of the units. In accordance with the classification of Wooldridge [74], the units are embedded into a dynamic and discrete production system affected by coincidences (e.g. machine breakdowns). In particular, the variability of traffic conditions in the lanes and the transfer cars is an important factor influencing the system state.

We have chosen an agent-based approach for the modeling of the units primarily because these agents automatically transfer their behavior to a new layout when the factory is restructured. This makes an agent-based approach very flexible and easy to handle. In contrast, a classical optimization approach must be formulated for a different setup anew, which is generally quite time-consuming.

Although our modeled units show both reactive and cooperative behavior, our agents do not act in a fully autonomous way. They also incorporate a certain degree of central steering, which allows one to integrate our distributed control concept into a hierarchical optimization and production planning. Note that it would, in principle, also be possible to implement a centrally controlled buffer-operating strategy steering the units [32, p. 494], based on hierarchical optimization. For this, one would usually start with the central determination of the optimal arrival sequence at the workstations, which requires us to solve a scheduling problem [49, 52, 50, 58]. Next, one could describe the movement of the units to a workstation as a vehicle routing problem (VRP) [21], which is another central optimization approach. As a result, fixed routes and time windows would be assigned to each unit (see the overview in [13, 55, 51, 20]). Intersecting flows (for example at turntables and chain crossovers) could be steered in the same way as in the control of traffic lights [54, pp. 128]. However, at least for online control, a purely central description of the entire system as

VRP would be unsuited due to the enormous complexity of the solution space, the numerically demanding search algorithm, and the considerable variability of real production processes.

In contrast to the classical optimization approach sketched above, we will in the following propose a distributed control of material flows [53, 42]. This decentralized approach fits an agent-based approach perfectly and has a greater flexibility, robustness, and performance under largely variable conditions such as the ones observed in many production systems with unexpected machine breakdowns, last minute orders, and other surprises.

Note that, besides the units, the material handling elements of the transport and buffer systems can also be treated as agents. Although the lanes do not perform independent actions, they are involved in the interaction processes with the units. The transfer cars are service agents, which react to requests from the units and relocate them. The transport systems, to which the transfer cars are assigned, can again be understood as VRP. An optimal driving strategy for the cars can be found for a given time window by solving this VRP [60, 45, 29, 5, 57].

2.3 Interactions as Basis for "Social" Behavior

In MAS, "social" behavior of the agents is of substantial importance. In the last few years, promising metaheuristics have been developed for the description of interactions leading to cooperative behavior [13, 9, 67]: Ant Colony Optimization (ACO) is an interesting multi-agent approach to the modeling of transportation problems [24, 11, 12, 48, 15, 22]. ACO is motivated by social, self-organizing insects [14, 71, 10, 68, 69] (see Chap. 2 for a detailed introduction). In ACO, the agents show the behavior of ants and move along the edges of a mathematical graph. The goal is the creation of efficient routes between the nodes with the help of distributed optimization [8, 66]. Indirect communication between the social insects is facilitated by *pheromones*, which are deposited along the edges [24, 23]. The feedback via the variable pheromone concentration can trigger an emergent collective behavior of the insects [59, 26].

Another approach is inspired by investigations of interactive pedestrian behavior. The basis of these considerations are models of self-driven many-particle systems [34]. A pedestrian regulates his or her speed and moves purposefully. In addition, all individuals react to other participants according to attractive or repulsive interaction effects ("social forces"), changing their actual speed and direction of motion. Investigations have shown that, for medium pedestrian densities, lanes consisting of pedestrians with the same desired walking direction are formed [35]. In "panic situations", however, the increased excitement of the pedestrians may generate intermittent mutual obstructions: Large noise amplitudes lead to a "freezing by heating effect" [38, 39, 40, 41].

2.4 Ingredients and Properties of the Modeled Transport and Buffer System

As indicated before, our model of the transport and buffer system is based on an agent-based concept. The agents represent units (or material handling elements) that can interact in direct and indirect ways with each other. The direct interaction takes place locally between the units in a lane and is described by the "interaction component" (see Sect. 3.4).

For the indirect interactions, a hindrance coefficient is formulated, which has a similar function as pheromones for social insects. It describes the effect of possible hindrances in a lane (see Sect. 3.2). For the path finding, we use a network algorithm that was developed to solve shortest-path problems [1]. Our algorithm considers the hindrance coefficients of the lanes and the temporal restriction given by the scheduled arrival at the workstation (see Sect. 3.3). The resulting indirect interactions between the units via the hindrance coefficients (stigmergy) support a decentralized optimization, steering the flow of units between the workstations similarly to self-organized traffic light control [53]. In connection with the indirect interaction principle, this path finding induces a movement based on the current situation in the plant. This procedure leads to proactive behavior of the units.

The problems of sorting and obstruction avoidance (see Sect. 7.3) are tackled by the combination of a distributed and a centralized approach (which may, however, be reformulated in terms of a decentralized approach as well): On the central level, all units heading for a single workstation are brought into the correct order (according to the production schedule) by a classical sorting procedure. The units receive their time of scheduled arrival from the assigned destination (e.g. workstation). On the local level, the sorting takes place via reactive local interactions (see Sect. 3.5). If a unit is blocked by another unit in the lane due to a wrong order, the blocking unit is informed and will often decide to leave the lane, thereby clearing the congestion (see "interaction component" in Sect. 3.4).

Thus, the model contains decentralized interactions that enable flexible adjustments to the current situation in the plant. By local interactions, hindrances can be successfully resolved or even avoided. However, the scheduled arrival at the workstation is centrally determined by the production program, which tries to reach a high throughput at low costs (i.e. little waste and few machine setups).

3 Overview of Model Ingredients

In the following subsections, we will describe our model, which can delineate arbitrary networks of material handling elements. The following questions must be answered:

- How do units find their paths in *arbitrary* networks, so that they arrive at the destination at the right time?
- How do the units *interact* with each other, so that they obstruct each other as little as possible and arrive in the *correct order* at the destination?
- How is the future action of the units determined by the goal of avoiding mutual hindrances?

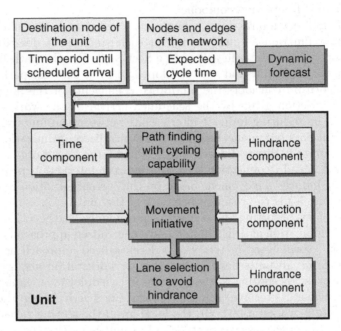

Fig. 3. Overview of modeling methods and their interdependencies.

Within our model, the plant layout is represented by a mathematical graph with nodes and directed edges, which is described in Sect. 4 in more detail.

In general, a transport and buffer system consists of lanes (e.g. conveying elements like roller conveyors) that can be loaded or unloaded by transport systems like automated guided vehicles or transfer cars on tracks. The lanes are usually equipped with engines and are automatically steered by photo sensors. Most of the lanes carry the material in only one direction ("first in, first out"). Therefore, the material flow of the lanes is assumed to be directed in our model.

As in real plants, several lanes that are connected to the same transport systems and follow the same direction of the material flow are combined into a buffer area. The buffer areas are represented by the nodes of the graph and are linked by the transfer systems. An edge of the graph corresponds to a transport connection of two buffer areas. Due to the directedness of the material flow within the lanes, the edges are directed as well.

The model of the transport and buffer system consists of the following procedures, whose interdependencies are illustrated in Fig. 3:

1. **Dynamic Forecast of the Expected Cycle Time and Estimation of the Possible Hindrances in Lanes**: After determining a fast and hindrance-minimal route in the system by path finding, both the expected cycle times, and the possible hindrances of each unit in the lanes are estimated.

2. **Path Finding with Cycling Capability and Automatic Determination of the Hindrance-Minimal Buffer**: A deviation from the fastest path is permitted by an informed search strategy [62], in order to allow moving to a hindrance-minimal buffer area. The path finding routine has the particular capability to generate cycles in the route.

3. **Movement Initiative**: The unit basically decides about its transport and buffering in the lane according to its own priority, but it considers requests of other units in the same lane to move away.

4. **Selection of the Next Lane to Avoid Hindrances**: During the relocation to the next buffer area, the following lane is selected, taking into account obstructions which, at a later time, may result from entering that lane.

Essential aspects and the results of the procedures are described in the following sections (for details see Appendix C in Ref. [65]).

3.1 Dynamic Forecast of the Expected Cycle Time of a Lane

In this section, we describe the operation of the lanes and the basic elements of the transport and buffer system. The lanes are supposed to transport the units and, at the same time, provide a buffering possibility. Therefore, it is *practically* impossible to distinguish between transport and buffering in the system. However, when *modeling* the entire cycle time, a distinction between transport and buffering time is necessary. Whether a unit in a lane is buffered or transported is decided on the basis of the superordinate buffer-operating strategy[7] or the temporal urgency of the unit to arrive at its destination on time (see Sect. 3.4).

Transport within a buffer lane takes place in four steps (see Fig. 4):

1. The transfer car selects a lane for the unit based on the current system conditions, the expected cycle time and the possible hindrances in the alternative lanes. The unit then *enters* that lane and is carried to its last free position to wait there.

---

[7] Buffer-operating strategies can be divided into allocation strategies and movement strategies [32]. The strategy of uniform distribution is a typical allocation strategy and distributes the units uniformly over all buffer areas. The strategy "first in, first out" is a movement strategy and performs the entrance and exit of units in the same order.

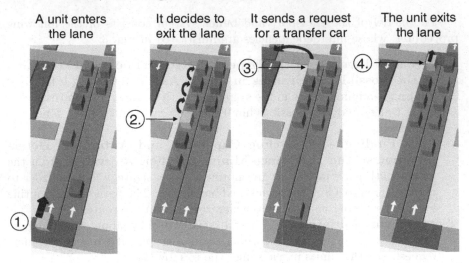

| A unit enters the lane | It decides to exit the lane | It sends a request for a transfer car | The unit exits the lane |

Fig. 4. Subsequent actions of a unit from the arrival in a lane to its removal. The information flow is symbolized by arrows.

2. At a certain time, the unit triggers the transport to the next buffer area by the "movement initiative". If the unit is blocked, it reports its priority to the blocking unit, which reports it to the next one, and so on until the priority message reaches the first unblocked unit. Then, all units will try to free up the lane, given their priorities are lower.
3. If the unit with the highest priority is no longer blocked after the exiting of the hindering units, it releases a *request* for the transfer car.
4. As soon as the transfer car is available, it loads the unit and transfers it to the next lane (*exit of the unit*).

In principle, the time period of a unit in a lane is determined by different influence factors, which include

- actual transport,
- buffering,
- the blockage by other units in the lane, and
- the time until the requested transport to the next lane takes place.

The forecast of the cycle time starts by assuming transport with no waiting (buffering). The actual duration of buffering as well as the best suitable buffering location are determined by a path finding algorithm with cycling capability.

The transport time $t(l)$ on a lane l depends on its level of occupancy. The expected cycle time of a unit entering the lane later on is estimated via the cycle times of the units that entered and exited previously. The estimation of the cycle time takes place by a forecast

- of the period during which the unit is blocked by other units in the lane (the duration from the second to the third step) and
- of the period beginning with the request for the transfer car (duration from the third step to the fourth step).

We have performed this forecast with an adjusted method of double exponential smoothing (ES2), which is an extension of the classical single exponential smoothing (ES1) [44, p. 60]. The ES2 can make a trend prediction (instead of a simple smoothing performed by the ES1).

In the exponential smoothing algorithms, the observed values t_n with $n \in \mathbb{N}$ and the smoothing factor $\alpha \in (0,1)$ are given. The ES1 $\tilde{t}_n^{(1)}$ calculates

$$\tilde{t}_n^{(1)} = \tilde{t}_{n-1}^{(1)} + \alpha \cdot \left(t_n - \tilde{t}_{n-1}^{(1)}\right) \tag{1}$$

with the forecast value $\tilde{t}_{n+1} = \tilde{t}_n^{(1)}$. By means of formula (1) the ES2 $\tilde{t}_n^{(2)}$ is calculated by

$$\tilde{t}_n^{(2)} = \tilde{t}_{n-1}^{(2)} + \alpha \cdot \left(\tilde{t}_n^{(1)} - \tilde{t}_{n-1}^{(2)}\right). \tag{2}$$

The forecast is then

$$\tilde{t}_{n+1} = \tilde{t}_n^{(1)} + \frac{1}{1-\alpha}\left(\tilde{t}_n^{(1)} - \tilde{t}_n^{(2)}\right).$$

Since ES2 is a trend function, it can predict unrealistically small or even negative values when the observed values decrease. We have, therefore, adjusted the procedure in a way that takes into account the minimum possible value $t_{\min}$ by

$$\tilde{t}_{n+1} \geq t_{\min}.$$

Therefore the adjustment of the ES2 has to fulfill the condition

$$\tilde{t}_n^{(1)} + \frac{1}{1-\alpha}\left(\tilde{t}_n^{(1)} - \tilde{t}_n^{\mathrm{korr}}\right) \geq t_{\min}.$$

Considering formula (2), this leads to

$$\tilde{t}_n^{\mathrm{korr}} = \min\left\{\tilde{t}_{n-1}^{\mathrm{korr}} + \alpha \cdot \left(\tilde{t}_n^{(1)} - \tilde{t}_{n-1}^{\mathrm{korr}}\right),\ \tilde{t}_n^{(1)} + (1-\alpha) \cdot \left(\tilde{t}_n^{(1)} - t_{\min}\right)\right\}.$$

Finally, the forecast value is determined by

$$\tilde{t}_{n+1} = \tilde{t}_n^{(1)} + \frac{1}{1-\alpha}\left(\tilde{t}_n^{(1)} - \tilde{t}_n^{\mathrm{korr}}\right). \tag{3}$$

The prediction (3) is performed every given time step or upon entering or exiting the lane.

3.2 Dynamic Forecast of Possible Hindrances in a Lane

A unit has to consider also the possible hindrances in the lanes while searching for a suitable path through the system. On the one hand, hindrances can be considered as blockage if units that are already buffered in the lane block the exit of another unit. On the other hand, the units in the lane can be removed by force, in order to allow for an unhindered transport of a newly entering unit. This will be often connected with relocation cycles.

For the estimation of possible hindrances from the point of view of a new unit entering a lane, the sequence of all units in the lane is compared with the scheduled order of exits. The so-called *removal index* $R(l)$ counts the number of undesired positions of units in a lane l resulting from a comparison of both sequences (see Fig. 5).

Let us assume n units $u_1, \ldots, u_n$ (with $n \in \mathbb{N}$) in a lane and that unit u_i will force R_i buffering units in the same lane to perform undesired removals. Then, the number R_{n+1} for a new entering unit u_{n+1} is determined by comparing the priorities $p(u_i)$ of the units u_i. The priority essentially reflects the urgency of a unit to be transported to its destination (see Sect. 3.4). With increasing priority, the necessity of transport to the next node becomes larger.

The number $R = R_{n+1}$ represents the removal index of the lane and is calculated by means of the following iterative formula:
For the first unit u_1 we have

$$R_1 = 0. \tag{4a}$$

If $p(u_i) < p(u_{n+1})$ for all $i = 1, \ldots, n$, then

$$R_{n+1} = n. \tag{4b}$$

If there exists an index $k \in \{1, \ldots, n\}$ with

$$p(u_k) \geq p(u_{n+1}) \text{ and } p(u_i) < p(u_{n+1}) \text{ for all } i = k+1, \ldots, n,$$

then we set

$$R_{n+1} = R_k + n - k. \tag{4c}$$

If a unit that will potentially enter a lane l has the smallest priority of the units in l and all units in l are descending order sorted by their priority, no relocation cycles will be necessary, and the removal index $R(l)$ for l is zero. However, the more the actual sequence of the units' priorities in a lane deviates from the scheduled sequence in which the units should arrive at their destination, the larger the number of undesired removals. Then the removal index increases. If the unit u_{n+1} that will potentially enter the lane l has the highest priority, then all n currently buffered units in the lane must exit, i.e. the removal index has its maximum value ($R(l) = n$).

Hindrances are not necessarily present in a highly occupied lane if the newly entered unit and the units buffered in the lane have the same destination without any overlap in the expected production times according to the

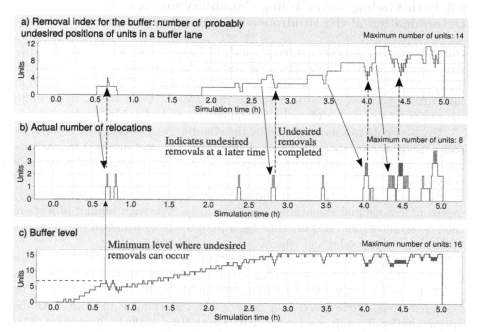

Fig. 5. Comparison of the relocation cycles with the removal index and number of units for a buffer area with two lanes and eight positions each. a) Removal index. b) Amount of transfer lane necessary for the relocation cycles. The causal influences are represented by arrows. c) Buffer level.

schedule. If the actual entrance and scheduled exit sequence differ from each other, there will be a direct correlation between the occupancy of the lane and the removal index. The higher the occupancy of the lane, the more the relocation cycles expected.

According to our approach, each unit tries to find a path as hindrance-free as possible. The possible hindrances within a lane l are described by the so-called *hindrance coefficient* $r(l)$, which is calculated from the removal index $R(l)$ according to formula (4) and the buffer level of l. The hindrance coefficients allow the unit to choose a favorable path to its destination. For this, it is necessary to facilitate an indirect communication between the units. Social insects establish this by the pheromone field [24, 23, 70]. In our model, the hindrance coefficient plays a similar role. It is determined by the occupancy and the removal index of the lane, which influence path finding. In this way, foresighted action of the units is possible.

3.3 Path Finding with Cycling Capability and Automatic Determination of the Hindrance-Minimal Buffer

Our path finding procedure is based on the A*-algorithm, which extends Dijkstra's algorithm by a destination-oriented heuristic (informed search strategy [62, p. 94]). The goal of the path finding is a hindrance-minimal path from the current position of the unit to the destination. Each search must fulfill the temporal restriction given by the scheduled arrival at the workstation.

The best path of a unit is found by the simultaneous consideration of a time and a hindrance component, which defines the temporal urgency to reach its goal on time. In this way, the *hindrance-minimal buffer area* is automatically determined.

All elements of the transport and buffer system predict their cycle times by means of exponential smoothing (see Sect. 3.1). The path finding procedure then determines the transport time for a path to the destination based on the cycle times of its nodes and edges.

In the original A*-algorithm the selection of the shortest path takes place on the basis of weighted edges. Since the estimation of possible hindrances is an integral part of our path finding procedure, the evaluation must be extended from a time-based to a more general assessment of utility: As the optimal buffer, if a unit is expected to arrive on time, is the hindrance-minimal buffer, the evaluation must also consider a hindrance component b_{hin} besides the time component b_{time} (see Fig. 3). Weighting the component b_{hin} with a parameter $\beta_{\text{path}} \geq 0$, the assessment of a node n is based on the function

$$b_{\text{time}}(n) + \beta_{\text{path}} \cdot b_{\text{hin}}(n). \tag{5}$$

Obviously, the orientation at arrival times and the effort to avoid hindrances can contradict each other. Therefore, a balance between both goals must be found. In principle, an urgent unit needs a fast path and will give little consideration to obstructions of other units. However, if a unit has sufficient time to reach its destination, the path finding selects a path minimizing hindrances of units with higher priority. The relative strength of time orientation and hindrance avoidance decides whether the units show cooperative or egoistic behavior.

Our path finding with *cycling capability* permits also paths with cycles. Therefore, a unit can potentially enter the track of a transfer car or a buffer area another time on its path to the destination. This is particularly meaningful if hindrances are considered in addition to the cycle time in the assessment of alternative paths.

The cycling capability allows for a deviation from the fastest path if a hindrance-minimal node can be reached. Sometimes, however, paths with cycles are even time-optimal if they bypass existing hindrances efficiently.

3.4 Movement Initiative

After completed path finding, the "movement initiative" decides about the transport to the next buffer area or buffering in the current lane (see Fig. 6). For this, the priority $p(u)$ of the unit u is determined (see Sect. 3.2), considering the following evaluation components:

1. *Pull component $p_{\text{pull}}(u)$:*
   - The attraction of the destination in order to be on time, considering the predicted cycle time to the unit's destination determined by the path finding procedure (see Sect. 3.3) and the scheduled arrival time (i.e. the temporal urgency as a function of transport time and arrival time),
   - the attraction of the hindrance-minimal buffer area on the path to the destination.
2. *Push component $p_{\text{push}}(u)$:* The repulsive force, if the unit is on a transfer lane that is exclusively intended for transport.
3. *Interaction component $p_{\text{inter}}(u)$:* The interaction with the other units in the lane, taking into account the removal priorities.

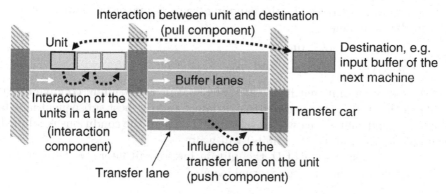

Fig. 6. Illustration of the movement initiative and of the interactions, considering decisions at the exit.

If a unit decides to exit a lane, but is blocked by other units, it informs these about its priority by means of the interaction component. Due to the interaction between the units, the blocking units will react, and exit the lane for the purpose of relocation cycles and a hindrance-free exit of the higher prioritized unit in the lane.

Let us assume n units $u_1, \ldots, u_n$ in the lane l with the unit u_1 at the exit of l. The exit of unit u_1 will be decided depending on the priority $p(u_1)$ and its components $p_{\text{pull}}(u_1)$, $p_{\text{push}}(u_1)$, and $p_{\text{inter}}(u_1)$.

The priority p_{lane} is a transferred priority. If there is a unit u in another lane that would like to relocate to lane l, then $p_{\text{lane}} = p(u) > 0$; otherwise

$p_{\text{lane}} = 0$. The interaction component $p_{\text{inter}}(u_1)$ is calculated by means of

$$p_{\text{inter}}(u_1) = \begin{cases} p_{\text{lane}} & \text{if } n = 1, \\ \max\{p_{\text{lane}}, p_{\text{pull}}(u_2), \ldots, p_{\text{pull}}(u_n)\} & \text{if } n \geq 2, \end{cases}$$

and represents the priorities of the blocked units. The priority $p(u_1)$ is determined by its components according to

$$p(u_1) = \max\{p_{\text{pull}}(u_1) + p_{\text{push}}(u_1), p_{\text{inter}}(u_1)\}.$$

The push and pull components are summarized, as they do not express the hindrance of other units. Since the interaction component $p_{\text{inter}}(u_1)$ represents the priorities of the blocked units, it independently influences the determination of the priority $p(u_1)$.

Depending on the priority and its components in comparison with a given *decision threshold* $D \geq 0$, the unit u_1 will exit, if one of the following conditions is fulfilled:

1. Unit u_1 decides upon its removal if

   $p(u_1) \geq D$ **and** u_1 can enter a subsequent lane.
2. A removal is enforced by the handling transfer lane or by obstructed units in the same lane if

   a) $p_{\text{push}}(u_1) \geq D$ **or**

   b) $p_{\text{inter}}(u_1) > p_{\text{pull}}(u_1)$ **and** $p_{\text{inter}}(u_1) \geq D$.

As soon as a unit that has requested removals is not hindered anymore, it will call the transfer car in order to exit lane l and the best subsequent lane to enter the next buffer area will be selected. If none of the conditions is fulfilled, the unit u_1 (and all blocked units) will wait.

With increasing value of D, the unit decides to exit later, i.e. D represents the reactivity of the unit to external events.[8]

3.5 Hindrance-Avoiding Selection of the Next Lane

If the movement initiative triggers the exit of a unit, a selection of the most suitable lane in the next buffer area is needed. Of course, it would be favorable if the units were buffered in the same sequence in which they are supposed to exit the lane according to the (optimized) production schedule. Thus, the best lane is the one whose sequence of units differs as little as possible from the scheduled order of units to exit. To occupy the transfer car as little as possible, a unit exits its lane only after the following lane has been chosen. This guarantees that there will be empty capacity for the unit in the selected

---

[8] More details about this decision-making process can be found in Appendix C.3.5 of [65, p. 188].

lane when it actually reaches this lane. Altogether, the selection process has to estimate the suitability of the lanes *and* to examine the availability of sufficient buffer capacity.

The selection process is made via *agent-based sorting*, i.e. each unit acts as an autonomous agent and selects the lane independently. There is no central decision maker who performs the sorting. From the point of view of an entering unit, the sequence of the units in the lane is compared with the scheduled order of exits (see Fig. 7). Note, however, that a binary interchange of units is not possible, in contrast to most conventional sorting procedures.

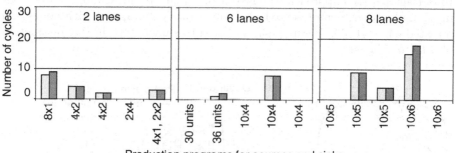

Production programs for sources and sinks

☐ Maximum calculated number of relocation cycles
■ Measured number of relocation cycles with agent-based sorting

Meaning of the program representation by the example:
The program 4x1, 2x2 consists of 4 jobs with always one unit and 2 jobs with 2 units.

Fig. 7. Comparison of the estimated number of necessary relocation cycles and the actually measured number generated by our agent-based sorting.

The selection process consists of the following steps:

1. If the buffer area consists of only a single lane, then this lane is selected.
2. If there is more than one lane in the area and if there exists a lane with a unit of the same job waiting at the last position of that lane for the same workstation, then decide on this lane.
3. Otherwise, the lane selection involves the following components:
   - *Hindrance component* b_{hin}: For all lanes, the suitability of entering is evaluated. This considers all hindrances that the unit has to expect *and* that it may cause. The smaller the evaluation $b_{hin}(l)$ of the lane l, the greater the suitability of that lane. The definition of the function b_{hin} determines the quality of the agent-based sorting. A feasible function b_{hin} is described in [65, p. 191].
   - *Resource component* b_{res}: Since the lanes can have different widths and the units can have different space requirements, the lanes are evaluated with respect to the utilization of the provided buffer capacity. A

possible evaluation function is $b_{\mathrm{res}}(l) = w(l)/w_{\mathrm{unit}}$ with the width $w(l)$ of the lane l and the width w_{unit} of the deciding unit.

Finally, $b(l) = b_{\mathrm{hin}}(l) + b_{\mathrm{res}}(l)$ is calculated for all lanes l of the buffer area. A lane l_0 is selected when it fulfills $b(l_0) = \min\limits_{l:\, w(l) \geq w_{\mathrm{unit}}} b(l)$.

4 Mathematical Abstraction of Interdependencies in the Transport and Buffer System

In our agent-based approach, the layout of a plant of, for example, a packaging manufacturer, is represented by a mathematical graph G with nodes and directed edges [1, 30]. Figure 9 shows the subsequent steps in the abstraction of a factory layout.

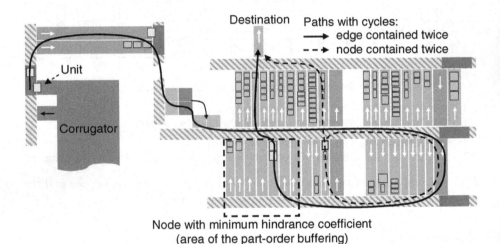

Fig. 8. Illustration of two paths containing the same edge twice (solid line) or the same node twice (dashed line). Cycles are generated by selecting a hindrance-minimal buffer area (i.e. the node with smallest hindrance coefficient).

Note that a unit can use *any* lane of a buffer area on its path (e.g. the area of the part-order buffering in Fig. 8): When searching for a path to the destination, the identification of an available link is more important than the determination of the respective lane. Therefore, homogeneous, parallel lanes are combined into sets of lanes, which form the nodes of the graph G (see Fig. 9b). Since turntables do not have a given direction, each of them is an individual node in the graph G.

The track of a transfer car connects different lanes. Similarly, the dispatch machines (for example pallet inserters) connect several lanes, since they have an input and an output buffer, and the number of units is not changed while

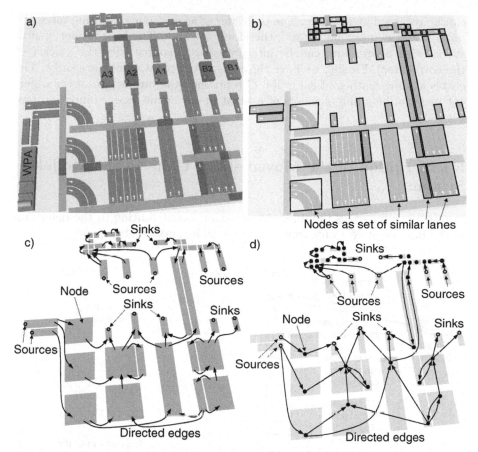

Fig. 9. Different steps in the abstraction of a factory layout. a) Example of a plant layout. b) Logical representation of the buffer system as a set of nodes of a mathematical graph. c) Representation of the transfer cars and the dispatch machines as directed edges of the graph (only a subset of edges is represented). d) Resulting mathematical graph representing the factory layout.

being processed, i.e. the machines transport an individual unit from the input buffer to the output buffer. Both, a dispatch machine and a transfer car track can be regarded as linkage of two nodes, i.e. they form the edges in the graph. Since only directed edges are modeled, a bidirectional transfer (e.g. between two neighboring turntables) is represented by two oppositely directed edges (see Fig. 9c).

The remaining workstations (e.g. the corrugator and the converting machines) form new units, so that the number of the incoming and outgoing units can be different. Therefore, these machines are modeled as sources and sinks respectively, which are typically interconnected in a certain way. The stacker of the corrugator forms stacks of raw boards for the output buffer and can be regarded as a source. The prefeeder of a converting machine takes the units from the input buffer and, therefore, is a sink. The load former creates stacks with boards and can be interpreted as a source (like the stacker of the corrugator). Finally, we have the exits of the plant, which are sinks. The corresponding mathematical graph G represents the interaction of the different material handling elements of a plant in our agent-based simulation (see Fig. 9d).

5 Description of the Movement of Units in the Modeled System

Figure 10 summarizes the different procedures contributing to the definition of the movement of a unit from its source to its sink (destination).

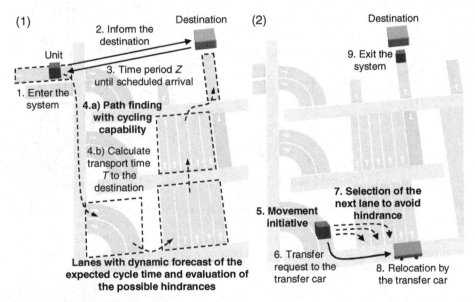

Fig. 10. Causal operational sequence of the movement of units in the modeled transport and buffer system.

1. The movement procedure starts with a unit's leaving a workstation and entering its output buffer. Consequently, the unit enters the transport and buffer system at this time.
2. The unit transfers the information that is has entered the system to the destination.
3. The destination schedules the sequence of the units, according to the production program, in which they can enter the input buffer of the machine.

From this, the destination derives their expected order of arrival, using the expected transport time T from their current position to the destination. The scheduled arrival time Z is then transferred back to the units. If $T \leq Z$, the unit has enough time to arrive at the destination on time and will buffer at a node for a time period $Z - T$. If, however $T < Z$, the unit will probably not arrive on time and will get high priority.

4. Considering time period Z until the scheduled arrival at its destination, the unit determines the best path from the current position to the input buffer of the machine. Since the nodes of the graph abstracting the factory network may represent *several* homogeneous lanes, the path does not specify the lane at this stage. The new lane is selected when the unit is transferred to the next node. After the unit has determined its path, it registers itself at the nodes and edges of the path. If the estimated (partial) cycle times at some nodes or edges change, the unit is informed about this. It may then adapt its expected cycle time or determine a new path.

5. The exit of the unit from the lane is decided according to different criteria (see Fig. 6). Blocked units report their priority to the next and eventually to the first unit in the respective lane, so that the blocking units consider exiting.

6. If the exit was decided and the unit is not blocked, the transfer car receives a request to relocate the unit to the next node.

7. If the next node consists of several lanes, the best lane is selected, considering the lane width and the possible hindrance of units buffered in those lanes.

8. As soon as the relocation is completed, the unit enters a lane of the next node, and the path finding procedure starts again.

9. If the unit arrives at its destination, it leaves the modeled system and finishes its movement procedure.

6 Implementation of the Model in a Simulation Environment

We have also developed a simulation software for our model of the transport and buffer system. Our software consists of different modules, which are controlled over a common program interface (see Fig. 11). The modules are developed as independent software units communicating with each other.

The simulator contains a library for the simulation of discrete events. During the simulation, the behavior of the objects is recorded and passed on to the statistics module, which automatically generates an HTML page with the simulation results. The units and their spatiotemporal dynamics are visualized with the help of a simulation player. Additionally, variables characterizing the units and the production machines can be displayed in separate windows.

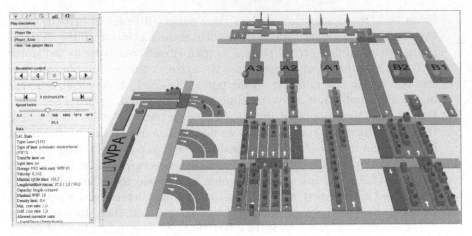

Fig. 11. Program interface representing a model of a packaging plant.

7 Path Finding as Basis for the Interaction of Units

For the following analysis, we will assume the plant layout shown in Fig. 12. Our goal is to study whether our path finding algorithm produces reasonable results. We will start investigating the influence of certain parameters on a *single* unit. Afterwards, we will continue with an analysis of the *interaction* of units belonging to *different* production jobs.

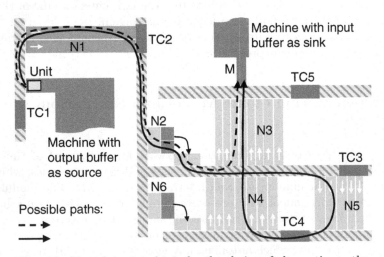

Fig. 12. Plant layout as basis for the choice of alternative paths.

7.1 Deviation from the Fastest Path by Variation of the Weight Parameter β_{path}

According to formula (5), the evaluation function of a node in the path finding algorithm consists of a time component and a hindrance component, which is weighted by the parameter β_{path}. With $\beta_{\text{path}} = 0$, only time-oriented goals are considered during the evaluation. The possibility of the deviation from the fastest path is reached by a parameter $\beta_{\text{path}} > 0$. Let us assume that the time until the scheduled arrival is $Z = 1400$ s and that the model parameters are specified as listed in the following table[9]:

Element n	Cycle time $t(n)$	Hindrance coefficient $r(n)$	Hindrance $c(n)$
Nodes N1, N2, N3, N6	70 s	1.5	105
Node N4	70 s	1.2	84
Node N5	70 s	1.0	70
Input buffer M	70 s	2.0	0
Edges	30 s	2.5	75

Figure 13a shows the simulation results as a function of $\beta_{\text{path}} \geq 0$. The upper diagram shows the evaluation function with its time and hindrance components. The values of these components are only determined by the respective path. They do not vary with the weight β_{path}, while the weighted sum (5), of course, does. Note, however, that the overall evaluation (5) is a smooth function, since the transition from one path to another occurs for a value of β_{path} at which their overall evaluations cross each other (i.e. where they are identical).

In Fig. 13, the transport time T and the cycle time for transport and buffering are represented in the diagram in the middle. The cycle time contains a buffering time if the unit arrives at the destination too early. Therefore, the cycle time is constant and equals $Z = 1400$ s as long as all three paths require lower transport times than given by this value. The transport time T changes only when transitions to another path occur.

The lower diagrams in Fig. 13 show the evaluation of the hindrance expected during the transport. Note that, although the transport time T becomes larger, the overall hindrance reflecting transport *and* buffering decreases at the transition points.

The simulation results presented in Fig. 13 illustrate that the path with the shortest transport time is chosen for small values of β_{path}. Note, however, that nodes N4 and N5 have lower hindrance coefficients than the other ones. Consequently, these nodes become more attractive for larger weights β_{path},

---

[9] The hindrance $c(n)$ of the elements n listed in the table is calculated according to $c(n) = t(n) \cdot r(n)$, except for the input buffer M, for which the hindrance is set to 0, as the units enter the input buffer already in a sorted way.

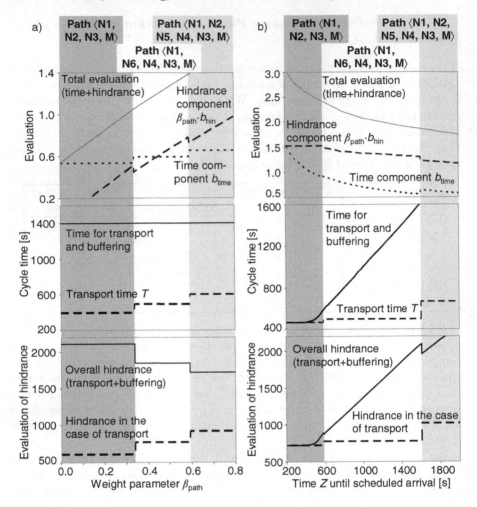

Fig. 13. Influence on the route choice of a) the weight parameter β_{path} and b) the time Z until the scheduled arrival of the unit at the input buffer M of the destination.

and the hindrance component becomes more influential on the path finding procedure as β_{path} increases. Therefore, nodes N4 and N5 are integrated into the path for sufficiently large values of β_{path}, although this leads to increased transport times (see the white and light grey areas).

7.2 Deviation from the Fastest Path by Variation of the Time Period Z Until the Scheduled Arrival at the Destination

Another important factor of path choice is the time period Z until the scheduled arrival at the destination. For small Z, the destination can possibly not be reached on time, so that a fast path is selected. The larger Z, the more likely a hindrance-avoiding path is chosen.[10]

For our analysis, we have assumed the weight $\beta_{path} = 1$ and model parameters according to the following table:

Element n	Cycle time $t(n)$	Hindrance coefficient $r(n)$
Nodes N1, N3, N6	70 s	1.5
Node N2	**140 s**	1.5
Node N4	70 s	**1.2**
Node N5	70 s	**1.0**
Input buffer M	70 s	2.0
Edges	30 s	2.5

Figure 13b shows the simulation results as a function of the time period Z until the scheduled arrival at the destination. For small values of Z, the fastest path is selected with a transport time of $T = 470$ s. A transport via the second path $\langle$N1, N6, N4, N3, M$\rangle$ requires $T = 500$ s. The latter path is only selected when the hindrance during buffering has become significant. The first signs of hindrance effects due to buffering can be seen for $Z > 470$ s.

7.3 Blockage

The blockage of a node is a further variable influencing path choice. We will show that path finding avoids nodes[11] at which the material flows are in danger of being blocked.

If the requested removals from a node are not processed, then the cycle time for transport increases even without any additional buffering at the node, just because of a temporary blockage of the units. Therefore, an emerging blockage due to delayed removals of units may be reflected by large transport cycle times.

If the units of a node are in a highly unsorted order, then the blockage can be caused by frequent relocation cycles binding possible removal capacity. Since the possible obstructions are described by the hindrance coefficient, a high value of this coefficient reflects the danger of blockages.

An *actual* blockage of a node develops either due to delayed removals or due to an increase in the number of hindrances. Therefore, the blockage of a node can be recognized by its large hindrance coefficient and the increasing cycle time for transport (even without additional buffering).

Let us now assume the scheduled arrival time $Z = 1050$ s, the weight $\beta_{path} = 1$ and the parameters listed in the following table:

---

[10] That applies only to the assumption of $\beta_{path} > 0$, so that the hindrance component is actually considered.

[11] A node can be avoided only, if there are alternative paths.

Element n	Cycle time $t(n)$	Hindrance coefficient $r(n)$
Nodes N1, N3	70 s	1.5
Node N2	variable	variable
Node N4	70 s	1.2
Node N5	70 s	1.0
Node N6	140 s	1.8
Input buffer M	70 s	2.0
Edges	30 s	2.5

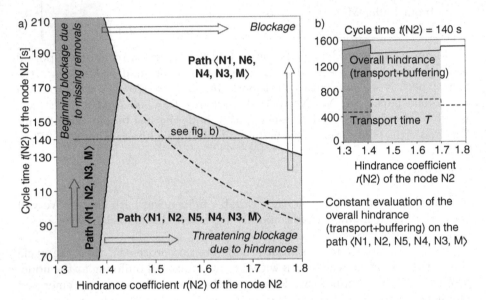

Fig. 14. a) Path finding as a function of cycle time $t(N2)$ and hindrance coefficient $r(N2)$ of node N2. The dashed line corresponds to the isocline of constant hindrance. b) Overall hindrance and transport time T along the isocline $t(N2) = 140$ s as a function of the hindrance coefficient $r(N2)$. One can see that the overall hindrance may drop, while the transport time T increases.

Figure 14a shows the route choice as a function of the cycle time $t(N2)$ and hindrance coefficient $r(N2)$ of node N2. For small hindrance coefficients, the path $\langle$ N1, N2, N3, M $\rangle$ is always selected, since it is fastest.

However, for cycle times $t(N2)$ smaller than 130 s, a transition to the path $\langle$ N1, N2, N5, N4, N3, M $\rangle$ containing the hindrance-minimal node N5 takes place when the coefficient $r(N2)$ grows.

If the regime of blockage of node N2 is finally reached, there is always a transition to the path $\langle$ N1, N6, N4, N3, M $\rangle$, which does not contain the node N2. Thus, the blockage of N2 is recognized and N2 is avoided. Whether or not a node is recognized as blocked, depends on the following:

- the intensity of the blockage, which is determined via the cycle time (without buffering) and the hindrance coefficient of the node,
- parameters such as cycle times or hindrance coefficients characterizing the efficiency of the infrastructure of the plant,
- the time period Z to the scheduled arrival at the destination, and
- the weight parameter β_{path}.

7.4 Cooperative Versus Egoistic Behavior

Let us now simplify the plant layout of Fig. 12 as depicted in Fig. 15. The machine "Corrugator" produces intermediate units (boards) of two jobs. Job j_1 has 100 units with a processing time of 1 min for one unit. The second job j_2 contains four units with a processing time of 25 min per unit. The machine "Conv M" converts[12] the boards to finished packages that leave the plant at the station "Exit". The converting machine is supposed to complete the units belonging to jobs j_1 first, and afterwards the units of job j_2.

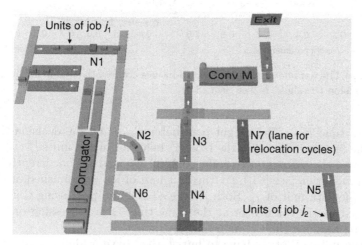

Fig. 15. Plant layout for the simulation of "social" (cooperative) behavior.

If the corrugator produces the jobs j_1 and j_2 in the same order as the one in which they are converted, the minimum cycle time is $T_{\mathrm{min}} = 116.25$ min for all units of job j_1 from the beginning at the corrugator to the finishing of the last unit at the converting machine.[13] The efficiency of the cooperative ("social") behavior can be quantified by comparing the actual cycle time with the

---

[12] Typical converting processes are cutting, printing, creasing, and gluing [63].

[13] A time period of about 16 min is needed for the transport of the units between the two machines, the waiting of the converting machine, and the setup of the machines.

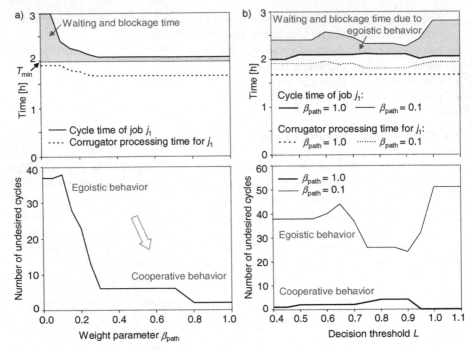

Fig. 16. a) The parameter β_{path} is used to induce cooperative behavior. b) Influence of the decision threshold L (see Sect. 3.4).

minimum time T_{min}. The weight parameter β_{path} and the decision threshold L are the factors influencing the (social) behavior of the units.[14]

Now, let the corrugator execute both jobs j_1 and j_2 concurrently through duplex production (see [64]), i.e. the first unit of job j_2 is finished at the same time as the 25th unit of j_1. Both jobs need 100 min processing time and will be finished by the corrugator at the same time. The processing order at the converting machine shall remain first j_1, then j_2.

As the units of job j_1 have to hurry, they have a high priority to get to their destination. Their "movement initiative" causes the units of job j_2 to consider this. Therefore, the decision threshold has only an influence on the temporal sequence of the decision-making process. Small variations in time can lead to large variations in the number of cycles. Obviously, this can occur only in the regime of egoistic behavior with many cycles.

For small values of β_{path}, the units of job j_2 select the fastest path to the destination without consideration of the possible obstruction of units belong-

---

[14] Actually there is another influencing parameter (see [65, p. 186]), which motivates the unit to move to the hindrance-minimal node of the path. Since only 4 out of 104 units behave egoistically, they are forced to cooperate in the interactions with the others (see Fig. 6).

ing to job j_1. This causes removals[15] of units belonging to job j_2 and leads to many relocation cycles (see Fig. 16a). As the units of j_2 hinder the units of j_1, they show egoistic behavior. However, with increasing β_{path}, the units of j_2 consider hindrances and decide for a buffering at node N5. By this cooperative behavior of the units of job j_2, the hindrance of the units of the more urgent job j_1 is avoided.

7.5 General Characteristics

Although our results were obtained for special plant layouts of a packaging manufacturer, our findings can be extended to more general settings: Under certain conditions, our model allows the units to diverge from the fastest path to the destination. For example, if a late arrival at the destination is expected, the unit possibly decides for a longer path if this facilitates buffering in an area with fewer hindrances.

Furthermore, our algorithm concept allows a unit to detect a substantial increase in the expected transport time for the decided path. If the anticipated transport time becomes too high, the unit possibly determines a better path to its destination and decides to bypass a congested buffer area.

In general, the characteristics of our approach reflect coordinated behavior as it is found in real plants operated by a central and goal-oriented planner considering reasonable prioritizations. The resulting transport of units ensures the feeding of each workstation with the right product in the right quantity at the right point in time [32, 43, 56].

8 Discussion

This chapter has described the modeling of transport and buffer systems based on an arbitrary layout and the movement of the units within that system, considering the scheduled arrival sequence at the workstations.

We have abstracted the material handling system as a mathematical graph with nodes and directed edges. Units (representing products or work in process) are treated as agents, which operate on the graph and interact in direct and indirect ways. The goal of their operation is the avoidance of blockages, which is achieved by indirect interactions minimizing the expected hindrances. So, the reduction of hindrances in the system is facilitated by cooperative behavior of the agents.

Our model contains decentralized decisions, which enable a flexible adjustment to the current situation in the plant. In particular, suitable local interactions can overcome mutual hindrances of the units. A combination of local and centralized procedures facilitates the arrival of the units at their destination in the right sequence. On the one hand, the units are arranged in

---

[15] The removals lead to relocation cycles via N7.

the right order by means of a classical sorting algorithm in accordance with the (optimized) production program. On the other hand, the units are sorted by relocations based on local interactions.

The high flexibility with respect to the restructuring of the layout or changes in the operation of a production system is a major advantage of the agent-based approach. Not only can new scenarios, such as effects of machine breakdowns or allocations of buffer areas to machines, be easily simulated and quickly evaluated, but also the effects of newly installed machines or relocated workstations on the operational procedures can be efficiently analyzed.

The developed simulation software can support planners in packing plants and other manufacturers in creating better production programs. Since the effects of the generated programs are simulated in advance, the planner can test which production programs are expected to cause operational hindrances in the material flows and consequential disturbances in the production, and which ones are not.

Note that the decentralized (local) control procedures of our agent-based approach could be also implemented by means of RFID tags attached to the units. Due to its flexibility in the layout and operation of production systems, this implementation would be applicable to many different plants. Then, various control strategies could be easily implemented by adjusting a few parameters only, thereby determining different operational programs.

Rather than using RFID tags to replace classical bar codes, our proposed implementation would enable more flexible, robust, and efficient decentralized control approaches in complex production systems. In our case, the units would search for a path through the production plant in an autonomous way, considering the scheduled completion times and hindrances in the system. While performing this task, our agents would use rudimentary intelligence and forecasting capabilities. This would both generate and use cooperative ("social") behavior of the individual units.

Acknowledgments

The authors wish to thank Andrew Riddell, John Williams, and Brian Miller of SCA Packaging Ltd. for their great support of this study.

References

1. R. K. Ahuja, T. L. Magnanti, and J. B. Orlin. *Network Flows: Theory, Algorithms, and Applications*. Prentice Hall, Inc. Upper Saddle River, NJ, USA, 1993.
2. T. Altiok. *Performance Analysis of Manufacturing Systems*. Springer, 1997.
3. D. Armbruster, K. Kaneko, and A. S. Mikhailov, editors. *Networks of Interacting Machines: Production Organization in Complex Industrial Systems and Biological Cells*. World Scientific Publishing, 2006.

4. D. Arnold and K. Furmans. *Materialfluss in Logistiksystemen*. Springer, 4th edition, 2005.
5. N. Ascheuer, S. O. Krumke, and J. Rambau. Online dial-a-ride problems: Minimizing the completion time. In: *Proceedings of the 17th Annual Symposium on Theoretical Aspects of Computer Science*, pages 639–650, 2000.
6. F. L. Baccelli, G. Cohen, and G. J. Olsder. *Synchronization and Linearity: An Algebra for Discrete Event Systems*. Wiley, 1992.
7. L. Ben-Naoum, R. Boel, L. Bongaerts, B. De Schutter, Y. Peng, P. Valckenaers, J. Vandewalle, and V. Wertz. Methodologies for discrete event dynamic systems: A survey. *Journal of the American Society for Horticultural Science*, 36(4):3–14, 1995.
8. C. Bertelle, A. Dutot, F. Guinand, and D. Olivier. Distribution of agent based simulation with colored ant algorithm. In: A. Verbraeck and W. Krug, editors, *Proceedings 14th European Simulation Symposium*, pages 766–771. SCS Europe BVBA, 2002.
9. C. Blum and A. Roli. Metaheuristics in combinatorial optimization: Overview and conceptual comparison. *ACM Computing Surveys*, 35:268–308, 2003.
10. E. Bonabeau. Social insect colonies as complex adaptive systems. *Ecosystems*, 1:437–443, 1998.
11. E. Bonabeau. Agent-based modeling: methods and techniques for simulating human systems. *Proceedings of the National Academy of Sciences of the United States of America (PNAS)*, 99 (Suppl. 3):7280–7287, 2002.
12. E. Bonabeau, M. Dorigo, and G. Theraulaz. Inspiration for optimization from social insect behaviour. *Nature*, 406:39–42, 2000.
13. O. Bräysy and M. Gendreau. Vehicle routing problem with time windows, Part I: Route construction and local search algorithms; vehicle routing problem with time windows, Part II: Metaheuristics. *Transportation Science*, 39:104–139, 2005.
14. S. Camazine, J.-L. Deneubourg, N. R. Franks, J. Sneyd, G. Theraulaz, and E. Bonabeau. *Self-Organization in Biological Systems*. Princeton University Press, 2001.
15. D. Corne, M. Dorigo, and F. Glover. *New Ideas in Optimization*. McGraw-Hill, 1999.
16. V. Darley. *Towards a Theory of Autonomous, Optimising Agents*. PhD thesis, Harvard University, Cambridge, 1999.
17. V. Darley, D. Sanders, and P. v. Tessin. An agent-based model of a corrugated box factory: The tradeoff between finished-goodstock and on-time-in-full delivery. In: H. Coelho and B. Espinasse, editors, *Proc. of the Fifth Workshop on Agent-Based Simulation*, pages 17–22, 2004.
18. B. de Schutter and T. van den Boom. Model predictive control for perturbed max-plus-linear systems. *Systems and Control Letters*, 45(1):21–33, 2002.
19. B. De Schutter and T. van den Boom. MPC for discrete-event systems with soft and hard synchronisation constraints. *International Journal of Control*, 76(1):82–94, 2003.
20. J. Desrosiers, F. Soumis, and M. Desrochers. Routing with time-windows by column generation. *Networks*, 14:545–565, 1984.
21. W. Domschke. *Logistik: Rundreisen und Touren*. Oldenbourg, 1997.
22. M. Dorigo. *Optimization, Learning and Natural Algorithms*. PhD thesis, Politecnico di Milano, 1992.

23. M. Dorigo, E. Bonabeau, and G. Theraulaz. Ant algorithms and stigmergy. *Future Generation Computer Systems*, 16:851–871, 2000.

24. M. Dorigo and T. Stützle. *Ant Colony Optimization*. MIT Press, 2004.

25. A. Drogoul and J. Ferber. Multi-agent simulation as a tool for modeling societies: Application to social differentiation in ant colonies. In: *Selected papers from the 4th European Workshop on Modelling Autonomous Agents in a Multi-Agent World, Artificial Social Systems*, pages 3–23. Springer, London, 1992.

26. A. Dussutour, V. Fourcassié, D. Helbing, and J.-L. Deneubourg. Optimal traffic organization in ants under crowded conditions. *Nature*, 428:70–73, 2004.

27. K. Furmans. *Ein Beitrag zur theoretischen Behandlung von Materialflußpuffern in Bediensystemnetzwerken.* Wissenschaftliche Berichte des Institutes für Fördertechnik und Logistiksysteme der Universität Karlsruhe (TH), 1992.

28. K. Furmans. *Bedientheoretische Methoden als Hilfsmittel der Materialflussplanung.* Wissenschaftliche Berichte des Institutes für Fördertechnik und Logistiksysteme der Universität Karlsruhe (TH), 2000.

29. M. Gendreau and J.-Y. Potvin. Dynamic vehicle routing and dispatching. In: T. Crainic and G. Laporte, editors, *Fleet Management and Logistics*, pages 115–126. Kluwer, 1998.

30. C. Godsil and G. Royle. *Algebraic Graph Theory*. Springer, New York, 2001.

31. W. Großeschallau. *Materialflussrechnung: Modelle und Verfahren zur Analyse und Berechnung von Materialflußsystemen.* Springer, Berlin, Heidelberg, 1984.

32. T. Gudehus. *Logistik 2*. Springer, 2000.

33. J. Hartwig. Photographs of despatch machines. 2006.

34. D. Helbing. Traffic and related self-driven many-particle systems. *Review of Modern Physics*, 73:1067–1141, 2001.

35. D. Helbing. The wonderful world of active many-particle systems. *Advances in Solid State Physics*, 41:357–368, 2001.

36. D. Helbing. Modelling supply networks and business cycles as unstable transport phenomena. *New Journal of Physics*, 5:90.1–90.28, 2003.

37. D. Helbing. Modeling and optimization of production processes: Lessons from traffic dynamics. In: G. Radons and R. Neugebauer, editors, *Nonlinear Dynamics of Production Systems*, pages 39–54. Wiley InterScience, 2004.

38. D. Helbing, L. Buzna, A. Johansson, and T. Werner. Self-organized pedestrian crowd dynamics: Experiments, simulations, and design solutions. *Transportation Science*, 39:1–34, 2005.

39. D. Helbing, I. Farkas, and T. Vicsek. Simulating dynamical features of escape panic. *Nature*, 407:487–490, 2000.

40. D. Helbing, I. J. Farkas, P. Molánr, and T. Vicsek. Simulation of pedestrian crowds in normal and evacuation situations. In: M. Schreckenberg and S. D. Sarma, editors, *Pedestrian and Evacuation Dynamics*, pages 21–58. Springer, Berlin, 2002.

41. D. Helbing, I. J. Farkas, and T. Vicsek. Freezing by heating in a driven mesoscopic system. *Phys. Rev. Lett.*, 84(6):1240–1243, 2000.

42. D. Helbing, S. Lämmer, and J.-P. Lebacque. Self-organized control of irregular or perturbed network traffic. In: C. Deissenberg and R. F. Hartl, editors, *Optimal Control and Dynamic Games*, pages 239–274. Springer, Dordrecht, 2005.

43. W. Hopp and M. Spearman. *Factory Physics*. McGraw-Hill/Irwin, 2000.

44. M. Hüttner. *Prognoseverfahren und ihre Anwendung*. Walter de Gruyter, Berlin, 1986.

45. J. J. Jaw, A. R. Odoni, H. N. Psaraftis, and N. H. M. Wilson. Heuristic algorithm for the multi-vehicle advance request dial-a-ride problem with time windows. *Transportation Research Part A: Policy and Practice*, 20B:243–257, 1986.

46. V. S. Kouikoglou and Y. A. Phillis. An exact discrete-event model and control policies for production lines with buffers. *IEEE Transactions on Automatic Control*, 36(5):515–527, 1991.

47. N. Krivulin. The max-plus algebra approach in modelling of queueing networks. In: *Summer Computer Simulation Conference*, pages 485–490. The Society for Computer Simulation, 1996.

48. C. Kube and E. Bonabeau. Cooperative transport by ants and robots. *Robotic and Autonomous Systems*, 30:85–101, 2000.

49. P. R. Kumar. Scheduling queueing networks: Stability, performance analysis and design. In: F. P. Kelly and R. J. Williams, editors, *IMA Volumes in Mathematic and its Applications*, pages 21–70. Springer, New York, 1995.

50. P. R. Kumar and T. I. Seidman. Dynamic instabilities and stabilization methods in distributed real-time scheduling of manufacturing systems. *IEEE Transactions on Automatic Control*, 35(3):289–298, 1990.

51. H. C. Lau, M. Sim, and K. M. Teo. Vehicle routing problem with time windows and a limited number of vehicles. *European Journal of Operational Research*, 148(3):559–569, 2003.

52. S. H. Lu and P. R. Kumar. Distributed scheduling based on due dates and buffer priorities. *IEEE Transactions on Automatic Control*, 36(12):1406–1416, 1991.

53. S. Lämmer, H. Korib, K. Peters, and D. Helbing. Decentralised control of material or traffic flows in networks using phase-synchronisation. *Physica A*, 363:39–47, 2006.

54. G. Mehlhorn, editor. *Der Ingenieurbau*. Ernst & Sohn, 1995.

55. R. Möhring, E. Köhler, E. Gawrilow, and B. Stenzel. Conflict-free real-time AGV routing. In: H. Fleuren, D. den Hertog, and P. Kort, editors, *Operations Research Proceedings 2004: Selected Papers of the Annual International Conference of the GOR*, pages 18–24. Springer, 2005.

56. S. Nahmias. *Production and Operations Analysis*. McGraw-Hill/Irwin, 2001.

57. G. Nikolakopoulou, S. Kortesis, A. Synefaki, and R. Kalfakakou. Real-time vehicle routing: Solution concepts, algorithms and parallel computing strategies. *European Journal of Operational Research*, 152(2):520–527, 2004.

58. J. R. Perkins and P. R. Kumar. Stable distributed, real-time scheduling of flexible manufacturing, assembly, disassembly systems. *IEEE Transactions on Automatic Control*, 34(2):139–148, 1989.

59. K. Peters, A. Johansson, and D. Helbing. Swarm intelligence beyond stigmergy: Traffic optimization in ants. *Künstliche Intelligenz*, 4:11–16, 2005.

60. H. N. Psaraftis. Dynamic programming solution to the single vehicle many-to-many immediate request dial-a-ride problem. *Transportation Science*, 14:130–154, 1980.

61. J. Quadrat, M. Akian, G. Cohen, S. Gaubert, and M. Viot. Max-plus algebra and applications to system theory and optimal control. In: *Proceedings of the International Congress of Mathematicians 1994*, pages 1502–1511. Birkhäuser, 1995.

62. S. J. Russell and P. Norvig. *Artificial Intelligence: A Modern Approach*. Prentice-Hall, 2002.

63. SCA Packaging Aylesford Ltd. *Technische Daten von Verarbeitungsmaschinen*, 2002-2005.

64. SCA Packaging UK & Ireland Ltd., Aylesford. *SCA Packaging Product Knowledge Programme*, 2003.

65. T. Seidel. *Modellierung von Produktionsnetzwerken aus der Perspektive interagierender Transportprozesse*. PhD thesis, Technische Universität Dresden, 2007.

66. D. J. Stilwell and J. S. Bay. Toward the development of a material transport system using swarms of ant-like robots. In: *Robotics and Automation, 1993. Proceedings., 1993 IEEE International Conference on*, pages 766–771, 1993.

67. E.-G. Talbi. A taxonomy of hybrid metaheuristics. *Journal of Heuristics*, 8:541–564, 2002.

68. G. Theraulaz and E. Bonabeau. Coordination in distributed buildings. *Science*, 269:686–688, 1995.

69. G. Theraulaz and E. Bonabeau. Modelling the collective building of complex architectures in social insects with lattice swarms. *Journal of Theoretical Biology*, 177:381–400, 1995.

70. G. Theraulaz and E. Bonabeau. A brief history of stigmergy. *Artificial Life*, 5(2):97–116, 1999.

71. G. Theraulaz, E. Bonabeau, and J.-L. Deneubourg. The mechanisms and rules of coordinated building in social insect. In: C. Detrain, J.-L. Deneubourg, and J. Pasteels, editors, *Information Processing in Social Insects*, pages 309–330. Birkhäuser, Basel, 1999.

72. G. Weiß. *Multiagent Systems: A Modern Approach to Distributed Artificial Intelligence*. MIT Press, 1999.

73. G. Weiß. Agent orientation in software engineering. *The Knowledge Engineering Review*, 16:349–373, 2001.

74. M. Wooldridge. *An Introduction to Multiagent Systems*. Wiley & Sons, 2002.

Organic Computing and Swarm Intelligence

Daniel Merkle, Martin Middendorf, and Alexander Scheidler

Department of Computer Science
University of Leipzig, Leipzig, Germany
{merkle,middendorf,scheidler}@informatik.uni-leipzig.de

Summary. The relations between swarm intelligence and organic computing are discussed in this chapter. The aim of organic computing is to design and study computing systems that consist of many autonomous components and show forms of collective behavior. Such organic computing systems (OC systems) should possess self-x properties (e.g., self-healing, self-managing, self-optimizing), have a decentralized control, and be adaptive to changing requirements of their user. Examples of OC systems are described in this chapter and two case studies are presented that show in detail that OC systems share important properties with social insect colonies and how methods of swarm intelligence can be used to solve problems in organic computing.

1 Introduction

Organic computing is a new field of computer science with the aim to design and understand computing systems that consist of many components and possess so-called self-x properties where "x" stands, for example, for "healing", "managing", "organizing", "optimizing" (e.g., [19, 37, 44]). One idea of organic computing is to use principles of self-organization in order to obtain systems with self-x properties. Computing systems that possess self-x properties and follow such design principles are called *organic computing systems* (*OC systems*).

Social insects, like ants and bees, are a particularly interesting source of inspiration for the design of OC systems. The main reason is that social insect colonies show a complex behavior even though the members of the colony are relatively simple individuals. Most of these behaviors can be called self-organized because there exists neither a central control nor a global work plan. A behavior of the colony which can be seen on a large scale (e.g., nest building or the formation of aggregations of thousands of individuals) but where the individuals act only according to simple rules that use input from their senses only about their local environment is called emergent. Examples of such emergent behavior are nest building of termites ([21]), the formation of bucket brigades during foraging of ants ([2]), the election of a new nest site by a swarm of bees ([26]), and the trail-laying behavior of ants that leads to short paths between their nest and food sources. The latter behavior has inspired the Ant Colony Optimization metaheuristic (see Chap. 2) that is used to solve combinatorial optimization problems ([15]). Another example is the behavior of ants to cluster larvae or corpses of dead ants, which has inspired the design of different clustering algorithms (e.g. [18, 23]).

Since self-organized systems can show emergent effects it is important to understand under which circumstances these effects might occur. Therefore, researchers have developed models that help explain the emergent behavior of social insects. Examples are threshold response models that have been proposed to explain the foraging behavior of ants ([4, 48, 50]); self-synchronization effects in the activity schedule of ants have been explained by models [12], and models have been used to describe the self-organized emergence of aggregations in social insects ([13, 14]). The emergent behavior of social insects and the related biological models have been used by researchers in swarm intelligence to build agent systems or swarm robots and to develop new optimization methods.

In this chapter we discuss connections between swarm intelligence and organic computing. Since organic computing is a relatively new research field it is too early to give an overview on the relations between organic computing and swarm intelligence. Therefore, we shortly present some example applications of organic computing methods from different areas (see Sec. 2). Then we present two case studies that show in some detail how methods of swarm intelligence are connected to problems in organic computing (see Sections 3 and 4).

The first study deals with the control of emergence effects in OC systems. In general, emergent behavior is considered to be an important aspect for OC systems (e.g., [42]). So far researchers have considered mostly the positive aspects of emergent behavior. They try to apply the principles of emergent behavior of natural systems to increase the capabilities of OC systems. Ideally, the autonomous components of an OC system should be able to create a complex emergent behavior without the knowledge of a global plan and without control information that they receive from a central controller. An example of such an emergent behavior that results from self-organization could be the task allocation between the components or the specialization of the components to different tasks by using reconfigurable hardware. Since a certain emergent behavior is typically seen as a desired property of an OC system there exist efforts to develop quantitative measures for emergence ([36]) in order to compare the strength of different OC systems. But, there exists also the potential danger that an OC system might show emergent properties that are unwanted and have not been foreseen when the system was designed. The question that is considered in the first part of this chapter is how an OC system can be controlled such that certain unwanted emergent behaviors can be prevented.

As an example model for this study the emergent clustering behavior of ants is used.

The second study in this chapter is related to the following observation. Typical for OC systems is that their components can adapt to environmental conditions. Hence, even if the components are all equal in principle they will show a slightly different behavior due to individual adaptations. Therefore, it is interesting to investigate what types of emergent effects might occur due to such slight differences in individual behavior. In this context, it is interesting that, as has been observed, ants with slightly different movement behavior can be found most often in different parts of the nest (see [45]). It is discussed in the second part of this chapter what patterns might occur in OC systems with moving components and what are their possible consequences on the behavior of OC systems.

The content of Section 3 and 4 is based on work that has been done within the project "Organization and Control of Self-Organizing Systems in Technical Compounds" within the German Research Foundation (DFG) priority programme on organic computing, and is based on publications [43] and [34].

2 Examples of Organic Computing Systems

In this section we shortly present some examples of the application of organic computing methods to different areas.

One application of organic computing in the field of hardware is the organic computing approach for very fast image processing that was proposed in [28, 16]. This approach is called Marching Pixels (MPs). The basic idea of MP is to use an embedded massively parallel array of pixel processor elements (PEs) to exploit emergent algorithms in order to solve difficult image-processing tasks. Marching pixels are seen as virtual organic units which are born, move, unite, are mutated, leave signatures on the ground, and die on the processor field. The task of the marching pixels is to carry out autonomously image preprocessing tasks, e.g., detection and tracking of moving objects. For the underlying technology there exist plans for future smart sensor chips which will integrate hundreds of millions of transistors. One idea for realizing the MPs approach is to use principles from the pheromone communication of ants to guide the pixels.

An Organic-Computing-inspired System-on-Chip (SoC) architecture which applies self-organization and self-calibration concepts to build reliable SoCs was proposed in [5]. This type of SoC architecture — called Autonomic SoC (ASoC) — provides lower overheads and a broader fault coverage than classical fault-tolerance techniques. The architecture essentially splits the SoC into two logical layers: the functional layer which contains the usual Intellectual Property components or Functional Elements (FEs), and the autonomous layer which consists of Autonomic Elements (AEs) and an interconnect structure among various AEs. FEs are either general-purpose CPUs, memories, on-chip busses, special-purpose processing units, or system and network interfaces as in a conventional design. AEs contain any extensions necessary to improve the reliability of the FE and convert the FE-AE pairs into autonomous units. The feasibility of this approach has been shown for the processing pipeline of a public-domain RISC CPU core.

Traffic systems are another application area of organic computing. It has been proposed to use self-organized inter-vehicle communication to recognize traffic jams [17]. One aim of this communication is to detect the front and the back of a traffic jam. Since the set of cars that forms the front or the back of the traffic jam changes, data about the traffic jam have to be transferred between the cars. Hence, a so-called "Hovering Data Cloud" is formed that is independent of the participating vehicles and stays with the beginning or the end of the traffic jam. This data is used to extract information for other cars to optimize the traffic flow.

Principles of organic computing are also applied to the design of controllers for traffic lights. Traffic flows in urban road traffic networks are changing constantly and on different time scales. Unfortunately, many such changes of traffic flow cannot be foreseen since the change might be due to public events, road works, or sudden incidents. Therefore, traffic light controllers need the ability to adjust quickly to changes in traffic situations and to react reasonably in situations that have not been anticipated in their design process. The Organic Traffic Control (OTC) project (see [41]) develops an adaptive traffic light controller architecture with learning capabilities. The overall architecture is self-optimizing because it is traffic-responsive and can adapt to larger changes in traffic due to a "simulation-based learning" approach.

Collaborating Traffic Lights (CTLs) is a project that tries to exploit the increasing amount of available sensor data about traffic to address the problem of global optimization of traffic flows (see [9]). The idea is to allow the controller or agent of the traffic lights at a junction to decide autonomously on the appropriate phase for the junction. The controller or agent would monitor the level of congestion at the junction under its control based on available sensor data and use this information to decide which action to take. Over time, the agent learns the appropriate action to take given the current level of congestion. In order to achieve optimal system-wide performance, the set of controllers or agents at traffic light junctions in the system should communicate their current status to controllers or agents at neighbouring upstream and downstream junctions.

Mobile robotics is also an area where organic computing methods are clearly useful. For example, the aim of the ORCA project ([35]) is to develop an architecture for mobile autonomous robots that is based on organic computing principles. The aim is to make the robots more reliable and robust. Also the design should eventually be easier compared to classical (fault-tolerant) approaches. The concepts that are used in the project are inspired by the functioning of the human autonomous nervous system and the human immune system. A robot shall be able to continuously monitor its own "health status" and ensure that it is stable and performing its task with optimum performance. In contrast to more classical approaches, error situations are not explicitly described in advance. If a new and unknown deviation from the healthy case is observed by a robot a counteraction is taken first. Based on success or failure, the robot will learn how to handle similar situations and to react faster and more appropriately (similarly to how the human immune system learns to fight against reoccurring infections).

The presented example applications of organic computing show that mobile components play a central role in many OC systems. This is one reason why methods that are inspired by the self-organized behavior of social insects have a great potential for future designs of OC systems. In each of the two following sections one such example of social-insect-inspired methods is discussed in detail.

3 Swarm-Controlled Emergence

The use of principles of self-organization from nature seems to have great potential for the design of OC systems. But recently there have been some concerns that self-organized computing systems might show an emergent behavior that is neither wanted nor intended or foreseen when they were designed. The term negative emergence has been used by some authors for such unwanted emergent behavior (see also [38] for a discussion of emergence). One important research question is how negative emergent behavior of OC systems can be prevented.

One possible approach to prevent negative emergence in OC systems that has been proposed by several researchers is to equip the systems with a so-called observer controller subsystem [40, 46] where a set of observers collects information about the system and based on this information the controllers send control information to the components to influence their behavior.

A potential disadvantage of this approach is that it relies fundamentally on (classical) controllers that send control messages to the components and thus directly restrict the autonomy of the single components of an OC system. Since this is against some central principles of organic computing like self-organization and self-autonomy it is interesting to search for alternative approaches.

In this section we present a new approach to prevent negative emergence in OC systems. This approach was proposed in [34] and is called swarm-controlled emergence. The general idea of swarm-controlled emergence is to add so-called anti-emergence components (anti-components) to an OC system which can prevent the occurrence of certain negative emergence effects. Ideally, the anti-components should not behave too differently from the normal components of the OC system. Then they can still do normal work in the OC system (eventually less efficient though because otherwise all components could become anti-components).

Characteristics of the swarm-controlled emergence approach that differ fundamentally from the observer controller approach are: i) the autonomy of the components (neither of the normal components nor of the anti-components) is not restricted, and ii) it is not necessary to have a corresponding communication structure for delivering control information.

In the following, we describe the principal ideas of the swarm-controlled emergence approach and give proof of concept for a test system and start investigations on special properties of the new approach. The chosen test case is one of the famous examples of emergent behavior of social insects which has several applications in computer science — the clustering behavior of ants (see [23, 24, 25, 27, 29, 30, 31, 32]). Ant clustering has been applied to solve combinatorial problems (e.g., clustering and sorting) and to study emergence in robotics (e.g., [10]).

3.1 Ant Clustering

Ant clustering refers to the behavior of ants to cluster their brood within the nest center (e.g., [18]) or to cluster dead corpses so that they form so-called ant cemeteries (e.g., [8, 10]). Both phenomena can be seen as emergent behavior and have been addressed by simple multi-agent models.

In the ant-clustering model that was proposed in [10] (see also Sect. 4.2 of Chap. 2), several items are distributed in a two-dimensional array of cells (at most one item per cell). Each agent walks randomly within the cell array, picks up an item that it finds with a certain probability, carries it around, and drops it with a certain probability. Formally, the probability p_p of an unladen agent picking up an item is $p_p = (k_1/(k_1 + f))^2$, where f is the fraction of cells in the neighborhood of the agent that are occupied with items and k_1 is a threshold value. Analogously, the probability that a laden agent drops an item if it is on a cell that is not occupied by an item is given by $p_d = (f/(k_2+f))^2$, with k_2 being a threshold value. Several methods for calculating the value f have been proposed. One method is to count how many items have been encountered by the agent during the last few time steps and define f as the fraction of time steps where the agent moved across cells that are occupied by an item. Another way to calculate f is to calculate the fraction of cells in the von Neumann neighborhood of the agent that are occupied with an item.

It was shown that the ant-clustering model fits the clustering behavior of real ants during the organization of cemeteries very well. More complicated patterns, as they occur, for example, in the clustering of brood of the ant *Lepthotorax* ([18]), where different types of brood are clustered in concentric rings, need more elaborated models. This is discussed in more detail in the second half of this chapter.

To illustrate the ant-clustering model and the concept of swarm-controlled emergence some experiments are described for a two-dimensional cell array. The size of the cell array that was used for the experiments is 500×500. In the initial state 2,500 cells (i.e., 1/100 of all cells) are occupied by an item.

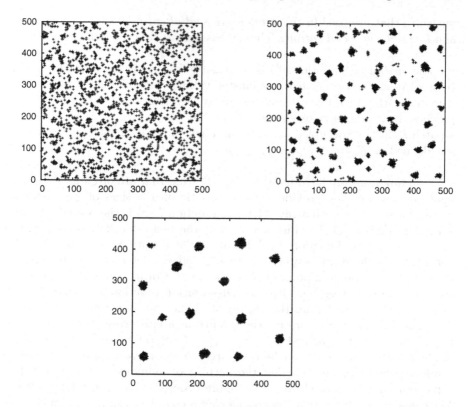

Fig. 1. Cell array with clustering agents: distribution of items after 100,000 (upper left), 1,000,000 (upper right), and 50,000,000 (lower) simulation steps

In all experiments 50 clustering agents were used that move on the cell array. The neighborhood of an agent is defined as the von Neumann neighborhood with radius 10, i.e., all cells for which $d_x + d_y \leq 10$ are said to be in the neighborhood of an agent, where d_x and d_y are the absolute distances of the considered cell to the cell of the agent in the two dimensions. The threshold parameters for the clustering agents were chosen as $k_1 = 0.05$ and $k_2 = 0.03$. The results of a typical test run are shown in Fig. 1. It can be seen that many small clusters have been formed after 100,000 simulation steps. With a growing number of simulation steps the number of clusters becomes smaller and the size of the clusters increases.

3.2 Cluster Validity and Clustering Measures

In order to study the effect of the swarm-controlled emergence approach to clustering it is necessary to measure the quality of a clustering. Since there exist several possibilities to define what a good clustering is several measures for the degree of clustering have been proposed in the literature. The measures that are used in this section are *spatial entropy*, *summary function*, and *hierarchical social entropy*. These measures are described in the following.

The *spatial entropy* ([6, 20]) is a measure for classifying spatial distributions of items according to their cluster validity on different spatial scales. Therefore, the two-dimensional cell array is partitioned into so-called s-patches, i.e., subarrays of size $s \times s$. Let p_I be the fraction of cells in an s-patch I that are occupied by an item. Then the spatial entropy E_s at scale s is defined as $E_s = -\sum_{I \in \{s-patches\}} p_I \log p_I$.

Two functions are introduced in the following that are often used for data analysis because they provide a good statistic on the sizes of gaps between items of a set $\mathcal{R}$ in a cell array. The first function $\hat{F}(r)$ is the probability that a random empty cell has distance r from the nearest cell that is occupied by an item of $\mathcal{R}$. Function $\hat{F}(r)$ is called the Empty Space Function and characterizes the gaps between clusters. Similarly, let $\hat{G}(r)$ be the average distance from a random point of $\mathcal{R}$ to the nearest other point of $\mathcal{R}$. Function $\hat{G}$ is the Nearest-Neighbor Distance Distribution function and characterizes how close the items within the cluster are. The so-called *summary function* $\hat{J}(r)$ (see [33]) is a measure for the quality of a clustering that is based on the $\hat{F}$ and $\hat{G}$ and is defined as $\hat{J}(r) = (1 - \hat{G}(r))/(1 - \hat{F}(r))$. The value of $\hat{J}$ for a pattern of items can be compared to the corresponding value for a random pattern to find whether the pattern of items can be interpreted as a pattern that is more clustered (or more ordered) than a random pattern. Mathematically this is done by considering a complete random point process with intensity λ, for which $F(r) = G(r) = 1 - exp(-\lambda \cdot \pi \cdot r^2)$ and $J(r) = 1$ hold (F, G, and J are the random functions that correspond to $\hat{F}$, $\hat{G}$, and $\hat{J}$). Therefore, a value of $\hat{J}(r) < 1$ indicates a clustered pattern, whereas a value of $\hat{J}(r) > 1$ can be interpreted an an ordered pattern. For the computation of $\hat{J}(r)$ the corresponding function in the R package spatstat ([1]) was used.

The *hierarchical social entropy* measure was proposed in [3]. This measure is based on a hierarchy of clusters that is computed in a bottom-up manner as follows. The bottom of the hierarchy is formed by assigning each item its own cluster (i.e., each cluster at the bottom contains only one item). Then iteratively the two nearest clusters are merged, until there is only one single cluster left. The distance between two clusters can be computed in different ways. In this section the so-called complete linkage method measure is used where the dissimilarity between two clusters is the maximal distance between two arbitrary items of both clusters (the values are between 0 and 1 where 1 means two items have maximal distance). The hierarchy of clusters can be visualized as a dendrogram that shows the agglomeration process of forming the hierarchy as a tree. The leaves of the tree are identified with the items to be clustered. Two nodes of the tree are siblings if their corresponding clusters are agglomerated during the hierarchical clustering. Note that each inner node of the dendrogram corresponds to a taxonomic level h, i.e., the two corresponding

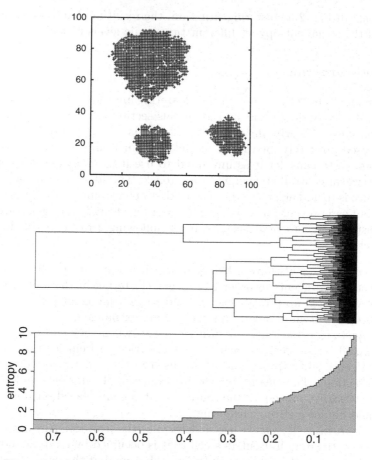

Fig. 2. Hierarchical social entropy; exemplary clustering situation on a 100×100 field (upper), resulting dendrogram (middle), and the value of the hierarchical social entropy $H(\mathcal{R}, h)$ at different taxonomic levels (lower)

clusters c_1 and c_2 have a dissimilarity of $d(c_1, c_2) = h$. For a given taxonomic level h the items are classified by the hierarchical clustering into clusters $\mathcal{C}(h) = \{c_1, \ldots, c_{M(h)}\}$. The hierarchical social entropy of a set of items $\mathcal{R}$ is defined as $\mathcal{S}(\mathcal{R}) = \int_0^\infty H(\mathcal{R}, h)dh$ where $H(\mathcal{R}, h) = -\sum_{i=1}^{M(h)} p_i \log_2(p_i)$ is the social entropy of $\mathcal{R}$ at level h (p_i is the fraction of items in cluster c_i). The hierarchical social entropy enables a total ordering according to the diversity of situations where items are distributed in a (two-dimensional) space. Note that the hierarchical social entropy is scale-invariant and allows us to address the extent of differences between clusters. In [3] the hierarchical social entropy was used to calculate the diversity of a group of robots in space. In this section it is used to distinguish between fine-grained and coarse-grained clustering situations. In Fig. 2 a clustering situation, the resulting dendrogram, and the value of the social entropy at different taxonomic levels is depicted.

3.3 Anti-clustering

The emergent clustering effect in the clustering ant model that was described in Sect. 3.1 is considered in the rest of this section as an unwanted negative emergent effect. Clearly, the clustering effect has a biological relevance and it can be used positively for various applications in computer science. But it is also possible to consider it as unwanted to use it as an example application for the swarm-controlled emergence approach. The idea of swarm-controlled emergence is to add agents to the system that behave similarly to the standard agents but can prevent (or reduce) the clustering effect. These agents are called anti-clustering agents, or $\mathcal{AC}$-agents. In the following, three types of $\mathcal{AC}$-agents are described.

i. *Reverse $\mathcal{AC}$-agents* have a behavior which is opposite to the behavior of the standard agents in order to prevent clustering in the following sense. The values of two probabilities p_p and p_d that an agent picks up an item or drops an item in a certain situation are exchanged.

ii. *Random $\mathcal{AC}$-agents* pick up an item always when they enter a cell that is occupied by an item. If such an agent carries an item it drops it with a fixed probability (probability 0.1 is used in the experiments described in this section). The idea is that the introduction of sufficient randomness in the clustering process in the sense that items are placed on random cells should hinder a strong clustering.

iii. *Deterministic $\mathcal{AC}$-agents* use a deterministic strategy. An agent always picks up the item if it enters a cell that is occupied by an item and always drops the item if no item is in the neighborhood of the current cell.

Experimental results are described in the following, where the influence of the different types of anti-clustering agents is described when they are added to a system with clustering agents.

Experiments with Reverse $\mathcal{AC}$-Agents

The influence of reverse $\mathcal{AC}$-agents is shown in Fig. 3. In the figure the distribution of the items after 1,000,000 simulation steps is shown for different numbers of reverse $\mathcal{AC}$-agents together with 50 clustering agents. It can be seen that it is not possible for the reverse $\mathcal{AC}$-agents to hinder the standard agents from performing a clustering. Even if 100 times more reverse $\mathcal{AC}$-agents are used the item distribution is similar for only standard agents after 1,000,000 simulation steps (see upper right part of Fig. 1). The differences between using 100 and 5000 reverse $\mathcal{AC}$-agents are relatively small. For the latter the clusters are more diffuse but the number of clusters is similar and nearly the same.

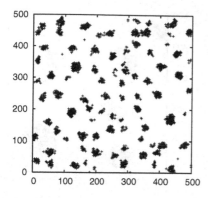 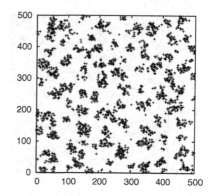

Fig. 3. Distribution of items with different numbers of reverse $\mathcal{AC}$-agents together with 50 standard agents after 1,000,000 steps; 100 reverse $\mathcal{AC}$-agents (left), 5000 reverse $\mathcal{AC}$-agents (right)

This results show that it is not a trivial task to find efficient anti-clustering agents.

The spatial entropy is $E_5 = 6.82$ and $E_5 = 7.39$ for 100 and 5000 reverse $\mathcal{AC}$-agents, together with 50 clustering agents. This is similar to the value of $E_5 = 6.53$ for the case with only clustering agents. The same holds for the hierarchical social entropy, which is $S = 11.67$ and $S = 12.60$ for 100 and 5000 reverse $\mathcal{AC}$-agents, together with 50 clustering agents. For the case with only clustering agents a similar value of $S = 11.88$ is obtained for the hierarchical social entropy.

Experiments with Random $\mathcal{AC}$-Agents

The distribution of items for different numbers of random $\mathcal{AC}$-agents together with 50 clustering agents after 50,000,000 simulation steps are depicted in Fig. 4. The figure shows that no strong clustering occurs for 100 random $\mathcal{AC}$-agents. Hence, a reasonable number of random $\mathcal{AC}$-agents is able to hinder the clustering and they might therefore be attractive candidates for use as anti-clustering agents.

But it can also be seen that using a medium number of 50 random $\mathcal{AC}$-agents even enhances the degree of clustering compared to the cases with fewer (0 or 10) or more 100 random $\mathcal{AC}$-agents. It is an interesting observation that the quality of the clustering can even improve with respect to using only clustering agents when a certain number of random $\mathcal{AC}$-agents is used. Even if the clusters that occur are slightly more diffuse their number is clearly smaller than for the case when no anti-clustering agents are used. In the former case it can be observed that the fraction of clustering agents that carry items is higher in later phases of the simulation runs. The reason is that the slightly diffuse clusters make it more likely that the clustering agents pick up an item. The result is that smaller clusters dissolve faster. It should be noted that this

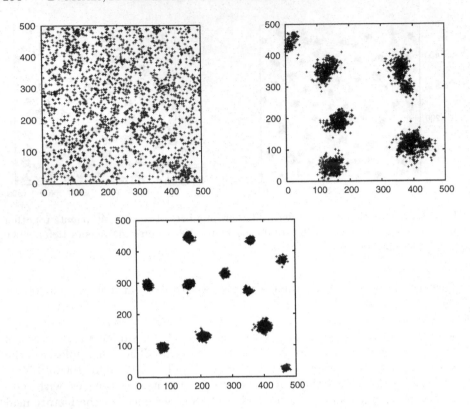

Fig. 4. Distribution of items with different numbers of random $\mathcal{AC}$-agents together with 50 clustering agents after 50,000,000 steps; 100 random $\mathcal{AC}$-agents (upper left), 50 random $\mathcal{AC}$-agents (upper right), 10 random $\mathcal{AC}$-agents (bottom)

finding is very interesting for ant-clustering algorithms in general because it shows the surprising fact that the addition of agents which have the effect of making clusters more diffuse can lead to improved clustering methods.

For the hierarchical social entropy a similar observation can be made. After 50,000,000 simulation steps the hierarchical social entropy is $\mathcal{S} = 13.64$, $\mathcal{S} = 8.07$, $\mathcal{S} = 6.91$, and $\mathcal{S} = 7.76$ for 100, 50, 10, and no random $\mathcal{AC}$-agents, together with 50 clustering agents. Hence, there is no good clustering for 100 and 50 random $\mathcal{AC}$-agents. But for a small number of 10 random $\mathcal{AC}$-agents it can be observed that the clustering quality is better than for the case with only clustering agents.

Clearly, what a good clustering method is depends on how clustering quality is defined and it can not be expected that for each quality measure there exist a suitable number of random $\mathcal{AC}$-agents that can improve the clustering quality when compared to the case with only clustering agents. This is illustrated for the Summary Function $J(r)$. It can be seen in Fig. 5 that the quality of the clustering always decreases with an increasing number of ran-

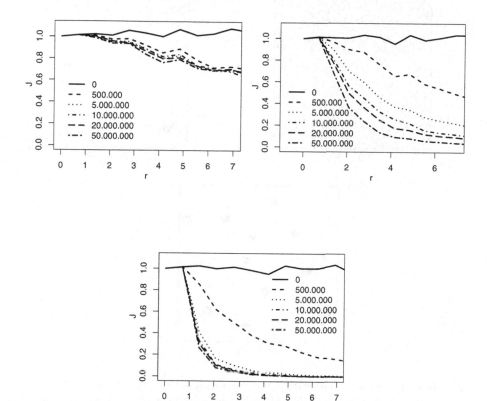

Fig. 5. Summary Function $J(r)$ for random $\mathcal{AC}$-agents together with 50 clustering agents at different time steps for the test runs for which the final item distribution is shown in Fig. 4; 100 random $\mathcal{AC}$-agents (upper left), 50 random $\mathcal{AC}$-agents (upper right), 10 random $\mathcal{AC}$-agents (bottom)

dom $\mathcal{AC}$-agents. The figure shows that no ordered item pattern occurs (the value $J(r)$ is always smaller than 1) and that, as expected, the more the random $\mathcal{AC}$-agents are used, the longer it takes until a certain degree of clustering occurs. This can be seen when comparing the curves for the same number of simulation steps in the three subfigures of Fig. 5. 100 random $\mathcal{AC}$-agents prevent a good clustering. The spatial entropy after 50,000,000 simulation steps is $E_5 = 7.56$, $E_5 = 6.73$, $E_5 = 5.78$, and $E_5 = 5.53$ for 100, 50, 10, and no random $\mathcal{AC}$-agents, together with 50 clustering agents. This shows that the clustering quality decreases with a growing number of random $\mathcal{AC}$-agents.

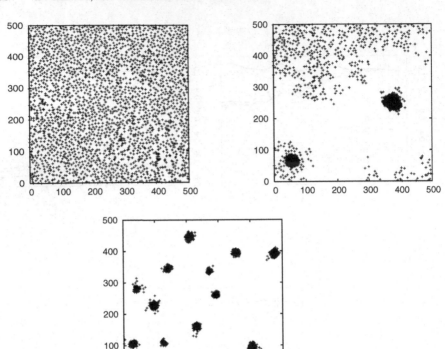

Fig. 6. Distribution of items with deterministic $\mathcal{AC}$-agents together with 50 standard agents after 50,000,000 steps; 50 deterministic $\mathcal{AC}$-agents (upper left), 35 deterministic $\mathcal{AC}$-agents (upper right), 10 deterministic $\mathcal{AC}$-agents (bottom)

Experiments with Deterministic $\mathcal{AC}$-Agents

The most interesting type of $\mathcal{AC}$-agents are the deterministic $\mathcal{AC}$-agents which have a deterministic picking and dropping behavior. Figure 6 shows the item distribution after 50,000,000 simulation steps for different numbers of anti-clustering agents. It can be seen that 50 deterministic $\mathcal{AC}$-agents clearly hinder the clustering agents from performing a successful clustering. This has been confirmed by tests where the initial item distribution was already clustered. In this case the deterministic $\mathcal{AC}$-agents destroyed the clustering successfully. The deterministic $\mathcal{AC}$-agents can be called efficient because they "win" against the same number of clustering agents.

A more detailed analysis is shown in Fig. 7, in which the Summary Function $J(r)$ is depicted. It can be seen in the upper-left subfigure that for 50 deterministic $\mathcal{AC}$-agents ordered patterns occur over time (i.e., no clustering occurs). Note that ordered patterns have been observed only for this type of

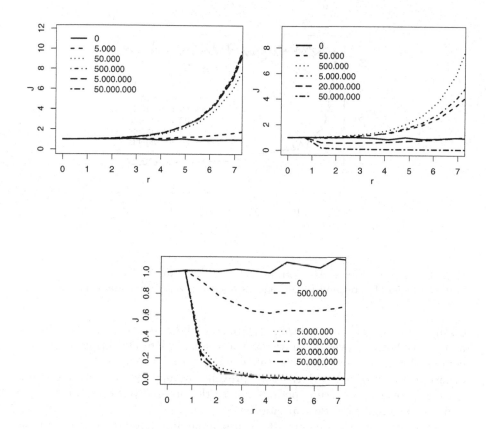

Fig. 7. Summary Function $J(r)$ after different number of time steps for different numbers of deterministic $\mathcal{AC}$-agents together with 50 standard agents for the test runs where the final clustering situation is shown in Fig. 6; 50 deterministic $\mathcal{AC}$-agents (upper left), 35 deterministic $\mathcal{AC}$-agents (upper right), 10 deterministic $\mathcal{AC}$-agents (bottom)

$\mathcal{AC}$-agent. Using only 10 $\mathcal{AC}$-agents cannot hinder the clustering agents from performing their task successfully (see Figs. 6 and 7).

An interesting behavior can be observed for a medium number of 35 deterministic $\mathcal{AC}$-agents. The upper-right subfigure of Figure 7 shows that after 500,000 steps ordered patterns occur, i.e., a clustering is prevented. But for an increasing number of steps, the ordered pattern disappears and a clustering occurs. After 20 million steps the value of the Summary Function at a radius of $r = 2$ is small ($J(2) \approx 0.5$) but for large radiuses the value becomes large (e.g., $J(7) \approx 7$). This is very interesting, because over time a situation

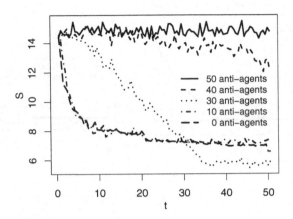

Fig. 8. Hierarchical social entropy S over time for different numbers $k \in$ $0, 10, 30, 40, 50$ of deterministic $\mathcal{AC}$-agents together with 50 standard agents

occurs where there is an ordering of items on a large scale but a clustering on a small scale. Moreover, it shows that a system where clustering agents are combined with anti-emergence agents can show a very complex behavior. The occurrence of ordered patterns that prevent the emergence behavior might exist only over certain time periods before the ordered patterns break down and the emergent behavior can appear.

Similarly as for the random $\mathcal{AC}$-agents, a medium number of deterministic $\mathcal{AC}$-agents even supports the clustering agents in their task. This can be seen in Fig. 8 where the values of the hierarchical social entropy S are shown over time for different numbers of deterministic $\mathcal{AC}$-agents. 50 agents can prevent a clustering ($S > 14.0$); when using no $\mathcal{AC}$-agents a clustering with $S \approx 7$ was achieved. But a medium number of 30 deterministic $\mathcal{AC}$-agents improves the final clustering with $S \approx 6$.

Summarizing Remarks

Altogether the experiments with different types of anti-clustering agents that are added to a system with clustering agents have shown that: i) a not very high number of random $\mathcal{AC}$-agents or deterministic $\mathcal{AC}$-agents is enough to prevent the emergence of clustering, ii) only few $\mathcal{AC}$-agents may help the clustering agents to cluster the items faster, iii) a combination of $\mathcal{AC}$-agents and clustering agents may lead to situations which have an ordered pattern on a large scale and a clustered pattern on a small scale.

4 OC Systems with Moving Components

Emergent patterns that occur when groups of simple agents move is obviously an interesting topic for biology (e.g., [13, 11, 47, 49]) and robotics (e.g., [10]) but it is also interesting for the design of OC systems that consist of moving components or where parts of the system are embedded into moving objects.

In the following we discuss the emergent sorting behavior of simple ant-like moving agents (see [45]). Sorting here means that agents with different behavior can be found most often in different parts of the nest area (or movement area). The starting point of this investigation is a study of Sendova-Franks and Lent [45] where the authors simulate the movement behavior of real ants in their nest. Using different models of moving behavior it was shown that sorting occurs in all models. In each model the moving behavior of the ants differs. The strongest sorting effect occurred when the ants' behavior differs by the strength of an attraction force towards the nest center (centripetal ant model). For the other three models the ants' behavior differs by the maximal turning angle during movement. Ants with a small turning angle tend to keep to the wall once they have collided with it. Thus it was concluded in [45] that the colony center or the wall can play the role of a pivot (or beacon) which appears to be necessary for the sorting.

First, the occurrence of high concentrations of agents in the center of the nest area is considered here as it was observed before in the simulations of [45]. Secondly, some changes to the movement models are described in order to obtain a behavior that is more realistic for organic computing applications. It is discussed which behavioral differences can lead to an emergent sorting behavior for these movement models. Thirdly, a movement model with attraction force is investigated for the case where there is more than one center of attraction. This scenario is interesting, for example, when there exist several service stations for the components of the OC system.

4.1 Cellular Automata Model

The experiments with the moving agents' models that are described in this part have been done with a probabilistic cellular automaton model. The cellular automaton was designed so that it is suitable for approximation of the behavior of the continuous model that was used in [45]. The latter model tries to reflect the situation and the dimensions in a real ant colony (see [43] for more details). The cellular automaton has an array R of cells where the length of a cell corresponds to a length of 0.4 mm. For most experiments $R = \{1, \ldots, 75\} \times \{1, \ldots, 50\}$ was used. The neighborhood of a cell (x_0, y_0) are all cells $(x_0 + x, y_0 + y)$ with $x^2 + y^2 \leq 13$. The body of an agent in the cellular automaton consists of 21 cells arranged in a circle. Formally, an agent at position (x_0, y_0) occupies all cells $(x_0 + x, y_0 + y)$ with $x^2 + y^2 \leq 5$. Each agent i has an internal parameter $0 \leq \mu_i \leq 1$ that influences its moving behavior and an actual direction of moving $0 \leq \alpha_i < 2\pi$. For each agent the probabilities of moving to one of the directly neighbored cells (i.e., the Moore neighborhood $n \in N_M = \{(-1, -1), \ldots, (1, 1)\}$) are calculated. In order for an agent to be able to move, all cells of the new place must be free (and within the array R).

In the cellular automaton model the agents move at each time step in random order so that each agent moves at most one cell per time step. Since an agent can move only to discrete positions it might not be possible for an agent to move exactly in direction α. Therefore, an agent has a probabilistic movement behavior where α is its expected direction. Similarly, the expected velocity is 0.3 mm/s $= 0.75$ cells/time step on average when considering a free run of an agent with no obstacles.

4.2 Movement Models

In this section we describe several movement models for agents. The first two models were proposed in [45] to model the movement behavior of real ants.

Avoiding Ant Model. In this model the agents do a correlated random walk. If unobstructed, i.e., an agent does not collide with a nestmate or the nest wall, the movement is as follows. The turning angle Θ_i is chosen randomly according to a uniform distribution between $-\Theta_i$ and Θ_i. The maximal turning angle Θ_i is different for all agents and depends on their individual parameter μ_i: $\Theta_i = (1 - \mu_i)\Theta^0 + \mu_i\Theta^1$. In [45] the standard values were $\Theta^0 = 60$ and $\Theta^1 = 15$. If the nest wall or a nestmate is in the sensing range, the agent will not move but only change its moving angle. In this case it avoids the obstacle explicitly by turning in one direction until it can move again. To define the turning direction assume that agent i collides with agent j. The sign of the scalar product between the vector that is perpendicular to the vector of the moving direction of agent i and the vector from the center of agent i to the center of agent j determines the direction of turning: $\Theta_i \leftarrow \text{sign}((- \sin \alpha_i, \cos \alpha_i) \cdot (x_j - x_i, y_j - y_i)) \cdot r$ where r is chosen randomly from a uniform distribution in the interval $(0, \Theta_i)$. A collision with the nest wall is handled analogously.

Centripetal Ant Model. It is very likely that agents have the ability to detect gradients in gas (CO_2) or pheromone concentrations [39]. Since the concentration of the gas is maximal in the center region of the nest where the brood is located [7], this could give the agent a chance to estimate the direction to the center. In the Centripetal Ant Model this is used to establish an attraction force towards the center of the nest. This attraction is different for different agents and depends on their internal parameter μ_i. For the calculation of the moving behavior a modified model from the clinotaxis model from [22] is used: $\Theta_i \leftarrow p_u\chi_i + p_b\tau_i \cdot (1 - \cos(\phi_i))/2$ where ϕ_i is the angle between the actual moving direction α_i and the vector towards the center of the nest. The values of p_u and p_b are randomly chosen from $\{-1, 1\}$ and they determine the direction of turning. The turning behavior depends on ϕ_i and the larger this angle

Fig. 9. Effect of different values of the internal parameter μ_i on the turning behavior in the Centripetal Ant Model; Z is the center of the nest; (left) for large μ_i there is only a slight difference between moving from or to the center; (right) for small μ_i the turning angle becomes significantly smaller the larger the angle between actual moving direction and the vector to the center

the more the agent will turn. The parameters χ_i and τ_i depend on the internal parameter μ_i of the agent: $\chi_i \leftarrow (1 - \mu_i)\chi^0 + \mu_i\chi^1$ and $\tau_i \leftarrow (1 - \mu_i)\tau^0 + \mu_i\tau^1$ with $\chi^0 = 0°$, $\chi^1 = 15°$, $\tau^0 = 30°$, and $\tau^1 = 0°$. Agents with larger μ_i will not be that much affected by their ϕ_i as agents with small μ_i (see Fig. 9). Therefore, for the agents with small μ_i the attraction to the colony center is larger than for agents with large μ_i.

It was shown in [43] that when the agents move according to the Centripetal Ant Model there occurs a cluster of non-moving agents in the center of the nest. Obviously, such a situation should not occur in OC systems. Therefore, the following variation of the movement models has been introduced. This model is also simple but the agents try to avoid a situation where they get stuck.

Model with Repulsive Behavior (Repulsive Model). The Centripetal Ant Model is slightly changed by modifying the agents' behavior in the case of a collision with a nestmate or the nest wall. In this case the agent turns according to the Avoiding Ant Model. Otherwise the moving behavior remains as in the Centripetal Ant Model. Hence, in the Repulsive Model the turning behavior of the agent is different for different situations.

For OC systems with moving components the question emerges whether small differences in speed or activity of the components can lead to a sorting effect. Differences in speed or activity can obviously occur when different types of components are used in such systems. But they might also occur when the components have power supplies of different quality or some components are loaded and carry items whereas other components are unloaded. An interesting question is whether differences in speed or activity between the agents can lead to a sorting effect even for very simple types of ant-like movement behavior. To investigate this two new movement models have been introduced that are described in the following.

Model with Speed Differences (Speed Model). In this model the agents have different velocities. The velocities are equidistantly distributed between ν_0,

$0 < \nu_0 < 1$, and 1, i.e., $\nu_i = \nu_0 + (i-1)(1-\nu_0)/(n-1)$ for $i \in \{1, \ldots, n\}$. Note that agent i moves ν_i times as fast as agent n. The position of the agents is now updated according to the formulas $x_i \leftarrow x_i + \nu_i \cdot \delta \cdot \cos \alpha_i$ and $y_i \leftarrow y_i + \nu_i \cdot \delta \cdot \sin \alpha_i$. The different velocities are realized in the stochastic cellular automaton such that agent i has expected velocity $\nu_i \cdot 0.3$ mm/s (recall that 0.3 mm/s is approximately the speed of a real ant). The movement and turning behavior are the same for all agents and do not depend on their internal parameter μ_i. All agents move and turn like an agent in the Centripetal Model with $\mu_i = 0$.

Model with Activity Differences (Activity Model). The Activity Model is similar to the Speed Model. The difference is that the agents have not only different velocities but also different turning behaviors. Similarly to the velocities, the turning angle is scaled by ν_i. Formally, if agent i can move, its turning angle is determined according to $\alpha_i \leftarrow \alpha_i + \nu_i \cdot \Theta_i$, with Θ_i calculated as in the Centripetal Ant Model. If the agent is obstructed, the turning behavior is defined as in the Avoiding Ant Model.

4.3 Experiments with Simple Environments

If not stated otherwise all experiments that are described in the following have been done over 100,000 time steps with 40 agents. The colony center is the point $Z = (35, 25)$. The distance of agent i from the colony center is computed as $r_i(t) = d((x_i \cdot 0.4 - 0.2, y_i \cdot 0.4 - 0.2), (15, 10))$ where (x_i, y_i) is the center of agent i in time step t. The distance r_i of agent i from the colony center is measured every 100 time steps. For a given time step t the mean distance $r_i^{\varnothing}(t)$ of agent i from the center is the average over all measured distances up to t. Let $r_i^{\varnothing} = r_i^{\varnothing}(100000)$ be the average distance measured over all time steps.

As in [45], Pearson's correlation coefficient k is used to measure the correlation of parameter μ_i and the mean distance of an agent from the nest center. A high value (≈ 1) for $k(t)$ indicates a strong correlation. As a second measure the slope of the linear relationship between the mean agent distance from the nest center and the internal parameter μ_i is measured. Given the mean distances $r_i^{\varnothing}(t)$ for all agents $i = 1, \ldots n$, the linear regression determines values $\alpha(t)$ and $\beta(t)$ such that the sum $\sum_{i=1\ldots n}(r_i^{\varnothing}(t) - (\alpha(t) + \beta(t)\mu_i))^2$ is minimized. The slope $\beta(t)$ can be seen as a measure for the degree of sortedness of the agents.

To depict the spatial distribution of the agents a similar strategy is used as proposed in [45]. The nest is divided into $15 \times 10 = 150$ squares and the agents are divided into five groups depending on their internal parameter $\mu_i : [0.0, 0.2), [0.2, 0.4), [0.4, 0.6), [0.6, 0.8), [0.8, 1.0)$. For each of these intervals the number of agents in every square was counted and divided by the total number of agents in the group. To investigate how active the agents are in different areas of the nest it was counted for every cell how often an agent enters that cell (measured over all time steps).

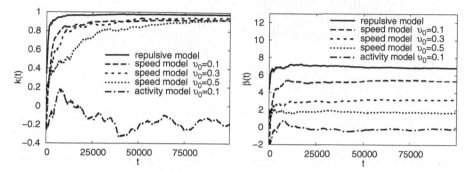

Fig. 10. Changes of correlation coefficient $k(t)$ (left) and slope $\beta(t)$ (right) for Repulsive Model, Speed Model, and Activity Model

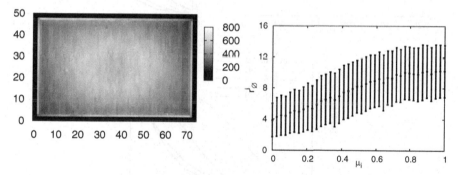

Fig. 11. Repulsive Model: number of time steps where an agent enters a cell (left) and average distance to nest center $r_i^\varnothing$ (right)

The results for different movement models with respect to the correlation coefficient $k(t)$ and changes of the slope $\beta(t)$ of the regression function are compared in Fig. 10. The figure shows that in the Repulsive Model and in the Speed Model the agents show a clear sorting behavior. In the Activity Model there is no clear indication for an agent sorting.

The motivation to introduce the Repulsive Model was to prevent the agents from getting stuck in the center of the nest. The left part of Fig. 11 shows that the agents in the center have a high movement activity and do not get stuck. The strength of the sorting behavior is shown in the right part of Fig. 11. Ants with $\mu \approx 0$ have an average distance of approximately 4 from the

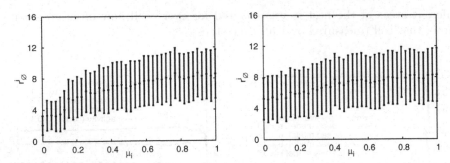

Fig. 12. Speed Model: average distance to nest center $r_i^{\varnothing}$; movement speed $\nu_0 = 0.1$ (left) and $\nu_0 = 0.3$ (right)

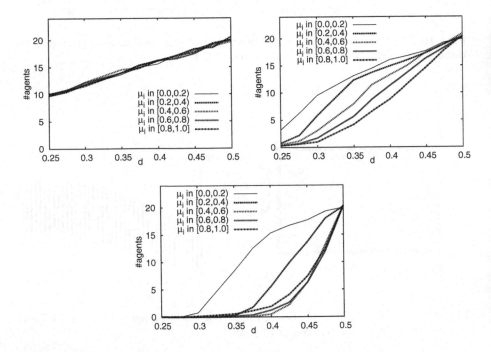

Fig. 13. Two nest centers; average number of agents for different classes $\mu_i \in [0.0, 0.2), [0.2, 0.4), [0.4, 0.6), [0.6, 0.8), [0.8, 1.0)$ in the smaller of the two areas at time step 0 (upper left), 2000 (upper right), and 50,000 (bottom); results are averaged over 100 test runs

center whereas agents with $\mu \approx 1$ have an average distance of approximately 10. The strength of the sorting behavior for the Speed Model is shown in Fig. 12. For large relative differences in movement speed ($\nu_0 = 0.1$) the sorting behavior is stronger than for smaller differences ($\nu_0 = 0.3$).

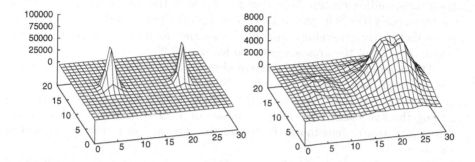

Fig. 14. Two nest centers; $d = 0.4$; $\mu_i \in [0, 0.2)$ (left) and $\mu_i \in [0.8, 1]$ (right); for both influence areas it is counted how often an agent enters a cell (measured over 50,000 time steps)

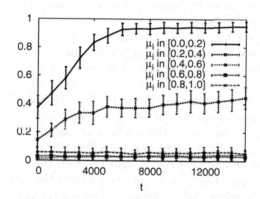

Fig. 15. Activation of the nest center in the larger area after $t \in \{0, 1000, \ldots, 15000\}$ simulation steps; $d = 0.4$; depicted is the fraction of agents for different classes $\mu_i \in [0.0, 0.2), [0.2, 0.4), [0.4, 0.6), [0.6, 0.8), [0.8, 1.0)$ that are in the smaller area after 50,000 simulation steps

4.4 Experiments with Complex Environments

For applications in organic computing it is interesting to consider complex environments with more than one focal point for the movement behavior. An example where this occurs is moving components which have several service stations they can visit. Therefore, a much larger environment of 600 cells × 400 cells with two centers (located at $(150, 200)$ and $(450, 200)$) is used for the experiments. Simulations were done with the Speed Model and 200 agents. The area is divided vertically at position $d \times 600$, $d \in [0, 1]$, such that in the left (right) part of area the turning behavior of the agents is influenced by the left (right) center. The effect is that the agents tend to turn toward the corresponding center. Note that for $d = 0.25$ the line dividing both areas passes exactly the left nest center. The agents are divided into five classes according to their μ-values. For all classes the number of agents in the left and right part of the area are counted for $d \in \{0.25, 0.30, \ldots, 0.50\}$.

The number of agents at different time steps is shown in Fig. 13 (results are averaged over 100 runs). A clear differentiation of the five agent classes is occurring over time. At the beginning the agents are equally distributed among the five classes for different values of d. Over time the faster agents (large μ_i values) occur more often than the slow agents in the larger part of the area.

The movement activity of agents with $\mu_i \in [0, 0.2)$ and agents with $\mu_i \in [0.8, 1)$ in different parts of the nest area for $d = 0.4$ is compared in Fig. 14. It can be seen that the slow agents occur next to both centers whereas the fast agents occur mainly in the part of the area with the center that has the larger influence region. These results show that moving agents with slightly different moving behavior can have very different spatial distributions in areas with several service stations.

A dynamic scenario is also considered where the service stations for the agents are not available starting from the same time. In the corresponding experiment it is assumed that the center in the larger part of the area becomes active several time steps later than the center in the smaller region. The results are given in Fig. 15 for the case where the center in the larger part of the area becomes active at time step $t \in \{0, 1000, \ldots, 15000\}$. Shown is the fraction of agents in each of the classes $\mu_i \in [0.0, 0.2), [0.2, 0.4), [0.4, 0.6), [0.6, 0.8), [0.8, 1.0)$ that are in the smaller area after 50,000 simulation steps. It can be seen, that the slow agents with $\mu_i \in [0.0, 0.2)$ are much more concentrated within the small part of the area if the center in the other part of the area becomes active late (more than 90% of these agents appear in the smaller part of the area if the service station appeared after $t > 6000$ simulation steps). The reason for this is that most of the slower agents are still fast enough to concentrate around the center in the small region during the first 6000 time steps. After that time they will leave the small area only with a very small probability. On the other hand it can be seen, that a large fraction of the faster agents ($\mu_i > 0.4$) can always be found in the larger area regardless of when the second center was added. This mechanism can possibly be used to implement a controlled separation process of agents with different properties in OC systems.

4.5 Summarizing Remarks

Emergent spatial sorting patterns for groups of randomly moving ant-like agents can be observed for simple movement models when there exist slight differences in the individual behavior of the agents. For different movement models the emergent spatial sorting effects have been described based on the results of simulation studies.

Scenarios with more complex environments where the movement of the agents can be influenced by several "center points" have also been considered. Such scenarios are relevant for applications in organic computing where the center points can be seen as service points for moving components of the system. It is interesting that the relative size of the influence area of the service points leads to an emergent effect that the spatial distribution of agents might differ strongly for agents with only slightly different moving behavior. A dynamic scenario where different times span between the activation times of two service stations for the agents was studied. Simulations have shown that the length of the time span has a significant influence on the distribution pattern of the agents and the type of influence is different for different moving behaviors.

5 Final Remarks

The connections between swarm intelligence and the new field of organic computing have been discussed in this chapter. Organic computing systems (OC systems) consist of many autonomous components that interact and show forms of collective behavior. OC systems are designed to possess different self-x properties (e.g., self-healing, self-managing, self-optimizing). Typically, OC systems will have a decentralized control and are able to adapt to their environment or the requirements of the user. Thus, OC systems share some important properties with social insect colonies. Clearly, there exist also many differences between technical systems like OC systems and biological systems like social insect colonies. Nevertheless, it was argued in this chapter that the similarities make methods of swarm intelligence that are often inspired by principles of collective behavior of social insects attractive for organic computing.

We have described some examples of OC systems and presented two case studies that show in detail how methods of swarm intelligence are connected to problems in organic computing. The topic of the first case study is a new approach to control emergent behavior in OC systems. This method is called swarm-controlled emergence and it was applied to control emergent clustering effects that can occur when a group of ant-like agents take up items, carry them around, and drop them. The starting point of the second study is an observation that biologists made with ants. Ants with slightly different movement behavior can be found on average in different parts of the nest. It is discussed what consequences this emergent effect might have on OC systems with moving components. Both case studies have shown that organic computing is strongly linked to swarm intelligence. There is a large potential for applications of swarm intelligence methods in organic computing and design problems for OC systems will provide new challenges for the development of new swarm intelligence methods.

References

1. A. Baddeley, R. Turner: spatstat website, URL: www.spatstat.org
2. C. Anderson, J. J. Boomsma, J. J. Bartholdi, III: Task partitioning in insect societies: bucket brigades. *Insectes Sociaux*, 49(2):171–180, 2002.
3. T. Balch: Hierarchical Social Entropy: An information theoretic measure of robot team diversity. *Autonomous Robots*, 8(3):209–237, 2000.
4. J.-C. de Biseau, J. M. Pasteels: Response thresholds to recruitment signals and the regulation of foraging intensity in the ant *Myrmica sabuleti* (Hymenoptera, Formicidae). *Behavioural Processes*, 48(3):137–148, 2000.
5. A. Bouajila, J. Zeppenfeld, W. Stechele, A. Herkersdorf, A. Bernauer, O. Bringmann, W. Rosenstiel: Organic computing at the system on chip level. In: G. De Micheli, S. Mir, and R. Reis (editors), *Proc. IFIP International Conference on Very Large Scale Integration of System on Chip (VLSI-SoC 2006)*. Springer, Berlin, Germany, 338–341, 2006.
6. E. Bonabeau, M. Dorigo, G. Theraulaz: Swarm Intelligence: ¿From Natural to Artificial Systems. Oxford University Press, New York, 1999.
7. M.D. Cox, G.B. Blanchard: Gaseous templates in ant nests. *Journal of Theoretic Biology*, 204:223–238, 2000.
8. L. Chrétien: Organisation spatiale du matériel provenant de l'excavation du nid chez Messor barbarus et des cadavres d'ouvrières chez Lasius niger (Hymenopterae: Formicidae). Université Libre de Bruxelles, Département de Biologie Animale, 1996.
9. R. Cunningham, J. Dowling, A. Harrington, V. Reynolds, R. Meier, V. Cahill: Self-Optimization in a Next-Generation Urban Traffic Control Environment. *ERCIM News*, 64:55–56, 2006.
10. J.-L. Deneubourg, S. Goss, N. Franks, A.B. Sendova-Franks, C. Detrain, L. Chretien: The dynamics of collective sorting: Robot-like ants and ant-like robots. In: J.-A. Meyer and S. Wilson, editors, *Proc. of the 1st Int. Conf., on Simulation of Adaptive Behavior: From Animals to Animats*, MIT Press/ Bradford Books, 356–363, 1991.
11. S. Depickère, D. Fresneau, J.-L. Deneubourg. A basis for spatial and social patterns in ant species: dynamics and mechanisms of aggregation. *Journal of Insect Behavior*, 17(1):81–97, 2004.
12. J. Delgado, R.V. Sole: Self-synchronization and task fulfilment in ant colonies, *Journal of Theoretical Biology*, 205(3):433–441, 2000.
13. J.L. Deneubourg, A. Lioni, C. Detrain: Dynamics of aggregation and emergence of cooperation. *The Biological Bulletin*, 202:262–267, 2002.
14. S. Depickère, D. Fresneau, J.-L. Deneubourg: Dynamics of aggregation in *Lasius niger* (Formicidae): influence of polyethism. *Insectes Sociaux*, 51(1):81–90, 2004.
15. M. Dorigo, T. Stützle: Ant Colony Optimization. MIT Press, 2004.
16. D. Fey, M. Komann, F. Schurz, A. Loos: An organic computing architecture for visual microprocessors based on marching pixels. *Proc. IEEE International Symposium on Circuits and Systems (ISCAS 2007)*, IEEE Press, 2689–2692, 2007.

17. S. P. Fekete, C. Schmidt, A. Wegener, S. Fischer: Hovering data clouds for recognizing traffic jams. In: T. Margaria, A. Philippou, and B. Steffen (editors), *Proc. 2nd International Symposium on Leveraging Applications of Formal Methods, Verification and Validation (IEEE-ISOLA 2006)*, IEEE Press, 213–218, 2006.

18. N.R. Franks, A. Sendova-Franks: Brood sorting by ants: distributing the workload over the work surface. *Behav. Ecol. Sociobiol.*, 30:109–123, 1992.

19. Gesellschaft für Informatik: Organic Computing / VDE, ITG, GI - Positionspapier. 2003.

20. H. Gutowitz: Complexity Seeking Ants. Unpublished report, 1993.

21. P. P. Grassé: La reconstruction du nid et les coordinations inter-individuelles chez Bellicositermes natalensis et Cubitermes sp. La theorie de la stigmergie: Essai d'interpretation des termites constructeurs. *Insectes Sociaux*, 6:41–83, 1959.

22. D. Grünbaum, Schooling as a strategy for taxis in a noisy environment. *Evolutionary Ecology*, 12:503–522, 1998.

23. J. Handl, B. Meyer: Improved ant-based clustering and sorting in a document retrieval interface. In: J.-J. Merelo Cuervós, P. Adamidis, H. G. Beyer, J. L. Fernández-Villacañas, and H.-P. Schwefel, editors, *Proc. Seventh Int. Conference on Parallel Problem Solving from Nature (PPSN VII)*. volume 2439 of Lecture Notes of Computer Science, Springer, Berlin, Germany, 913–923, 2002.

24. J. Handl, J. Knowles, M. Dorigo: Strategies for the increased robustness of ant-based clustering. In: G. Di Marzo Serugendo, A. Karageorgos, O.F. Rana, and F. Zambonelli, editors, *Postproc. of the First International Workshop on Engineering Self-Organising Applications (ESOA 2003)*, volume 2977 of Lecture Notes of Computer Science, Springer, Berlin, Germany, 90–104, 2003.

25. J. Handl, J. Knowles, M. Dorigo: On the performance of ant-based clustering. In: A. Abraham, M. Köppen, and K. Franke, editors, *Proc. 3rd Int. Conf. on Hybrid Intell. Systems (HIS 2003)*, IOS Press, 2003.

26. S. Janson, M. Middendorf, M. Beekman: Searching for a new home — scouting behavior of honeybee swarms. *Behavioral Ecology*, 18:384–392, 2007.

27. P. M. Kanade, L. O. Hall: Fuzzy Ants as a clustering concept. In: E. Walker, editor, *Proceedings 22nd International Conference of the North American Fuzzy Information Processing Society NAFIPS 2003*, IEEE Systems Man and Cybernetics Society, 227–232, 2003.

28. M. Komann, D. Fey: Marching pixels computing principles in embedded systems using organic parallel dardware. *Proc. International Symposium on Parallel Computing in Electrical Engineering (PARELEC'06)*, 369–373, IEEE CS Press, 2006.

29. P. Kuntz, D. Snyers: Emergent colonization and graph partitioning. In: D. Cliff et al., editors, *Proc. of the 1st Int. Conf, on Simulation of Adaptive Behavior: From Animals to Animats*, MIT Press, 494–500, 1994.

30. N. Labroche, N. Monmarche, G. Venturini: A new clustering algorithm based on the chemical recognition system of ants. In: F. van Harmelen, editor, *Proc. of the 15th European Conference on Artificial Intelligence, ECAI 2002*, IOS Press, 345–349, 2002.

31. N. Labroche, N. Monmarche, G. Venturini: AntClust: Ant clustering and web usage mining. In: E. Cantú-Paz et al., editors, *Proc. of the 2003 Genetic and Evolutionary Computation Conference*, volume 2723 of Lecture Notes of Computer Science, Springer, Berlin, Germany, 25–36, 2003.

32. N. Labroche, N. Monmarche, G. Venturini: Visual clustering with artificial ants colonies. In: V. Palade, R.J. Howlett, L.C. Jain, editors, *Proc. of the 7th International Conference on Knowledge-Based Intelligent Information & Engineering Systems (KES 2003)*, volume 2773 of Lecture Notes of Computer Science, Springer, Berlin, Germany, 332–338, 2003.

33. M.N.M. van Lieshout, A.J. Baddeley: A nonparametric measure of spatial interaction in point patterns. *Statistica Neerlandica*, 50:344–361, 1996.

34. D. Merkle, M. Middendorf, A. Scheidler: Swarm controlled emergence — Designing an anti-clustering ant system. *Proc. of the 2007 IEEE Swarm Intelligence Symposium*, IEEE Press, 10 pages, 2007.

35. F. Mösch, M. Litza, A. El Sayed Auf, E. Maehle, K.-E. Großpietsch, W. Brockmann: ORCA – Towards an organic robotic control. In: H. de Meer and J.P.G. Sterbenz, editors, *Proc. 1st International Workshop on Self-Organizing Systems (IWSOS 2006) and 3rd International Workshop on New Trends in Network Architectures and Services (EuroNGI 2006)*, volume 4124 of Lecture Notes of Computer Science, Springer, Berlin, Germany, 251–253, 2006.

36. M. Mnif, C. Müller-Schloer: Quantitative emergence. In: *Proc. of the 2006 IEEE Mountain Workshop on Adaptive and Learning Systems (IEEE SMCals 2006)*, 2006.

37. C. Müller-Schloer, C. von der Malsburg, R. P. Würtz: Organic computing. *Informatik Spektrum*, 27(4):332–336, 2004.

38. C. Müller-Schloer, B. Sick: Emergence in organic computing systems: Discussion of a controversial concept. In: L.T. Yang, H. Jin, J. Ma, and T. Ungerer, editors, *Proc. 3rd International Conference on Autonomic and Trusted Computing*, volume 4158 of Lecture Notes of Computer Science, Springer, Berlin, Germany, 1–16, 2006.

39. G. Nicolas and D. Sillans: Immediate and latent effects of carbon dioxide on insects. *Annual Review of Entomology*, 34:97–116, 1989.

40. U. Richter, M. Mnif, J. Branke, C. Müller-Schloer, H. Schmeck: Towards a generic observer/controller architecture for organic computing. Köllen Verlag, LNI P-93, 112–119, 2006.

41. F. Rochner, H. Prothmann, J. Branke, C. Müller-Schloer, H. Schmeck: An organic architecture for traffic light controllers. *Proc. GI Jahrestagung 2006*, Lecture Notes in Informatics 93:120–127, 2006.

42. F. Rochner, C. Müller-Schloer: Emergence in technical systems. *it — Special Issue on Organic Computing*, 47:188–200, 2005.

43. A. Scheidler, D. Merkle, M. Middendorf: Emergent sorting patterns and individual differences of randomly moving ant like agents. In: S. Artmann and P. Dittrich, editors, *Proc. 7th German Workshop on Artificial Life (GWAL-7)*, IOS Press, 11 pp., 2006.

44. H. Schmeck: Organic computing – A new vision for distributed embedded systems. *Proc. of the Eighth IEEE International Symposium on Object-Oriented Real-Time Distributed Computing (ISORC 2005)*, IEEE CS Press, 201–203, 2005.

45. A.B. Sendova-Franks, J.V. Lent: Random walk models of worker sorting in ant colonies. *Journal of Theoretical Biology*, 217:255–274, 2002.

46. T. Schöler, C. Müller-Schloer: An observer/controller architecture for adaptive reconfigurable stacks. In: M. Beigl and P. Lukowicz, editors, *Proc. of the 18th International Conference on Architecture of Computing Systems (ARCS 05)*,

volume 3432 of Lecture Notes of Computer Science, Springer, Berlin, Germany, 139–153, 2006.

47. R.V. Sole, E. Bonabeau, J. Delgado, P. Fernandez, J. Marin: Pattern formation and optimization in army ant raids. *Artificial Life*, 6(3):219–226, 2000.

48. G. Theraulaz, E. Bonabeau, J.L. Deneubourg: Response threshold reinforcement and division of labour in insect colonies. *Proc. Roy. Soc. London B*, 265:327–335, 1998.

49. G. Theraulaz, E. Bonabeau, S.C. Nicolis, R.V. Sole, V. Fourcassie, S. Blanco, R. Fournier, J.L. Joly, P. Fernandez, A. Grimal, P. Dalle, J.L. Deneubourg: Spatial patterns in ant colonies. *Proc. Natl. Acad. Sci.*, 99(15):9645–9649, 2002.

50. G. Theraulaz, E. Bonabeau, R.V. Sole, B. Schatz, J.L. Deneubourg: Task Partitioning in a ponerine ant. *Journal of Theoretical Biology*, 215(4):481–489, 2002.

Natural Computing Series

H.-P. Schwefel, I. Wegener, K. Weinert (Eds.): **Advances in Computational Intelligence. Theory and Practice.** VIII, 325 pages. 2003

A. Ghosh, S. Tsutsui (Eds.): **Advances in Evolutionary Computing. Theory and Applications.** XVI, 1006 pages. 2003

L.F. Landweber, E. Winfree (Eds.): **Evolution as Computation.** DIMACS Workshop, Princeton, January 1999. XV, 332 pages. 2002

M. Hirvensalo: **Quantum Computing.** 2nd ed., XI, 214 pages. 2004 (first edition published in the series)

A.E. Eiben, J.E. Smith: **Introduction to Evolutionary Computing.** XV, 299 pages. 2003

A. Ehrenfeucht, T. Harju, I. Petre, D.M. Prescott, G. Rozenberg: **Computation in Living Cells. Gene Assembly in Ciliates.** XIV, 202 pages. 2004

L. Sekanina: **Evolvable Components. From Theory to Hardware Implementations.** XVI, 194 pages. 2004

G. Ciobanu, G. Rozenberg (Eds.): **Modelling in Molecular Biology.** X, 310 pages. 2004

R.W. Morrison: **Designing Evolutionary Algorithms for Dynamic Environments.** XII, 148 pages, 78 figs. 2004

R. Paton[†], H. Bolouri, M. Holcombe, J.H. Parish, R. Tateson (Eds.): **Computation in Cells and Tissues. Perspectives and Tools of Thought.** XIV, 358 pages, 134 figs. 2004

M. Amos: **Theoretical and Experimental DNA Computation.** XIV, 170 pages, 78 figs. 2005

M. Tomassini: **Spatially Structured Evolutionary Algorithms.** XIV, 192 pages, 91 figs., 21 tables. 2005

G. Ciobanu, G. Păun, M.J. Pérez-Jiménez (Eds.): **Applications of Membrane Computing.** X, 441 pages, 99 figs., 24 tables. 2006

K.V. Price, R.M. Storn, J.A. Lampinen: **Differential Evolution.** XX, 538 pages, 292 figs., 48 tables and CD-ROM. 2006

J. Chen, N. Jonoska, G. Rozenberg: **Nanotechnology: Science and Computation.** XII, 385 pages, 126 figs., 10 tables. 2006

A. Brabazon, M. O'Neill: **Biologically Inspired Algorithms for Financial Modelling.** XVI, 275 pages, 92 figs., 39 tables. 2006

T. Bartz-Beielstein: **Experimental Research in Evolutionary Computation.** XIV, 214 pages, 66 figs., 36 tables. 2006

S. Bandyopadhyay, S.K. Pal: **Classification and Learning Using Genetic Algorithms.** XVI, 314 pages, 87 figs., 43 tables. 2007

H.-J. Böckenhauer, D. Bongartz: **Algorithmic Aspects of Bioinformatics.** X, 396 pages, 118 figs., 9 tables. 2007

P. Siarry, Z. Michalewicz (Eds.): **Advances in Metaheuristics for Hard Optimization.** XVI, 481 pages, 66 figs., 83 tables. 2008

J. Knowles, D. Corne, K. Deb (Eds.): **Multiobjective Problem Solving from Nature. From Concepts to Applications.** XVI, 412 pages, 178 figs., 53 tables. 2008

P.F. Hingston, L.C. Barone, Z. Michalewicz (Eds.). **Design by Evolution.** XII, 362 pages, 143 figs., 20 tables. 2008

C. Blum, D. Merkle (Eds.): **Swarm Intelligence: Introduction and Applications.** X, 284 pages, 70 figs., 14 tables. 2008